OCR Chemistry for AS

Graham Hill
Andrew Hunt

The Publishers would like to thank the following for permission to reproduce copyright material:

Photo credits: **p.1** Photodisc; **p.2** © lookGaleria/Alamy (inset: AJ Photo/Science Photo Library); **p.3** *t* Gregory Ochocki/Science Photo Library, *c* © IBM; **p.4** *t* Colin Cuthbert/Science Photo Library, *c* Geoff Tompkinson/Science Photo Library, *b* © Scott Camazine; **p.7** Science Photo Library/Science Source; **p.8** Science Photo Library/Sheila Terry; **p.11** Rex Features; **p.19** Science Photo Library/US Dept. of Agriculture; **p.20** Martyn Chillmaid; **p21** Martyn Chillmaid; **p.24** © Christina Kennedy/Brand X/Corbis; **p.25** Tim Graham/Getty Images; **p.27** © Bettmann/Corbis; **p.32** Science Photo Library/Simon Fraser; **p.34** Science Photo Library/David Parker; **p.35** Martin Sookias photography; **p.35** Hodder & Stoughton; **p.36** *t* Roger Scruton, *c* James Davis Photography; **p.39** GSF Picture Library; **p.42** Geoff Tompkinson/Science Photo Library; **p.43** Roger Scruton; **p.59** Alamy/ImageState, **p.71** *t* Philippe Plailly/Eurelios/Science Photo Library, *cl* © Alan Goldsmith/Corbis, *cr* © David Samuel Robbins/Corbis; **p.72** *t* © Peter Arnold, Inc./Alamy, *c* Andrew Lambert/Science Photo Library, *b* Andrew Lambert/Science Photo Library; **p.75** David Noton Photography/Alamy; **p.76** *t* Andrew Lambert/Science Photo Library, *b* © PjrFoto/Studio/Alamy; **p.79** © Owaki-Kulla/Corbis; **p.81** © AGStockUSA, Inc./Alamy; **p.87** Charles D. Winters/Science Photo Library; **p.88** Volker Steger/Science Photo Library, **p.90** Peter Scoones/Science Photo Library; **p.94** Science Photo Library/Jean-Loup Charmet; **p.95** *t* Science Photo Library/Russ Lappa, *b* Science Photo Library/Andrew Lambert; **p.97** *t* both Science Photo Library, *b* Science Photo Library/US Dept. of Energy; **p.102** *tl* Russ Lappa/Science Photo Library, *c* Bob Gibbons/Science Photo Library, *bl* Andrew Lambert/Science Photo Library; **p.103** Science Photo Library/Andrew Lambert; **p.105** *l* Martin Land/Science Photo Library, *r* Natural History Museum, London; **p.106** Martin Bond/Science Photo Library; **p.107** Science Photo Library; **p.109** both Andrew Lambert/Science Photo Library; **p.110** Andrew Lambert/Science Photo Library; **p.112** Andrew Lambert/Science Photo Library; **p.115** © Randy Faris/Corbis; **p117** Photodisc; **p.118** *c* © Bob Krist/Corbis, *l* © Nordicphotos/Alamy; **p.121** Geoff Tompkinson/Science Photo Library; **p.127** © Car Culture/ Corbis; **p.128** Science Photo Library/Kaj R. Svensson; **p.131** *tr* Saturn Stills/Science Photo Library, *b* Alamy/Hideo Kurihara; **p.133** Ingram; **p.138** both Science Photo Library/Agstockusa; **p.139** Sue Cunningham Photographic; **p.140** *t* © Justin Kase zfourz/Alamy, *b* Alamy/Paul Glendell; **p.141** Andrew Lambert/Science Photo Library; **p.143** Roger Scruton; **p.148** reproduced by kind permission of Unilever PLC (from an original in Unilever Archives); **p.151** Emma Lee/Life File; **p.152** Andrew Lambert/Science Photo Library; **p.154** Ecoscene/Martin Jones; **p.159** © Amit Bhargava/Corbis; **p.160** Corbis; **p.161** Andrew Lambert Photography/Science Photo Library; **p.165** Bill Barksdale/ AGStockUSA/Science Photo Library; **p.168** Science Photo Library/BSIP/Keene; **p.169** Science Photo Library/Dr Jeremy Burgess; **p.170** Science Photo Library/Jerry Mason; **p.173** PA Photos/AP Photo/Amy Sinisterra; **p.175** © SHOUT/Alamy; **p.177** Phoenix Mission/University of Arizona/ Corby Waste; **p.178** © Gareth Price; **p.181** © Hodder Education; **p.182** Saturn Stills/Science Photo Library; **p.185** www.ctpimaging.co.uk; **p.198** *t* Geoff Tompkinson/Science Photo Library, *c* © Israel Sanchez/epa/Corbis; **p.202** Science Photo Library/Philippe Plailly/Eurelios; **p.203** Science Photo Library; **p.208** *t* Leonard Lessin/Science Photo Library, *c* © Corbis, All Rights Reserved, *b* Andrew Lambert/Science Photo Library; **p.216** E. Wolff/BAS; **p.219** Science Photo Library/NASA; **p.221** © Reuters/Corbis; **p.222** Alamy/Colin Palmer; **p.223** Science Photo Library/Tony Craddock; **p224** Science Photo Library/Biosym Technologies, Inc.; **p.226** Maximilian Stock Ltd/Science Photo Library; **p.228** courtesy of Monsanto; **p.230** © Bill Stormont/Corbis; **p.234** David Zutler, Founder and Developer of World's First Planet Friendly Bottle; **p.235** *t* Alex Bartel/Science Photo Library, *c* Tek Image/Science Photo Library

b = bottom, *c* = centre, *l* = left, *r* = right, *t* = top

Artwork credits: Philip Allan Updates: Dr Eric Wolff, graph of estimated temperature difference from today and carbon dioxide and methan in ppmv against year before, from ***Chemistry Review***, volume 15, number 1 (September, 2005); **Carbon Footprint Ltd**: Redrawn pie chart showing components of a typical footprint from (www.carbonfootprint.com/carbon_footprint.html); **Chemicals Industries Association**: Pie chart of products produced by the chemical industry in the UK from **www.cia.org.uk** (Chemicals Industries Association, 2006); **www.greener.industry.org**: Diagram of manufacture of ethanoic acid from carbon monoxide and ethanol from diagram of Reaction mechanism on screen 6 of the series of web pages on ethanoic acid; **IPCC**: Graph of temperature differences from 1961-1990 average against year (redrawn) from Presentation of the Working Group 1 report (www.ipcc.ch/present/presentations.htm); **Royal Society of Chemistry**: Graph of carbon dioxide in ppmv against year AD from *Education in Chemistry*, volume 41, number 5 (September, 2004); **Transport for London**: Flow diagram of hydrogen-fuelled London bus and supply of hydrogen (slightly modified) from www.tfl.gov.uk.

Acknowledgements: Every effort has been made to trace all copyright holders, but if any have been inadvertently overlooked the Publishers will be pleased to make the necessary arrangements at the first opportunity.

Contents

Unit 1 Atoms, bonds and groups 1

1 Atomic structure 2
2 Chemical quantities 19
3 Acids, bases and salts 32
4 Redox 50
5 Electronic structure 59
6 Bonding and structure 70
7 Periodicity 94
8 Group 2 – the alkaline earth metals 102
9 Group 7 – the halogens 109

Unit 2 Chains, energy and resources 117

10 Introduction to organic chemistry 118
11 Alkanes 133
12 Alkenes 143
13 Alcohols 157
14 Halogenoalkanes 165
15 Instrumental analysis 173
16 Enthalpy changes and energetics 181
17 Rates and equilibria 198
18 Chemistry of the air 215
19 Green chemistry 226
The periodic table of elements 239
Index 241

Introduction

Welcome to *OCR Chemistry for AS*. This book covers everything in the OCR specification with 6 topics for Unit 1 and 13 topics for Unit 2. Test yourself questions throughout the book will help you to think about what you are studying while the Activities give you the chance to apply what you have learnt in a range of modern contexts. At the end of each topic you will find exam-style Review questions to help you check your progress.

There is also a CD-ROM inside the back cover, which contains an interactive copy of the book. Wherever the CD-ROM icon appears in the book, this links directly to an external resource (requires a web connection). These resources include:

- Data tables, for use when answering questions
- Tutorials, which work through selected problems and concepts using a voiceover and animated diagrams
- Practical guidance to support experimental and investigative skills.

All diagrams and photographs can be launched and enlarged directly from the page. There are also Learning outcomes available at the beginning of every topic and answer files to all Test yourself and Activity questions. Extension questions, covering some ideas in greater depth, are available at the end of each topic. All resources can be saved to your local hard drive.

With the powerful Search tool on the CD-ROM, key words can be found in an instant, leading you to the relevant page or alternatively to resources associated with each key word.

Acknowledgements

We would like to acknowledge the suggestions from teachers who commented on our initial plans: Ian Davis, Neil Dixon and Tim Joliffe.

The team at Hodder Education, led by Gillian Lindsey and supported by Anne Trevillion, has made an extremely valuable contribution to the development of the book and the CD resources.

Graham Hill and Andrew Hunt

Unit 1

Atoms, bonds and groups

1 Atomic structure
2 Chemical quantities
3 Acids, bases and salts
4 Redox
5 Electronic structure
6 Bonding and structure
7 Periodicity
8 Group 2 – the alkaline earth metals
9 Group 7 – the halogens

1 Atomic structure

Modern chemistry began when scientists understood the difference between elements and compounds. Soon afterwards, chemists began to explain chemical changes in terms of atoms and molecules. This led to experiments and theories, which opened up our understanding of atomic structure.

1.1 Why study chemistry?

Chemistry is one of the sciences that helps to us understand ourselves and all the materials and living things in the Universe.

Some of the reasons for studying chemistry are outlined below. How well do these reasons match your expectations as you start this advanced chemistry course?

Looking for patterns in chemical behaviour

Part of being a chemist involves getting a feel for the way in which chemicals behave. Chemists get to know chemicals just as people get to know their friends and family. They look for patterns in the chemical reactions between substances and recognise that some of the patterns are familiar. For example, the elements sodium and potassium are soft and stored under oil because they react so readily with air and water, and copper sulfate is blue, like many other copper compounds. By understanding patterns, chemists can identify and make plastics like polythene and medicines like aspirin (Figure 1.1).

Figure 1.1 ▶ Aspirin is probably the most common medicine we use. The bark of willow trees was used to ease pain for more than 2000 years. Early in the twentieth century, chemists extracted the active ingredient from willow bark. Their understanding of patterns in the behaviour of similar compounds enabled them to synthesise aspirin.

Definitions

Elements are the simplest chemicals – they cannot be broken down into simpler chemicals.

Compounds are chemicals that contain two or more elements chemically combined together.

An **atom** is the smallest particle of an element.

A **molecule** is a particle containing two or more atoms joined together chemically.

An **ion** is an atom, or a group of atoms, that has become electrically charged by the loss or gain of one or more electrons.

Working out the composition and structure of materials

New materials exist because chemists understand how atoms, ions and molecules are arranged in different materials and the forces that hold these particles together. Thanks to this knowledge, we can enjoy fibres that breathe but are waterproof, plastic ropes that are 20 times stronger than similar ropes of steel (Figure 1.2) and metal alloys that can remember their shape. Understanding the structure and bonding of materials is a central theme in modern chemistry.

Controlling chemical changes

Four simple questions are at the heart of most chemical investigations.

- *How much?* How much of the reactants do we need to make the product and how much of the product will we produce?
- *How fast?* How can we make sure that a reaction goes at the right speed: not too fast and not too slow? How can we control the speed of reactions?
- *How far?* Will the chemicals we use react completely and make the product or will the reaction stop before we get all we want? If it does, what can we do to get as much of the product as possible?
- *How do reactions occur?* When chemical reactions take place, the atoms in substances are rearranged. This means that some bonds must break and others must form. But, which bonds between atoms are breaking and which new bonds are forming during a reaction?

Figure 1.2 ▲
Plastic ropes made of polymers 20 times stronger than steel have made rock climbing and ice climbing safer and easier.

Developing new skills

Chemistry involves doing things as well as gaining knowledge and understanding about materials. Chemists use their thinking skills and practical skills to solve practical problems. One of the frontiers of today's chemistry involves nanotechnology, in which chemists work with particles as small as individual atoms (Figure 1.3).

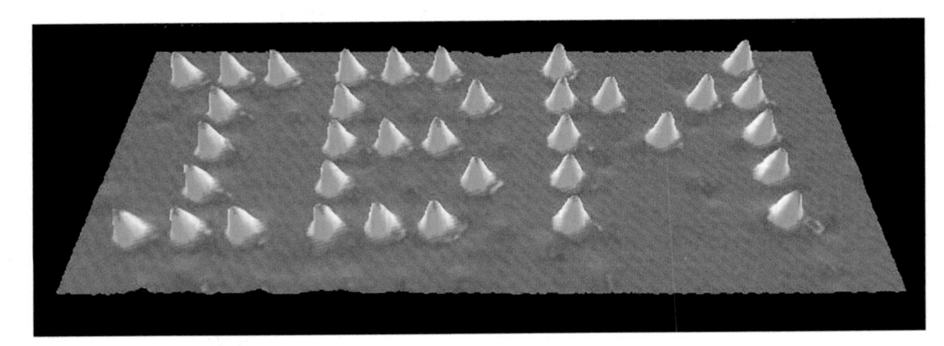

Figure 1.3 ◄
In the 1990s, two scientists working for IBM cooled a nickel surface to −269 °C in a vacuum chamber. They then introduced a tiny amount of xenon so that some of the xenon atoms stuck to the surface of the nickel. Using a special instrument called a scanning tunnelling microscope, the scientists moved individual xenon atoms around on the nickel surface to construct the IBM logo – each blue blob is the image of a single xenon atom.

Chemists increasingly rely on modern instruments to explore structures and chemical changes. They also use information technology to store data, search for information and publish their findings.

Recognising the value of chemistry to society

Chemists study materials and try to change raw materials into more useful substances. All the new materials that we have are made from natural raw resources by chemical reactions. On a large scale, the chemical industry converts raw materials from the earth, sea and air into valuable new products for society. A good example of this is the Haber process, which turns natural gas and air into ammonia – the chemical used to make fertilisers, dyes and explosives.

A vital task for chemists is to analyse materials and find out what they are made of. When chemists have analysed a substance, they use symbols and formulae to show the elements it contains. Symbols are used to represent the atoms in elements and formulae are used to represent the ions and molecules in compounds. Analysis is involved in checking that our drinking water is pure and that our food is safe to eat. People worry about pollution of the environment, but we would not understand the causes or the scale of this pollution without chemical analysis.

Linking theories and experiments

Scientists test their theories by doing experiments. In chemistry, experiments often begin with careful observation of what happens as chemicals react and change. Theories are more likely to be accepted if their predictions turn out to be correct when tested by experiments.

One of the reasons Mendeleev's periodic table was so successful was that his predictions about the properties of missing elements turned out to be so accurate.

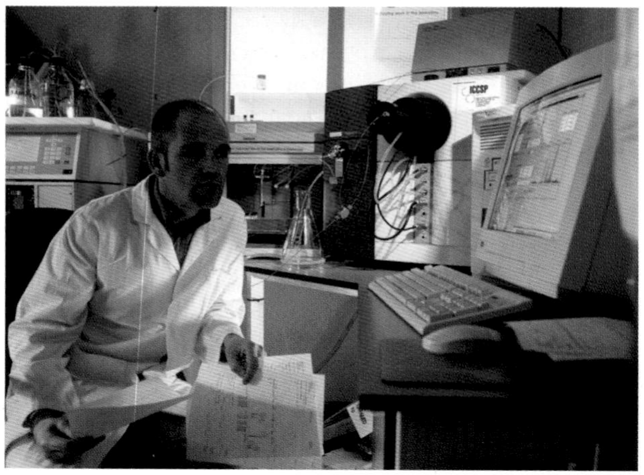

Figure 1.4 ▶
The scientist is operating a machine which uses liquid chromatography followed by mass spectrometry (LCMS) to research anti-cancer (chemotherapy) drugs.

The study of chemistry is about more than 'what we know'. It is also about 'how we know'. For example, the study of atomic structure has provided evidence about the nature and properties of protons, neutrons and electrons. This has led to an explanation of the properties of elements and the patterns in the periodic table in terms of the electron structures of atoms.

Definitions

Symbols are used to represent elements. For example, Fe for iron and C for carbon.

Formulae are used to represent compounds. For example, H_2O for water and NaCl for sodium chloride.

Learning to enjoy and take an interest in chemistry

As a schoolgirl, Dorothy Hodgkin became intensely interested in crystals and their structures. In 1964, she won a Nobel Prize for her use of X-rays to determine the structures of complex molecules such as penicillin and vitamin B12. In 1996, Harry Kroto shared the Nobel Prize in Chemistry for his part in the discovery of a new form of carbon called 'buckyballs'. According to Sir Harry, 'Science is to do with fun and solving puzzles.'

Figure 1.5 ▶
Professor Harry Kroto with models of his buckyballs.

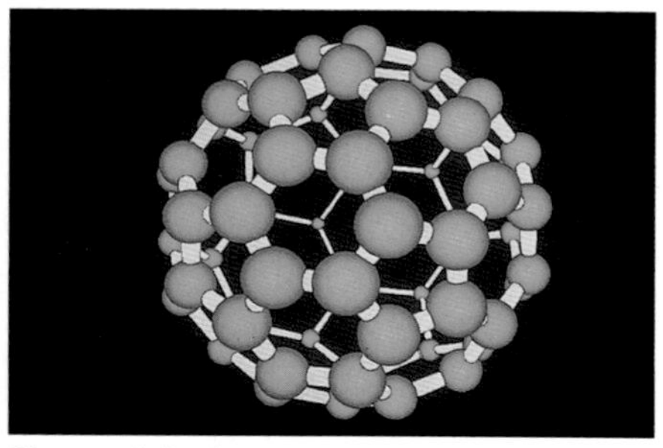

Figure 1.6 ▲
A computer model of a C_{60} buckyball – 60 carbon atoms arranged like a cage.

Some people particularly enjoy the practical side of chemistry. They get pleasure from working with chemicals, obtaining high yields of products and getting accurate results. Others are fascinated by the theory of chemistry and by using models to explain how materials react and change. Yet others are interested in chemistry because of the way in which it can improve our lives – especially through medicine, nutrition, pharmacy, dentistry and the science of materials.

Test yourself

1 Remind yourself of some patterns in the ways that chemicals behave.
   a) What happens when a more reactive metal (such as zinc) is added to a solution of a compound of a less reactive metal (such as copper sulfate) in water?
   b) What forms at the negative electrode (cathode) during the electrolysis of a metal compound?
   c) What happens when an acid (such as hydrochloric acid) is added to a carbonate (such as calcium carbonate)?
   d) What do sodium chloride, sodium bromide and sodium iodide look like?

2 This question concerns the substances ice, salt, sugar, copper, steel and limestone. Which of these substances contain:
   a) uncombined atoms
   b) ions
   c) molecules?

3 Write down the three most important reasons why you chose to study AS Chemistry?

1.2 From elements to compounds

Compounds form when two or more elements combine. Apart from the atoms of helium and neon, all atoms combine with other atoms.

Compounds of non-metals with non-metals

Water, carbon dioxide, methane in natural gas, sugar and ethanol (alcohol) are examples of compounds of two or more non-metals. Most of these non-metal compounds melt and vaporise easily. They may be gases, liquids or solids at room temperature, and they do not conduct electricity.

In these compounds of non-metals, the atoms combine in small groups to form molecules. For example, methane contains one carbon atom bonded to four hydrogen atoms. The formula of the molecule is CH_4. Figure 1.7 shows three ways of representing a methane molecule. Models like that shown in the first diagram represent atoms as little spheres.

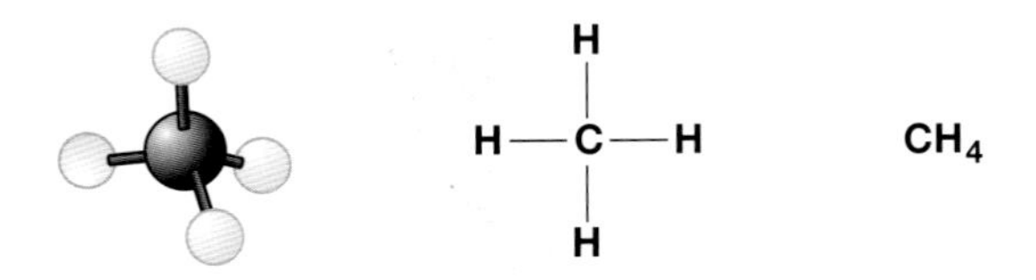

Figure 1.7 ▲
Ways of representing a molecule of methane.

It is possible to work out the formula of most non-metal compounds if you know how many bonds the atoms normally form (Table 1.1).

Element	Symbol	Number of bonds formed	Colour in molecular models
Carbon	C	4	Black
Nitrogen	N	3	Blue
Oxygen	O	2	Red
Sulfur	S	2	Yellow
Hydrogen	H	1	White
Chlorine	Cl	1	Green

Table 1.1 ▲
Symbols, number of bonds and colour codes of some non-metals.

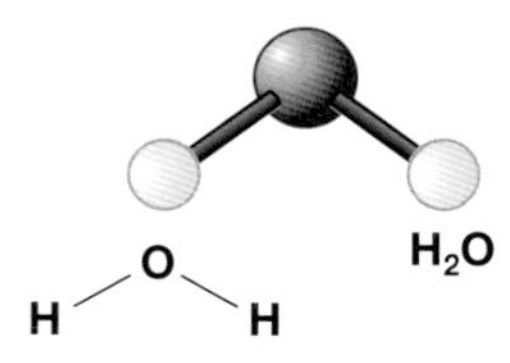

Figure 1.8 ▲
Ways of representing a molecule of water.

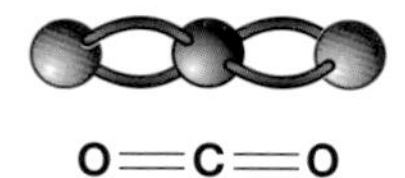

Figure 1.9 ▲
Bonding in carbon dioxide showing the double bonds between atoms.

Water is a compound of oxygen and hydrogen. Oxygen atoms form two bonds and hydrogen atoms form one bond. Two hydrogen atoms can therefore bond to one oxygen atom (Figure 1.8), so the formula of water is H_2O.

In some non-metal compounds, there are double or even triple bonds between the atoms (Figure 1.9). Notice also that there is a strict colour code for the atoms of different elements in the molecular models – these colours are also shown in Table 1.1.

In practice, it is not possible to predict the formulae of all non-metal compounds because the bonding rules in Table 1.1 cannot account for the formulae of carbon monoxide, CO, or sulfur dioxide, SO_2.

Test yourself

4 Draw the various ways of representing the following compounds (like those for methane in Figure 1.7):
   a) hydrogen chloride
   b) carbon disulfide.

5 Name the elements present and work out the formulae of the following compounds:
   a) hydrogen sulfide
   b) dichlorine oxide
   c) hydrogen nitride (ammonia).

Compounds of metals with non-metals

Common salt (sodium chloride), limestone (calcium carbonate) and copper sulfate are all examples of compounds of metals with non-metals. These metal/non-metal compounds melt at much higher temperatures than compounds of non-metals and are solids at room temperature. They conduct electricity when molten but not as solids. Metal/non-metal compounds conduct electricity as liquids because they consist of ions. For example, sodium chloride consists of sodium ions (Na^+) and chloride ions (Cl^-).

The formula of sodium chloride is NaCl (or Na^+Cl^-) because the positive charge on one Na^+ ion is balanced by the negative charge on one Cl^- ion. A crystal of sodium chloride contains equal numbers of sodium ions and chloride ions (Figure 1.10).

Figure 1.10 ▶
A space-filling model and a ball-and-stick model showing the structure of sodium chloride.

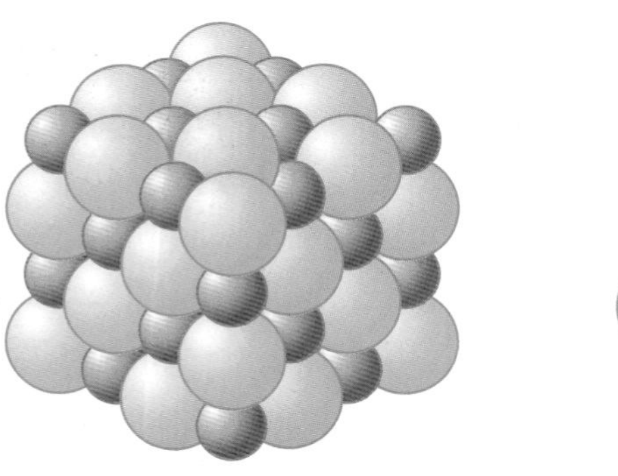

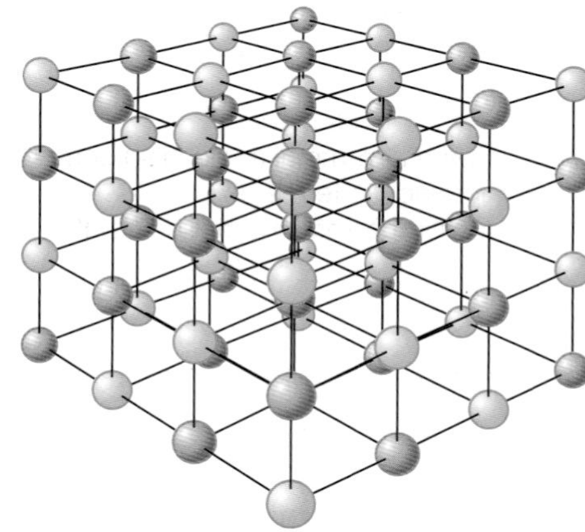

The formulae of all metal/non-metal (ionic) compounds can be obtained by balancing the charges on positive and negative ions. For example, the formula of potassium oxide is K_2O (or $(K^+)_2O^{2-}$). Here, two K^+ ions balance the charge on one O^{2-} ion.

Elements such as iron, which have two different ions (Fe^{2+} and Fe^{3+}), have two sets of compounds – iron(II) compounds such as iron(II) chloride, $FeCl_2$, and iron(III) compounds such as iron(III) chloride, $FeCl_3$.

Table 1.2 shows the names and formulae of some more ionic compounds. Notice that the formula of magnesium nitrate is $Mg(NO_3)_2$, or $Mg^{2+}(NO_3^-)_2$.

The brackets around NO_3^- show that it is a single unit containing one nitrogen and three oxygen atoms bonded together and with one negative charge. Other ions, like OH^-, SO_4^{2-} and CO_3^{2-} must also be treated as single units and put in brackets when there are two or three of them in a formula.

Data

Name of compound	Formula
Magnesium nitrate	$Mg^{2+}(NO_3^-)_2$ or $Mg(NO_3)_2$
Aluminium hydroxide	$Al^{3+}(OH^-)_3$ or $Al(OH)_3$
Zinc bromide	$Zn^{2+}(Br^-)_2$ or $ZnBr_2$
Lead nitrate	$Pb^{2+}(NO_3^-)_2$ or $Pb(NO_3)_2$
Calcium iodide	$Ca^{2+}(I^-)_2$ or CaI_2
Copper(II) carbonate	$Cu^{2+}CO_3^{2-}$ or $CuCO_3$
Silver sulfate	$(Ag^+)_2SO_4^{2-}$ or Ag_2SO_4

Table 1.2 ▲
Names and formulae of some ionic compounds.

Test yourself

Data

6 The formula of aluminium hydroxide must be written as $Al(OH)_3$. Why is $AlOH_3$ wrong?
7 Write the formulae of the following ionic compounds:
   a) potassium sulfate
   b) aluminium oxide
   c) lead carbonate
   d) zinc hydroxide
   e) iron(III) sulfate.
8 Which of the following compounds consist of molecules and which consist of ions?
   a) Octane (C_8H_{18}) in petrol
   b) Copper(I) oxide
   c) Concentrated sulfuric acid
   d) Lithium fluoride
   e) Phosphorus trichloride
9 Compare non-metal (molecular) compounds with metal/non-metal (ionic) compounds in terms of:
   a) melting points and boiling points
   b) conduction of electricity as liquids.

1.3 Early ideas about atoms

The idea that all substances could be made of atoms was suggested by the Greek philosopher Democritus more than 2400 years ago.

Democritus believed that if a lump of metal, such as iron, was cut into smaller and smaller pieces, you would end up with miniscule and invisible pieces of iron that could not be cut any smaller. He called these smallest particles of a substance 'atomos', which means 'indivisible'.

Although the Greeks were great thinkers, they did no experiments. So, Democritus' ideas about atoms were never tested. People had to try to understand his ideas about invisible particles without any evidence. It was not surprising that his ideas about atoms were eventually forgotten.

About 2000 years after Democritus, chemists in Europe began to purify substances and to carry out experiments with them. They found that many

Figure 1.11 ▲
The Greek philosopher Democritus, who lived from 460 BC to 370 BC.

Figure 1.12 ▲
John Dalton was born in 1766 in the village of Eaglesfield in Cumbria. His father was a weaver. Dalton was always curious and liked to study. When he was only 12 years old, he started to teach children in the village school. For most of his life, he taught science and carried out experiments at the Presbyterian College in Manchester.

substances could be broken down (decomposed) into simpler substances, which they called elements. These elements could be combined to make compounds.

By the eighteenth century, chemists had started to carry out accurate experiments. To their surprise, they found that the weights of elements that reacted were always in the same proportions. For example, water always contained 1 part by weight of hydrogen to 8 parts by weight of oxygen, and black copper oxide always contained 1 part by weight of oxygen to 4 parts by weight of copper.

At the start of the nineteenth century, John Dalton puzzled over these results and concluded that everything would make sense if elements were made of indivisible particles. He believed that compounds like copper oxide were made of particles of copper and oxygen with different masses and that these always combined in the same ratios. Dalton called the indivisible particles **atoms** in recognition of the ideas first proposed by Democritus.

Dalton began to publish his atomic theory in 1808. The main points in his theory are:

- all elements are made up of indivisible particles called atoms
- all the atoms of a given element are identical and have the same mass
- the atoms of different elements have different masses
- atoms can combine to form molecules in compounds
- all the molecules of a given compound are identical.

Although some scientists were reluctant to accept Dalton's ideas, his atomic theory caught on because it could explain the results of most experiments.

Even today, Dalton's atomic theory is still useful and very helpful. However, research during the last 110 years has shown that atoms are **not** indivisible and that all atoms of the same element are **not** identical.

Test yourself

10 Why did Democritus' idea about atoms never catch on?
11 Why was the idea of atoms accepted by scientists when Dalton re-introduced it about 200 years ago?
12 Look at the five main points in Dalton's atomic theory. Which of these points:
   a) are still correct
   b) are now incorrect?
13 Look at the formulae in Figure 1.13, which Dalton used for water, carbon dioxide and black copper oxide.

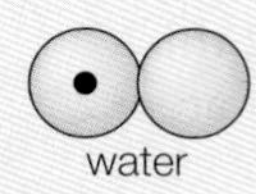

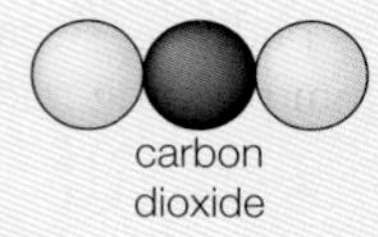

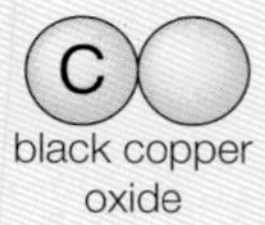

Figure 1.13 ▲
Dalton's formulae for three compounds.

   a) Write the formulae that we use today for these compounds.
   b) What symbols did Dalton use for carbon, oxygen, hydrogen and copper?
   c) Which one of the formulae did Dalton get wrong?

1.4 Inside atoms

Just more than a century ago, scientists still thought that atoms were just like Dalton had described them – solid, indestructible particles like tiny snooker balls. Then, between 1897 and 1932, experiments showed that atoms contained three smaller particles – electrons, protons and neutrons. But, how were they discovered and how did this change our ideas about atoms?

1897: Thomson discovers electrons

In 1897, JJ Thomson was investigating the conduction of electricity by gases. When he passed 15 000 volts across the terminals of a tube containing air, the glass walls glowed bright green. Rays, travelling in straight lines from the negative terminal, hit the glass and made it glow. Experiments showed that a narrow beam of the rays could be deflected by an electric field (Figure 1.14). When passed between charged plates, the rays always bent towards the positive plate. This showed they were negatively charged.

Further study showed that the rays consisted of tiny negative particles that were about 2000 times lighter than hydrogen atoms. This surprised Thomson. He had discovered particles smaller than atoms. Thomson called the tiny negative particles electrons.

Thomson obtained electrons with different gases in the tube and when the terminals were made of different substances. This suggested that all substances contained electrons.

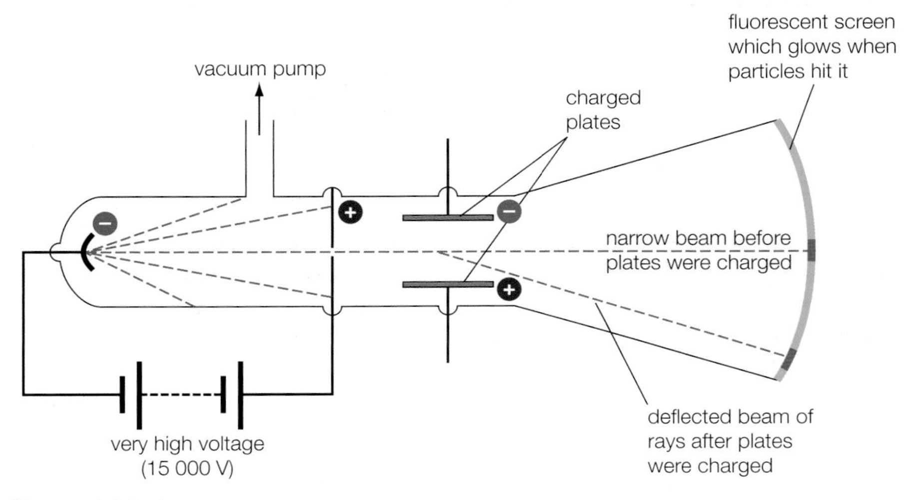

Figure 1.14 ▲
The effect of charged plates on a beam of electrons.

Thomson knew that atoms have no electrical charge overall. So, the rest of the atom must have a positive charge to balance the negative charge of the electrons.

In 1904, Thomson put forward a model for the structure of atoms. He suggested that atoms were tiny balls of positive material with electrons embedded in them like fruit in a Christmas pudding. Because of this, Thomson's ideas became known as the 'plum pudding' model of atomic structure (Figure 1.15).

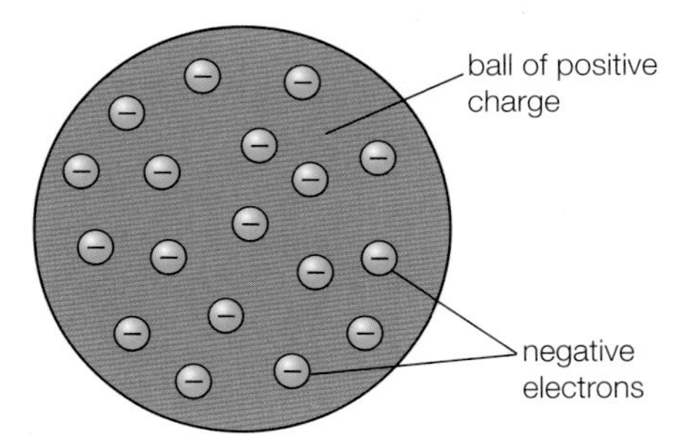

Figure 1.15 ▲
Thomson's plum pudding model for the structure of atoms.

Activity

Rutherford and his colleagues find the positive nucleus

In 1898, Ernest Rutherford showed that radioactive materials give out at least two types of radiation. He called these alpha 'rays' and beta 'rays'.

At the time, Rutherford and his colleagues did not know exactly what alpha rays were, but they did know that alpha rays contained particles. These alpha particles were small, heavy and positively charged. Rutherford and his colleagues used the alpha particles from radioactive substances as tiny 'bullets', which they fired at atoms.

In 1909, Rutherford designed an experiment in which his colleagues Hans Geiger and Ernest Marsden directed narrow beams of positive alpha particles at very thin gold foil only a few atoms thick (Figure 1.16). They expected the particles to pass straight through the foil or to be deflected slightly.

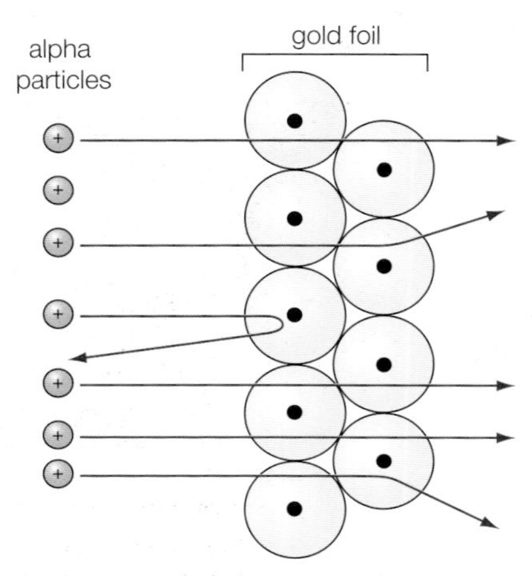

Figure 1.16 ▲
When positive alpha particles are directed at a very thin sheet of gold foil, they emerge at different angles. Most pass straight through the foil, some are deflected and a few seem to rebound from the foil.

The results showed that:

- most of the alpha particles went straight through the foil
- some of the alpha particles were scattered (deflected) by the foil
- a few alpha particles rebounded from the foil.

1. Why do you think most of the alpha particles pass straight through the foil?
2. What conclusions can you draw about the size of any positive and negative particles in the gold atoms if most alpha particles pass straight through the foil?
3. Why were some alpha particles deflected?
4. How did these results cast doubts on Thomson's plum pudding model of atomic structure?
5. Why do you think a few alpha particles seemed to rebound from the foil?
6. In 1911, Rutherford interpreted these results and put forward a new model for the structure of atoms. How do you think Rutherford described atoms?

1911: Rutherford's nuclear model of atomic structure

Rutherford summarised the results of Geiger and Marsden's experiment by saying that atoms have a very small positive nucleus surrounded by a much larger region of empty space in which electrons orbit the nucleus like planets orbiting the Sun (Figure 1.17).

Rutherford's model of atomic structure was called the nuclear model. This model quickly replaced Thomson's plum pudding model and is still the basic model of atomic structure that we use today.

Through the work of Thomson, Rutherford and their colleagues, we now know that:

- atoms have a small positive nucleus surrounded by a much larger region of empty space in which there are tiny negative electrons
- the positive charge of the nucleus is due to positive particles that Rutherford called protons
- protons are about 2000 times heavier than electrons
- the positive charge on one proton is equal in size but opposite in sign to the negative charge on one electron
- atoms have equal numbers of protons and electrons, so the positive charges on the protons cancel out the negative charges on the electrons
- the smallest atoms are those of hydrogen, which contain one proton and one electron. The next smallest atoms are those of helium, which have two protons and two electrons, then lithium atoms, which have three protons and three electrons, and so on.

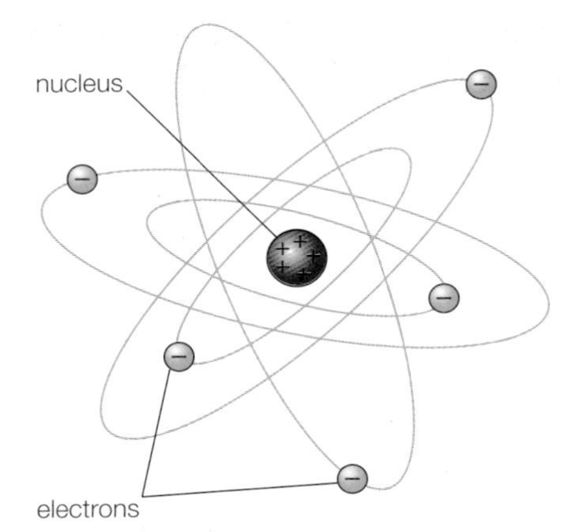

Figure 1.17 ▲
Rutherford's nuclear model for the structure of atoms. Rutherford pictured atoms as miniature solar systems with electrons orbiting the nucleus like planets around the Sun.

Although Rutherford was successful in explaining many aspects of atomic structure, one big problem remained. If hydrogen atoms contain one proton and helium atoms contain two protons, then the relative masses of hydrogen and helium atoms should be one and two, respectively. However, the mass of helium atoms relative to hydrogen atoms is four and not two.

In 1932, James Chadwick, one of Rutherford's colleagues, discovered the source of the extra mass in helium. He showed that the nuclei of atoms contain uncharged particles as well as positively charged protons. Chadwick called these uncharged particles neutrons. Experiments showed that neutrons have the same mass as protons. This helped Chadwick to explain the problem concerning the relative masses of hydrogen and helium atoms (Table 1.3).

Hydrogen atoms have one proton and no neutrons, so a hydrogen atom has a relative mass of one unit. Helium atoms have two protons and two neutrons, so a helium atom has a relative mass of four units. This makes a helium atom four times as heavy as a hydrogen atom (Figure 1.19).

Figure 1.18 ▲
If the nucleus of a hydrogen atom was enlarged to the size of a marble and placed in the centre of the pitch of Wembley stadium, the atom's one electron would be whizzing around somewhere in the stands.

	Hydrogen atoms	Helium atoms
Number of protons	1	2
Number of neutrons	0	2
Relative mass	1	4

Table 1.3 ▲
The relative masses of hydrogen and helium atoms.

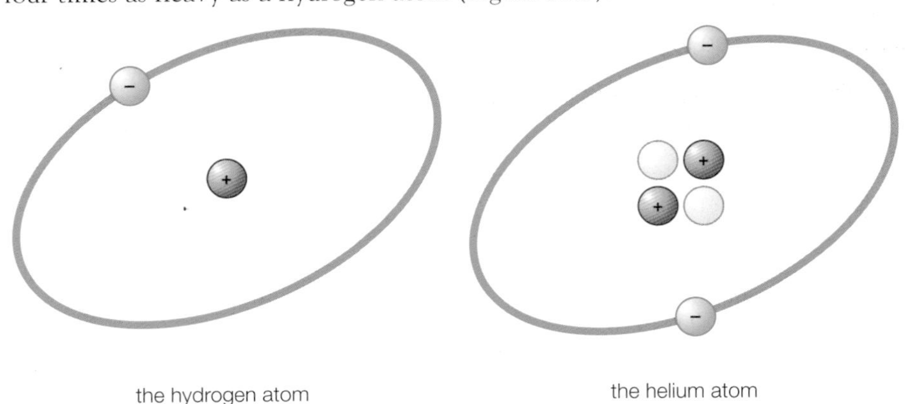

Figure 1.19 ▲
Protons, neutrons and electrons in a hydrogen atom and a helium atom.

We now know that all atoms are made up from protons, neutrons and electrons. The relative masses, relative charges and positions within atoms of these sub-atomic particles are summarised in Table 1.4.

Particle	Relative mass (atomic mass units)	Relative charge	Position within atoms
Proton	1	+1	Nucleus
Neutron	1	0	Nucleus
Electron	$\frac{1}{2000}$	−1	In space outside the nucleus

Table 1.4 ▲
Relative masses, relative charges and positions within atoms of protons, neutrons and electrons.

Atoms and ions

The structure of atoms helps us understand how ions are formed from atoms by the loss or gain of electrons. A helium atom (He) has two protons (each with one positive charge), two neutrons and two electrons (each with one negative charge). If an electron is removed from a helium atom, it leaves an ion (charged particle) with two positive charges and only one negative charge. This gives an overall charge on the ion of one positive charge. We can therefore write the symbol He^+ for this particle.

If two electrons are removed from a helium atom, the remaining particle has two positive charges and no negative charges. This can be represented by the symbol He^{2+} (Figure 1.20).

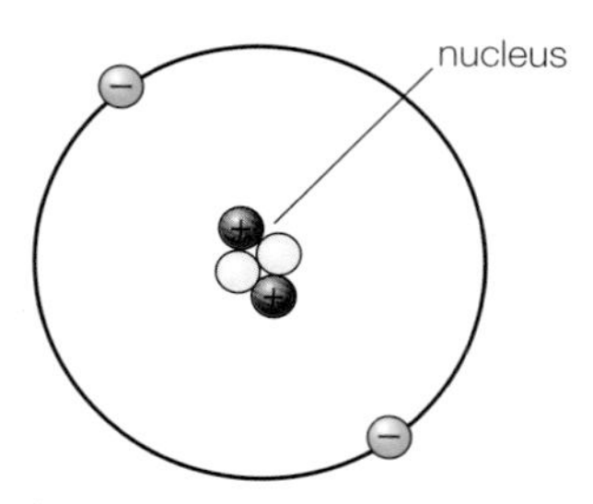

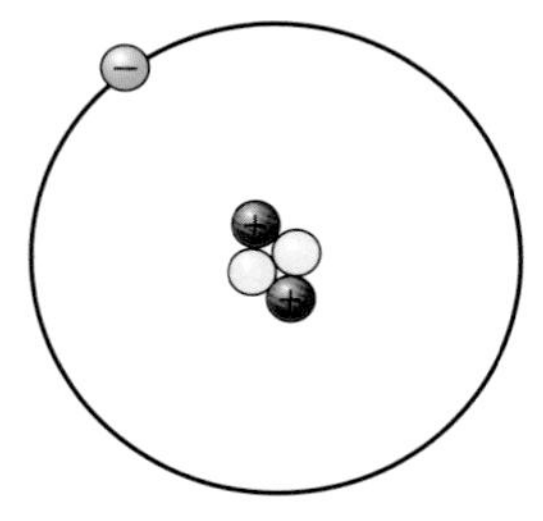

Figure 1.20 ▶
Protons, neutrons and electrons in a helium atom and helium ions.

1.5 Atomic number and mass number

All the atoms of a particular element have the same number of protons, and atoms of different elements have different numbers of protons.
The only atoms with one proton are hydrogen atoms; the only atoms with two protons are helium atoms, the only atoms with three protons are lithium atoms, and so on. This means that the number of protons in an atom determines which element it is. Because of this, scientists have a special name for the number of protons in the nucleus of an atom. They call it the atomic number, and they use the symbol Z to represent atomic number. So, hydrogen has an atomic number of 1 (Z = 1), helium has an atomic number of 2 (Z = 2), and so on.

Protons do not account for all the mass of an atom. Neutrons in the nucleus also contribute to the atom's mass. The mass of an atom therefore depends on the number of protons plus the number of neutrons. This number is called the mass number of the atom (symbol A).

Hydrogen atoms, with one proton and no neutrons, have a mass number of 1 (A = 1). Lithium atoms, with 3 protons and 4 neutrons, have a mass number of 7 (A = 7). Aluminium atoms, with 13 protons and 14 neutrons, have a mass number of 27 (A = 27).

Definitions

The **atomic number** of an atom is the number of protons in its nucleus. The term proton number is sometimes used instead of atomic number.

The **mass number** of an atom is the number of protons plus the number of neutrons in its nucleus. Protons and neutrons are sometimes called nucleons, so the term nucleon number is an alternative to mass number.

There is an agreed shorthand system for showing the mass number and atomic number of an atom. Figure 1.21 shows this for an aluminium atom: $^{27}_{13}Al$. Using this shorthand system, we can also represent ions. For example, an aluminium ion is written as $^{27}_{13}Al^{3+}$.

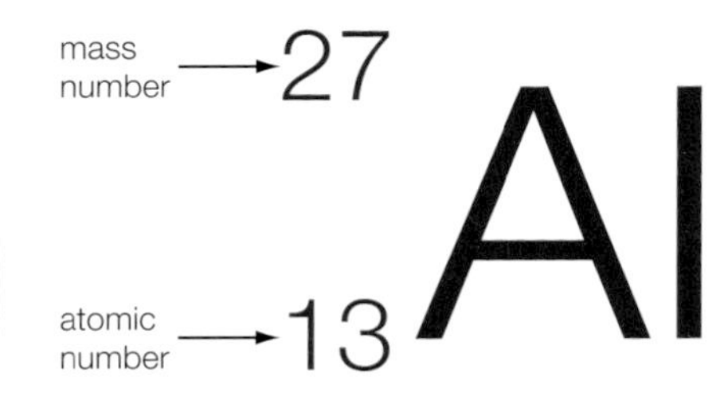

Figure 1.21 ▲
The mass number and atomic number can be shown with the symbol of an atom.

Test yourself

14 How many protons, neutrons and electrons are there in the following atoms and ions?
a) $^{9}_{4}Be$
b) $^{39}_{19}K$
c) $^{235}_{92}U$
d) $^{19}_{9}F^{-}$
e) $^{40}_{20}Ca^{2+}$

15 Write symbols to show the mass number and atomic number for the following atoms and ions:
a) an oxygen atom with 8 protons, 8 neutrons and 8 electrons
b) an argon atom with 18 protons, 22 neutrons and 18 electrons
c) a sodium ion with a charge of +1 and a nucleus of 11 protons and 12 neutrons
d) a sulfur ion with a charge of −2 and a nucleus with 16 protons and 16 neutrons

1.6 Comparing the masses of atoms

Individual atoms are far too small to be weighed on a balance. However, the mass of one atom can be compared with that of another atom using an instrument called a mass spectrometer (Figure 1.22). The relative masses of atoms are called relative atomic masses. The measurement of relative atomic masses using mass spectrometry is covered more fully in Topic 15.

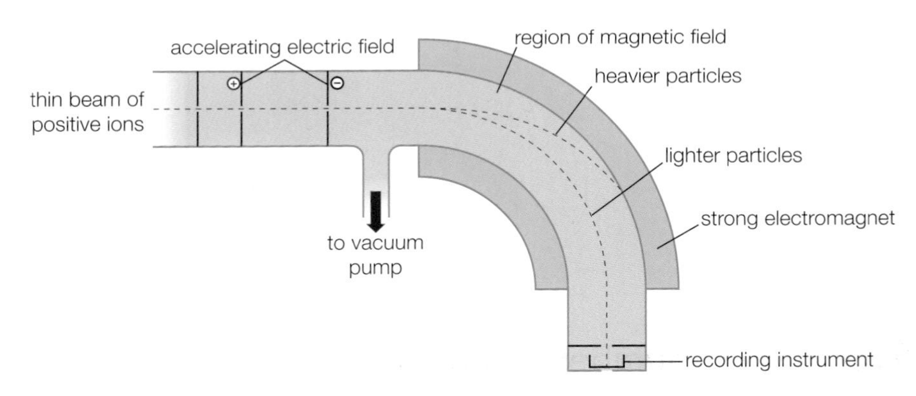

Figure 1.22 ◄
A mass spectrometer produces positive ions of elements to be analysed. A thin beam of the ions is first accelerated through an electric field. The ions then pass through a magnetic field, which deflects them. The extent of deflection depends on the mass of the ions – lighter ions are deflected more than heavier ions. From the extent of deflection, it is possible to compare the masses of different atoms and to make a list of their relative masses.

Chemists originally measured relative atomic masses using hydrogen as the standard. Then, because of the existence of isotopes it became necessary to choose one particular isotope as the standard. Today, the isotope carbon-12 ($^{12}_{6}C$) is chosen as the standard and given a relative mass of exactly 12. The relative atomic mass of an element is therefore the average mass of an atom of the element relative to one twelfth of the mass of an atom of the isotope carbon-12.

$$\text{Relative atomic mass} = \frac{\text{average mass of an atom of the element}}{\frac{1}{12} \times \text{the mass of one atom of carbon-12}}$$

The symbol for relative atomic mass is A_r, where r stands for relative. Using the carbon-12 scale, the relative atomic mass of hydrogen is 1.0 and that of chlorine is 35.5. So, we can write $A_r(H) = 1$ and $A_r(Cl) = 35.5$ or H = 1 and Cl = 35.5 for short. Notice that the values of relative atomic masses have no units because they are relative. Table 1.5 lists a few relative atomic masses. The full complement of the relative atomic masses of all stable elements is shown in the periodic table on page 239.

Definition

The **relative atomic mass** of an element is the average mass of an atom of the element relative to one twelfth the mass of an atom of carbon-12.

Element	Symbol	Relative atomic mass
Carbon	C	12.0
Hydrogen	H	1.0
Helium	He	4.0
Nitrogen	N	14.0
Oxygen	O	16.0
Magnesium	Mg	24.3
Sulfur	S	32.1
Iron	Fe	55.8
Copper	Cu	63.5

Table 1.5 ▲
Relative atomic masses of a few elements.

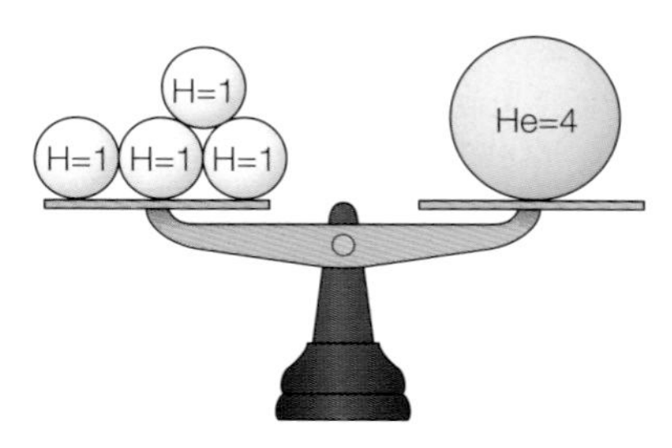

Figure 1.23 ▲
If atoms could be weighed, the scales would show that helium atoms are four times as heavy as hydrogen atoms.

Test yourself

16 How many times heavier (to the nearest whole number) are:
 a) C atoms than H atoms
 b) Mg atoms than C atoms
 c) S atoms than He atoms
 d) C atoms then He atoms
 e) Fe atoms than N atoms?

Definition

The **relative molecular mass** of an element or a compound is the sum of the relative atomic masses of all the atoms in its molecular formula.

Relative molecular mass and relative formula mass, M_r

Relative atomic masses can also be used to compare the masses of different molecules. The relative masses of molecules are called relative molecular masses (symbol M_r).

$$\text{For oxygen, } O_2,\ M_r(O_2) = 2 \times A_r(O) = 2 \times 16 = 32$$

$$\text{For sulfuric acid, } M_r(H_2SO_4) = 2 \times A_r(H) + A_r(S) + 4 \times A_r(O) = (2 \times 1) + 32.1 + (4 \times 16) = 98.1$$

Definition

The **relative formula mass** of a compound is the sum of the relative atomic masses of all the atoms in its formula.

Metal compounds consist of giant structures of ions and not molecules. To avoid the suggestion that their formulae represent molecules, chemists use the term relative formula mass not relative molecular mass for ionic compounds and for other compounds with giant structures such as silicon dioxide, SiO_2.

$$\text{For magnesium nitrate, } M_r(Mg(NO_3)_2) = A_r(Mg) + 2 \times A_r(N) + 6 \times A_r(O) = 24.3 + (2 \times 14) + (6 \times 16) = 148.3$$

Test yourself

17 What is the relative molecular mass of:
 a) chlorine, Cl_2
 b) sulfur, S_8
 c) ethanol, C_2H_5OH
 d) tetrachloromethane, CCl_4?

18 What is the relative formula mass of:
 a) magnesium chloride, $MgCl_2$
 b) iron(III) oxide, Fe_2O_3
 c) hydrated copper(II) sulfate, $CuSO_4.5H_2O$?

1.7 Isotopes and relative isotopic masses

Look closely at Figure 1.24, which shows a mass spectrometer print-out (mass spectrum) for chlorine. When atoms of chlorine are ionised in a mass spectrometer, the beam of ions separates into two paths and produces two peaks on the mass spectrum. This shows that the atoms from which the ions formed must have different masses. These atoms of the same element with different masses are called isotopes.

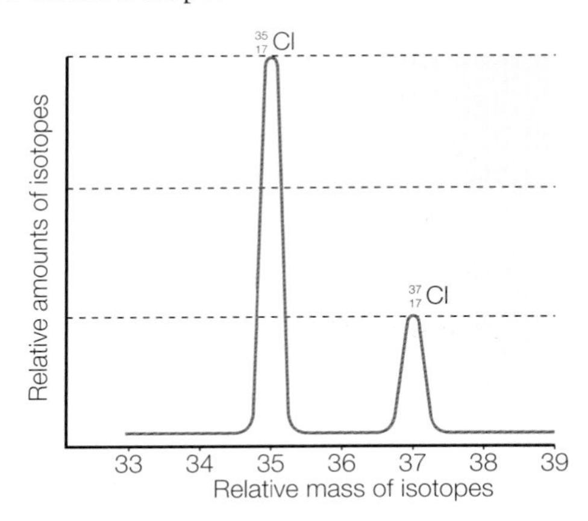

Figure 1.24 ◀
A mass spectrum for chlorine.

From the two peaks in Figure 1.24, we can deduce that chlorine consists of two isotopes with relative masses of 35 and 37. These relative masses are best described as relative isotopic masses, because they give the relative mass of particular isotopes relative to atoms of carbon-12. Notice that the relative masses of pure isotopes are called relative isotopic masses, whereas the relative masses of the atoms in an element, which often contain a mixture of isotopes, are called relative atomic masses.

The heights of the two peaks in Figure 1.24 show that the relative proportions of chlorine-35 to chlorine-37 are 3:1. This means that for every four chlorine atoms, three are chlorine-35 and one is chlorine-37.

The isotope chlorine-35 has a mass number of 35 and contains 17 protons and 18 neutrons, whereas chlorine-37 has a mass number of 37 and contains 17 protons and 20 neutrons. Table 1.6 summarises the important similarities and differences in isotopes.

Isotopes have the same:	Isotopes have different:
number of protons	numbers of neutrons
number of electrons	mass numbers
atomic number	physical properties
chemical properties	

Table 1.6 ▲
Important similarities and differences in isotopes.

Calculating relative atomic masses

Using the print-outs (mass spectra) from mass spectrometers like that in Figure 1.24, it is possible to calculate the relative atomic masses of elements from the relative abundance of their isotopes. The mass spectrum in Figure 1.24 shows that chlorine consists of two isotopes, $^{35}_{17}Cl$ and $^{37}_{17}Cl$, with relative atomic masses of 35 and 37.

The relative atomic mass of an element is the average mass of one atom of the element. This average must take into account the relative amounts of the different isotopes. For example, if chlorine contained 100% $^{35}_{17}Cl$, its relative atomic mass would be 35.0. If it contained 100% $^{37}_{17}Cl$, its relative atomic mass would be 37.0. A 50:50 mixture of the isotopes $^{35}_{17}Cl$ and $^{37}_{17}Cl$ would have a relative atomic mass of 36.0. Figure 1.24 shows that naturally occurring chlorine contains three times as many $^{35}_{17}Cl$ atoms as $^{37}_{17}Cl$ atoms – that is, $\frac{3}{4}$ or 75% is $^{35}_{17}Cl$ and $\frac{1}{4}$ or 25% is $^{37}_{17}Cl$.

The relative atomic mass of chlorine therefore can be calculated as:

75% chlorine-35 + 25% chlorine-37

$$= \frac{75}{100} \times 35.0 \quad + \frac{25}{100} \times 37$$

$$= 26.25 \quad + 9.25$$

$$= 35.5$$

Definitions

Isotopes are atoms of the same element with the same atomic number but different mass numbers.

Relative isotopic mass is the mass of one atom of an isotope relative to the mass of one atom of carbon-12, for which the relative mass is defined as exactly 12.

Test yourself

19 Look carefully at the mass spectrometer print-out for tungsten in Figure 1.25.

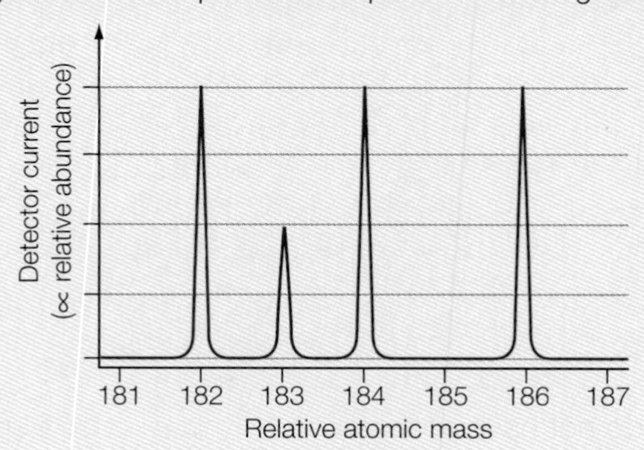

Figure 1.25 ▲
A mass spectrum of the element tungsten.

a) How many different ions are detected in the mass spectrum of tungsten?
b) What are the relative masses of these different ions?
c) What are the relative proportions of the different ions?

20 Magnesium contains three naturally occurring isotopes: $^{24}_{12}Mg$, $^{25}_{12}Mg$ and $^{26}_{12}Mg$.

a) How many protons and neutrons are present in the nuclei of each of these isotopes?
b) What is the relative atomic mass of a sample of magnesium that contains 80% $^{24}_{12}Mg$, 10% $^{25}_{12}Mg$ and 10% $^{26}_{12}Mg$?

Activity

Finding the relative atomic mass of neon

Look carefully at the mass spectrum of neon in Figure 1.26.

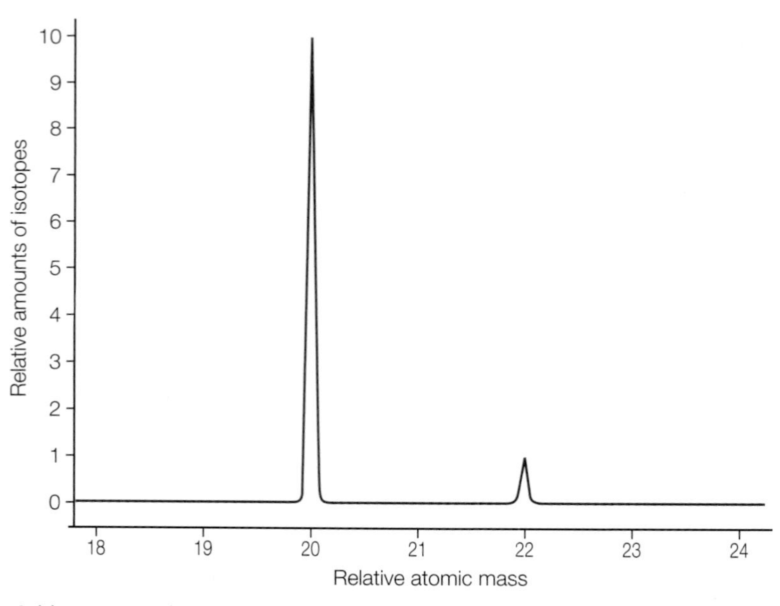

Figure 1.26 ◄
The mass spectrum of neon.

1 How many isotopes does neon contain?

2 What are the relative isotopic masses of the isotopes of neon?

3 What are the relative amounts of the isotopes of neon?

4 Start a spreadsheet program (e.g. Excel) on your computer and open up a new spreadsheet for your results.

a) Enter the relative isotopic masses of the isotopes of neon in column 1 and the relative amounts of these isotopes in column 2.

b) Enter a formula in column 3 to work out the contribution of each isotope to the overall relative atomic mass of neon. (Hint: Look at the calculation of the relative atomic mass of chlorine in Section 1.7.)

c) Finally, enter a formula in column 4 to calculate the relative atomic mass of neon.

(If you do not have access to a computer with a spreadsheet program, calculate the relative atomic mass of neon yourself.)

5 Pure samples of the isotopes of neon have different densities and slightly different boiling points.

a) Why is this?

b) Which isotope of neon will have the highest density?

c) Which isotope of neon will have the highest boiling point?

6 Isotopes of the same element have different physical properties but the same chemical properties. Why is this?

7 Samples of neon collected in different parts of the world have slightly different relative atomic masses. Why is this?

Tutorial

REVIEW QUESTIONS

1 Antimony has two main isotopes – antimony-121 and antimony-123. A forensic scientist was asked to help a crime investigation by analysing the antimony in a bullet. This was found to contain 57.3% of ^{121}Sb and 42.7% of ^{123}Sb.

a) Explain what you understand by the term relative atomic mass. **(3)**

b) Calculate the relative atomic mass of the sample of antimony from the bullet. Write your answer to an appropriate number of significant figures. **(3)**

c) State one similarity and one difference between the isotopes in terms of subatomic particles. **(2)**

2 a) During mass spectrometry, the following processes occur: acceleration, deflection, detection and ionisation. Write these processes in the order in which they occur in a mass spectrometer. **(2)**

b) The isotopes in naturally occurring silicon – ^{28}Si, ^{29}Si and ^{30}Si – can be separated by mass spectrometry.

i) Explain what you understand by the term isotope. **(2)**

ii) Copy and complete the table below to show the composition of isotopes ^{28}Si and ^{30}Si. **(2)**

Isotope	Protons	Neutrons	Electrons
^{28}Si			
^{30}Si			

iii) Why do samples of silicon extracted from different samples of clay have slightly different relative atomic masses? **(1)**

3 This question concerns the following five species:

$^{16}_{8}O^{2-}$ $^{19}_{9}F^{-}$ $^{20}_{10}Ne$ $^{23}_{11}Na$ $^{25}_{12}Mg^{2+}$

a) Which two species have the same number of neutrons? **(2)**

b) Which two species have the same ratio of neutrons to protons? **(2)**

c) Which one species does not have 10 electrons? **(1)**

d) What do the numbers 16, 8 and 2– represent in the symbol $^{16}_{8}O^{2-}$? **(3)**

4 Chlorine has two isotopes: chlorine-35 and chlorine-37.

a) One molecule of chlorine can be written as $^{35}_{17}Cl_2$. Write the formulae for the two other molecules of chlorine. **(1)**

b) What are the relative formula masses of the three different molecules of chlorine? **(1)**

c) How do you think the boiling point of chlorine-35 will compare with that of chlorine-37? Explain your answer. **(3)**

d) Explain how mass and charge are distributed within a chlorine atom in terms of protons, neutrons and electrons. **(4)**

5 a) Why is it correct to use the term relative molecular mass for water but incorrect to use this term for sodium chloride? **(2)**

b) What term should be used with sodium chloride in place of relative molecular mass? **(1)**

c) Explain the terms relative atomic mass and relative isotopic mass using the following data: Naturally occurring chlorine contains two isotopes – chlorine-35 and chlorine-37 – and its relative atomic mass is 35.5. **(6)**

d) When would the relative atomic mass and relative isotopic mass of an element have the same value? **(1)**

2 Chemical quantities

Chemists often need to measure how much they have of a particular chemical. Analysts in pharmaceutical companies regularly test samples of tablets and medicines to check that they contain the right amount of a drug. Food manufacturers check the purity and amounts of different chemicals in the raw materials they buy and in the foods they produce. Industrial chemists measure the amounts of substances they need for chemical processes.

Figure 2.1 ◄
This chemist is pipetting a solution of one reactant into beakers containing solutions, of varying concentrations, of a second reactant.

2.1 Chemical amounts – the mole

When chemists are determining formulae or working with equations, they need to measure amounts that contain equal numbers of atoms, molecules or ions. Chemists have balances to measure masses and graduated containers to measure volumes, but there is no instrument to measure chemical amounts directly. Instead, chemists must first calculate the chemical amounts they need and then measure out the masses or volumes of these substances.

Chemists use the term 'mole' to describe an amount of substance that contains a standard number of atoms, molecules or ions. The word 'mole' entered the language of chemistry at the end of the nineteenth century and is based on a Latin word that means a heap or a pile. The chemist's symbol for mole is 'mol'.

The key to working with chemical amounts in moles is to know the relative atomic masses of different atoms (Section 1.6). The relative atomic masses of all stable elements are shown in the periodic table on page 239.

Definition

The term **mole** (symbol 'mol') is the unit in which chemists measure an amount of a substance that contains a standard number of atoms, ions or molecules.

Molar mass

In chemistry, the amount of a substance is measured in moles, and 1 mole of an element is equal to its relative atomic mass in grams. So, 1 mole (1 mol) of carbon is 12 g and 1 mole (1 mol) of copper is 63.5 g. These masses of one mole are usually called molar masses (symbol M). So, the molar mass of carbon, M(C), is $12\,g\,mol^{-1}$ and the molar mass of copper, M(Cu), is $63.5\,g\,mol^{-1}$.

Similarly, the molar mass of the molecules of an element or a compound is numerically equal to its relative molecular mass. So, the molar mass of one oxygen molecule, $M(O_2)$, is:

$$2 \times A(O) = 2 \times 16\,g\,mol^{-1}$$
$$= 32\,g\,mol^{-1}$$

and the molar mass of sulfuric acid, $M(H_2SO_4)$, is:

$$2 \times A(H) + A(S) + 4 \times A(O) = (2 \times 1) + 32.1 + (4 \times 16)\,g\,mol^{-1}$$
$$= 98.1\,g\,mol^{-1}.$$

Definition

Molar mass is the mass of one mole (i.e. the mass per mole) of a chemical. The unit is $g\,mol^{-1}$. As always with molar amounts, the symbol or formula of the chemical must be specified.

Likewise, the molar mass of an ionic compound is numerically equal to its relative formula mass. Therefore, the molar mass of magnesium nitrate, $M(Mg(NO_3)_2)$, is:

$A(Mg) + 2 \times A(N) + 6 \times A(O)$
$= 24.3 + (2 \times 14) + (6 \times 16)\ g\,mol^{-1}$
$= 148.3\ mol^{-1}$.

Tutorial

Amount in moles

The mole is the standard (SI) unit for amount of substance. So, the amount of any substance is measured in moles, which is normally abbreviated to mol. Therefore:

- 12 g of carbon contains 1 mol of carbon atoms
- 24 g of carbon contains 2 mol of carbon atoms
- 240 g of carbon contains 20 mol of carbon atoms.

Notice that: amount of substance (mol) $= \dfrac{\text{mass of substance/g}}{\text{molar mass/g mol}^{-1}}$

Figure 2.2 ▶ One mole amounts of copper (Cu), carbon (C), iron (Fe), aluminium (Al), mercury (Hg) and sulfur (S).

Tutorial

Avogadro constant

Relative atomic masses show that one atom of carbon is 12 times as heavy as one atom of hydrogen. This means that 12 g of carbon will contain the same number of atoms as 1 g of hydrogen. Similarly, an atom of oxygen is 16 times as heavy as an atom of hydrogen, so 16 g of oxygen will also contain the same number of atoms as 1 g of hydrogen.

In fact, the relative atomic mass in grams (that is, 1 mole) of every element (1 g of hydrogen, 12 g carbon, 16 g oxygen, etc) will contain the same number of atoms. This number is called the Avogadro constant in honour of the Italian scientist Amedeo Avogadro. The symbol used for the Avogadro constant is either L or N_A. Experiments show that the Avogadro constant, N_A, is equal to $6.02 \times 10^{23}\ mol^{-1}$. Written out in full, this is 602,000,000,000,000,000,000,000 per mole. The value of N_A is usually simplified to $6 \times 10^{23}\ mol^{-1}$.

The Avogadro constant is not simply the number of atoms in 1 mole of an element. It is also the number of molecules or formula units in 1 mole of the relevant substances. Therefore, 1 mole of oxygen (O_2) contains 6×10^{23} O_2 molecules, and 2 moles of oxygen (O_2) contains $2 \times 6 \times 10^{23}$ O_2 molecules.

Therefore, the number of atoms, molecules or formula units
= amount of the chemical in moles × the Avogadro constant.

It is vital, of course, to specify the particles concerned when calculating the amount of a substance or the number of particles of a substance. For example, 2 g of hydrogen will contain 2 mol of hydrogen (H) atoms (12×10^{23} atoms) but only 1 mol of hydrogen (H_2) molecules (6×10^{23} molecules).

Figure 2.3 ◀ One mole amounts of some ionic compounds.

Test yourself

1 What is the amount in moles of:
 a) 20.05 g of calcium atoms
 b) 3.995 g of bromine atoms
 c) 159.8 g of bromine molecules
 d) 6.41 g of sulfur dioxide molecules
 e) 10 g of sodium hydroxide, NaOH?

2 What is the mass of:
 a) 0.1 mol of iodine atoms
 b) 0.25 mol of chlorine molecules
 c) 2 mol of water molecules
 d) 0.01 mol ammonium chloride, NH_4Cl
 e) 0.125 mol of sulfate ions, SO_4^{2-}?

3 How many moles of:
 a) sodium ions are there in 1 mol of sodium carbonate, Na_2CO_3
 b) bromide ions are there in 0.5 mol of barium bromide, $BaBr_2$
 c) nitrogen atoms are there in 2 mol of ammonium nitrate, NH_4NO_3?

4 Use the Avogadro constant (6×10^{23}) to calculate:
 a) the number of chloride ions in 0.5 mol of sodium chloride, NaCl
 b) the number of oxygen atoms in 2 mol of oxygen molecules, O_2
 c) the number of sulfate ions in 3 mol of aluminium sulfate, $Al_2(SO_4)_3$.

5 One cubic decimetre of tap water was found to contain 0.1116 mg of iron(III) ions (Fe^{3+}) and 12.40 mg of nitrate ions (NO_3^-).
 a) What are these masses of Fe^{3+} and NO_3^- in grams?
 b) What are the amounts of Fe^{3+} and NO_3^- in moles?
 c) What are the numbers of Fe^{3+} and NO_3^- ions?

2.2 Empirical and molecular formulae

Although the formulae of most compounds can be predicted, the only sure way of knowing a formula is by experiment. This has been done for all common compounds and their formulae can be checked in tables of data.

Experimental (empirical) formulae

Empirical evidence is information based on experience or experiment. Chemists use the term 'empirical formulae' to describe formulae calculated from the results of experiments.

An experiment to find an empirical formula involves measuring the masses of elements that combine in the compound. From these masses, it is possible to calculate the number of moles of atoms that react and hence the ratio of atoms that react. This gives an empirical formula that shows the simplest whole number ratio for the atoms of different elements in a compound.

Worked example

Analysis of 20.1 g of iron bromide showed that it contained 3.8 g of iron and 16.3 g of bromine. What is its formula?

Answer

	Iron		Bromine
Combined masses	3.8 g		16.3 g
Molar mass	55.8 g mol^{-1}		79.9 g mol^{-1}
Combined moles of atoms	$\frac{3.8\,g}{55.8\,g\,mol^{-1}}$ = 0.068 mol		$\frac{16.3\,g}{79.9\,g\,mol^{-1}}$ = 0.204 mol
Ratio of combined atoms	0.068 / 0.068 = 1		0.204 / 0.068 = 3
Simplest ratio of atoms	1	:	3
Empirical formula	$FeBr_3$		

Sometimes, the analysis of a compound shows the percentages of the different elements rather than their masses. These percentages can, of course, be regarded as combining masses and the empirical formula can be worked out in the same way. Table 2.1 shows an example of this. Notice how the calculation to find the empirical formula of ethene can be summarised in a table.

This empirical formula for ethene shows the simplest ratio of carbon and hydrogen atoms. However, the actual formula, which shows the correct number of carbon and hydrogen atoms in one molecule of ethene, could be CH_2, C_2H_4, C_3H_6, C_4H_8, etc, because all of these formulae give CH_2 as the simplest ratio.

Experiments show that the relative molecular mass of ethene is 28. As the relative molecular mass of CH_2 is only 14, the actual molecules of ethene must be twice as heavy as CH_2. This means that the actual formula of ethene must be C_2H_4.

Formulae such as C_2H_4, which show the actual number of atoms of each element in one molecule of a compound, are called molecular formulae.

	C	H
Combined percentage by mass	85.7%	14.3%
Combined masses	85.7 g	14.3 g
Molar mass	12 g mol^{-1}	1 g mol^{-1}
Combined moles	$\frac{85.7\,g}{12\,g\,mol^{-1}}$ = 7.14 mol	$\frac{14.3\,g}{1\,g\,mol^{-1}}$ = 14.3 mol
Ratio of atoms	7.14 / 7.14 = 1.0	14.3 / 7.14 = 2.0
Simplest ratio of atoms	1	2
Empirical formula	CH_2	

Table 2.1 ▲
Finding the empirical formula of ethene.

Definitions

An **empirical formula** shows the simplest whole number ratio for the atoms of each element present in a compound.

A **molecular formula** shows the actual number of atoms of each element in one molecule of a compound.

Test yourself

6 What is the empirical formula of the compound in which:
   a) 0.6 g carbon combines with 0.2 g hydrogen
   b) 1.02 g vanadium combines with 2.84 g chlorine
   c) 1.38 g sodium combines with 0.96 g sulfur and 1.92 g oxygen?
7 What is the empirical formula of the compound in which the percentages of the elements present are:
   a) 34.6% copper, 30.5% iron and 34.9% sulfur
   b) 2.04% hydrogen, 32.65% sulfur and 65.31% oxygen
   c) 52.18% carbon, 13.04% hydrogen and 34.78% oxygen.
8 Ethanoic acid (acetic acid) is the main constituent of vinegar. It contains only carbon, hydrogen and oxygen. 0.90 g of pure ethanoic acid was found to contain 0.36 g of carbon and 0.06 g of hydrogen.
   a) What is the mass of oxygen in 0.90 g of ethanoic acid?
   b) Calculate the empirical formula of ethanoic acid. Set out your calculation like the calculation for ethene in Table 2.1.
   c) The relative molecular mass of ethanoic acid is 60. What is its molecular formula?

Activity

Finding the formula of red copper oxide

A student investigated the formula of red copper oxide by reducing it to copper using natural gas, as shown in Figure 2.4.

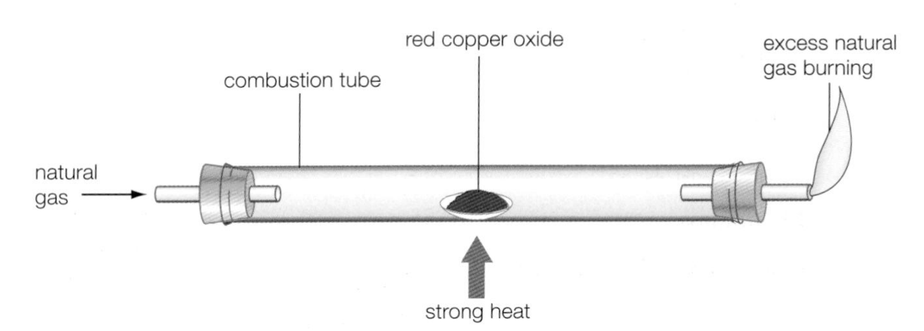

Figure 2.4 ◀ Reducing red copper oxide by heating in natural gas.

The experiment was carried out five times, each time starting with different masses of red copper oxide. The results are shown in Table 2.2.

Table 2.2 ◀ Experimental results.

Experiment number	Mass of red copper oxide/g	Mass of copper in the oxide/g
1	1.43	1.27
2	2.14	1.90
3	2.86	2.54
4	3.55	3.27
5	4.29	3.81

1 Look at Figure 2.4. What safety precautions should the student take during the experiments?

2 What steps should the student take to ensure that all the copper oxide is reduced to copper?

3 Start a spreadsheet program (e.g. Excel) on a computer and open up a new spreadsheet for your results. Enter the experiment numbers and the masses of copper oxide and copper in the first three columns of your spreadsheet, as in Table 2.2.

4 a) Enter a formula in column 4 to work out the mass of oxygen in the red copper oxide that was used.

b) Enter a formula in column 5 to find the amount of copper in moles in the oxide.

c) Enter a formula in column 6 to find the amount of oxygen in moles in the oxide.

5 From the spreadsheet, plot a line graph of amount of copper (*y*-axis) against amount of oxygen (*x*-axis). Print out your graph. (If you cannot plot graphs directly from the spreadsheet, draw the graph by hand.)

6 Which of the points should be disregarded in drawing the line of best fit?

7 a) What is the average value of $\frac{\text{amount of copper/mol}}{\text{amount of oxygen/mol}}$ from your graph?

b) How many moles of copper combine with one mole of oxygen in red copper oxide?

c) What is the empirical formula of red copper oxide?

8 How did the student improve the reliability of the results?

9 Is the result for the formula of red copper oxide valid? Explain your opinion.

(Hint: Valid results are reliable, obtained by fair tests in which only one variable changes and are free of observer or experimenter bias.)

10 Look at the nutritional data on the packaging of a chocolate bar.

a) Is there any evidence that the data is:
i) accurate
ii) reliable
iii) valid?

b) What could producers do to provide evidence that the data about their product is:
i) accurate
ii) reliable
iii) valid?

Figure 2.5 ▲
When sparklers burn, bits of magnesium react with oxygen in the air.

2.3 Chemical equations

Burning and rusting are good examples of chemical reactions. When they occur, chemical bonds in the reactants break and then new bonds form in the products. The photo in Figure 2.5 shows sparks from a sparkler. These sparks are bits of burning magnesium. When magnesium burns in air, it reacts with oxygen to form white magnesium oxide. A word equation for the reaction is:

magnesium + oxygen → magnesium oxide

Balancing chemical equations

When chemists write equations, they use symbols and formulae rather than words. However, writing a word equation is a useful first step towards a balanced chemical equation with symbols. There are four key steps in writing an equation.

Step 1: Write a word equation for the reaction.

magnesium + oxygen → magnesium oxide

Step 2: Write symbols for the elements and formulae for the compounds in the word equation.

$$Mg + O_2 \rightarrow MgO$$

Remember that oxygen, hydrogen, nitrogen and the halogens are diatomic molecules with two atoms in their molecules, so they are written as O_2, H_2, N_2, Cl_2, Br_2 and I_2. All other elements are shown as single atoms.

Step 3 Balance the equation by putting numbers in front of the symbols and formulae, so that the numbers of each type of atom are the same on both sides of the equation.

$$2Mg + O_2 \rightarrow 2MgO$$

Never change a formula to make an equation balance. The formula of magnesium oxide is always MgO and never MgO_2 or Mg_2O.

Step 4: Add state symbols to show the state of each substance in the equation – (s) for solid, (l) for liquid, (g) for gas and (aq) for an aqueous solution (a substance dissolved in water).

$$2Mg(s) + O_2(g) \rightarrow 2MgO(s)$$

Balanced chemical equations are more useful than word equations because they show:

- the symbols and formulae of the reactants and products
- the relative numbers of atoms and molecules in the reaction.

Definition

A **balanced chemical equation** describes a chemical reaction using symbols for the reactants and products. The numbers of each kind of atom are the same on both sides of the equation.

Practical guidance

Modelling equations

Chemists often use models in order to understand what is happening to the different atoms, molecules and ions during a reaction. Figure 2.6 shows how molecular models can give a picture of the reaction between hydrogen and oxygen at an atomic level.

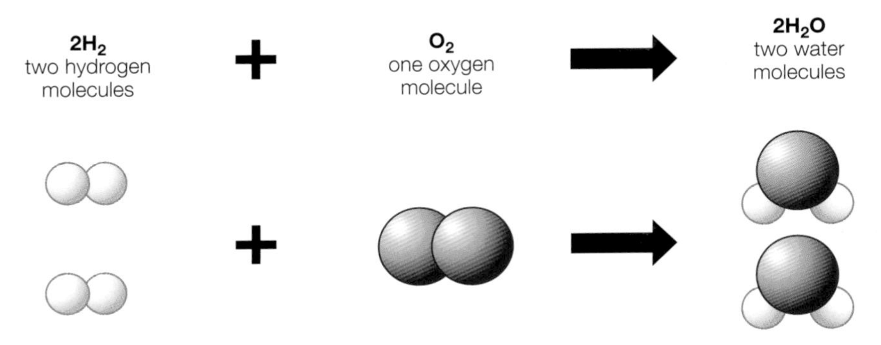

Figure 2.6

Using molecular models like those in Figure 2.6, it is possible to see which bonds break in the reactants and which bonds are formed in the products. The models also confirm that the atoms have simply been rearranged during the reaction, as there are exactly the same atoms on both sides of the equation.

Test yourself

9 Write balanced equations for the following word equations.

a) sodium + oxygen ⟶ sodium oxide
b) sodium oxide + water ⟶ sodium hydroxide
c) hydrogen + chlorine ⟶ hydrogen chloride
d) zinc + hydrochloric acid (HCl) ⟶ zinc chloride + hydrogen
e) methane (CH_4) + oxygen ⟶ carbon dioxide + water
f) iron + chlorine ⟶ iron(III) chloride

Data

10 a) Use molecular models to show what happens when methane burns in oxygen (Figure 2.7) and write an equation for the reaction.

b) Draw a diagram for this reaction similar to that for hydrogen reacting with oxygen in Figure 2.6.

Figure 2.7 ▲
When methane (natural gas) burns on a hob, it reacts with oxygen in the air to form carbon dioxide and water.

2.4 Calculations from equations

An equation is not just a useful shorthand method for describing what happens during a reaction. In industry, in medicine and whenever chemists make products from reactants, it is vitally important to know the amounts of reactants that are needed for a chemical process and the amount of product that can be obtained. Chemists can calculate these amounts using equations.

Calculating the masses of reactants and products

There are four key steps in solving problems using equations.

Step 1: Write the balanced equation for the reaction.

Step 2: Write down the amounts in moles of the relevant reactants and products.

Step 3: Convert these amounts in moles to masses.

Step 4: Scale the masses to the quantities required.

Test yourself

11 What mass of calcium oxide, CaO, forms when 25 g calcium carbonate, $CaCO_3$, decomposes on heating?

12 What mass of sulfur combines with 8.0 g copper to form copper(I) sulfide, Cu_2S?

13 What mass of sulfur is needed to produce 1 kg of sulfuric acid, H_2SO_4?

Worked example

What mass of iron can be obtained from 1 kg of iron(III) oxide (iron ore)?

Step 1: $Fe_2O_3(s) + 3CO(g) \rightarrow 2Fe(s) + 3CO_2(g)$

Step 2: $1 \text{ mol } Fe_2O_3 \rightarrow 2 \text{ mol Fe}$

Step 3: $M(Fe_2O_3) = (2 \times 55.8) + (3 \times 16)$
$= 111.6 + 48$
$= 159.6 \text{ g mol}^{-1}$

So, $159.6 \text{ g } Fe_2O_3 \rightarrow 2 \times 55.8 \text{ g Fe} = 111.6 \text{ g Fe}$

Step 4: $159.6 \text{ g } Fe_2O_3 \rightarrow 111.6 \text{ g Fe}$

$$1 \text{ g } Fe_2O_3 \rightarrow \frac{111.6 \text{ g}}{159.6} \text{ Fe} = 0.7 \text{ g Fe}$$

1 kg of iron(III) oxide produces 0.7 kg of iron.

Calculating the volumes of gases in reactions

The volume of a gas depends on three things:

- the amount of gas in moles
- the temperature
- the pressure.

At a fixed temperature and fixed pressure, the volume of a gas depends only on the amount of gas in moles. The particular kind of gas and the formula of the gas do not matter.

After thousands of measurements, chemists have found that the volume of 1 mole of every gas occupies about 24 dm^3 (24 000 cm^3) at room temperature (25 °C) and atmospheric pressure. This volume of 1 mole of gas is called the molar volume.

At 0 °C (273 K) and 1 atmosphere pressure (100 kPa), the molar volume is 22.4 $dm^3 \, mol^{-1}$ (22 400 $cm^3 \, mol^{-1}$).

So, 1 mole of oxygen, O_2, and 1 mole of carbon dioxide, CO_2, each occupy 24 dm^3 at room temperature, 2 moles of O_2 will therefore occupy 48 dm^3 and 0.5 moles of O_2 will occupy 12 dm^3 at room temperature. Notice from these simple calculations that:

$$\text{volume of gas/cm}^3 = \text{amount of gas/mol} \times \text{molar volume/cm}^3 \text{ mol}^{-1}$$

So, under laboratory conditions,

volume of gas/cm^3 = amount of gas/mol × 24 000 $cm^3\,mol^{-1}$

Avogadro was the first to suggest that equal volumes of all gases at the same temperature and pressure contain the same number of molecules. Later, scientists found that 22.4 dm^3 of all gases at s.t.p. contained 6×10^{23} molecules, and this number is now known as the Avogadro constant.

Measuring the volumes of gases in reactions

The apparatus in Figure 2.9 can be used to measure the reacting volumes of dry ammonia, NH_3, and dry hydrogen chloride, HCl.

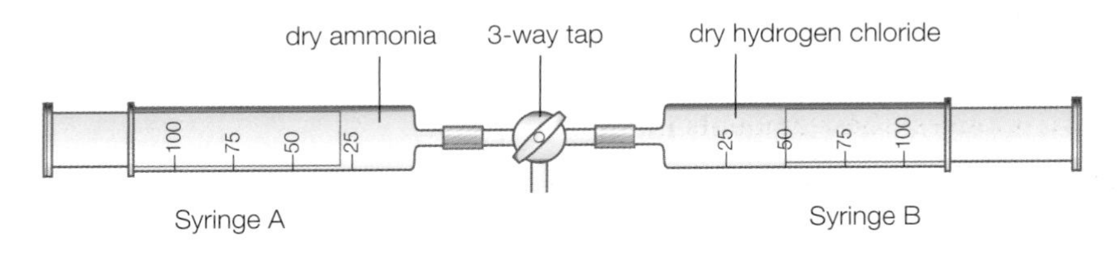

Figure 2.9 ▲
The apparatus for measuring the reacting volumes of ammonia and hydrogen chloride.

When 30 cm^3 of ammonia and 50 cm^3 of hydrogen chloride are mixed, ammonium chloride, NH_4Cl, forms as a white solid. The volume of this solid is insignificant. The volume of gas remaining is 20 cm^3, which turns out to be excess hydrogen chloride.

30 cm^3 of NH_3 reacts with 30 cm^3 of HCl

∴ 1 cm^3 of NH_3 reacts with 1 cm^3 of HCl

24 dm^3 of NH_3 reacts with 24 dm^3 of HCl

→ 1 mole of NH_3 reacts with 1 mole of HCl

Notice from this experiment that the ratio of the reacting volumes is the same as the ratio of the reacting amounts in moles. This is always the case when gases react.

$NH_3(g)$ +	$HCl(g)$	→	$NH_4Cl(s)$
	1 mol		1 mol
	1 volume		1 volume

Test yourself

14 What is the amount, in moles, of gas at room temperature and pressure in:
- a) 240 000 cm^3 chlorine
- b) 48 cm^3 hydrogen
- c) 3 dm^3 ammonia?

15 What are the volumes of the following amounts of gas at room temperature and pressure?
- a) 2 mol nitrogen
- b) 0.0002 mol neon
- c) 0.125 mol carbon dioxide?

Definition

The **molar volume** of a gas is the volume of 1 mole. At room temperature and atmospheric pressure, the molar volume of all gases is 24 $dm^3\,mol^{-1}$.

Figure 2.8 ▲
The Italian scientist Amedeo Avogadro (1776–1856).

Temperature scales

Two temperature scales are in common use – the Kelvin scale and the Celsius scale. On the Kelvin scale, temperatures are measured in units called Kelvin, symbol K. On the Celsius scale, temperatures are measured in degrees Celsius (or centigrade), symbol °C. Strictly speaking, the standard (SI) unit of temperature is the Kelvin.

Fortunately, temperatures on both scales use the same size units. Therefore:

absolute zero, −273 °C, is 0 K

0 °C is 273 K

25 °C is 298 K

100 °C is 373 K

and, in general, t °C is (273 + t) K

Gas volume calculations

Gas volume calculations are fairly easy when all the relevant substances are gases. When this is the case, the ratio of the gas volumes in the reaction is the same as the ratio of the numbers of moles in the equation.

Worked example

What volume of oxygen reacts with 60 cm^3 methane and what volume of carbon dioxide is produced if all volumes are measured at room temperature and pressure?

The equation for the reaction is:

$CH_4(g)$	+	$2O_2(g)$	→	$CO_2(g)$	+	$2H_2O(l)$
1 mol		2 mol		1 mol		insignificant volume
1 volume		2 volumes		1 volume		of liquid below 100 °C

So, 60 cm^3 methane reacts with 120 cm^3 oxygen to produce 60 cm^3 carbon dioxide.

Test yourself

16 Assuming that all gas volumes are measured at the same temperature and pressure, what volume of:

a) nitrogen forms when 2 dm^3 ammonia, NH_3, decomposes into its elements

b) oxygen is needed to react with 50 cm^3 ethane, C_2H_6, when it burns and what volume of carbon dioxide forms?

By measuring the volumes of gases involved in reactions and then converting these volumes to amounts of gas in moles, it is possible to deduce equations for reactions. This is illustrated in the next activity.

Activity

Finding an equation for the reaction of magnesium with hydrochloric acid

A small piece of magnesium was weighed and then added to excess dilute hydrochloric acid, HCl(aq), in the apparatus shown in Figure 2.10. A vigorous reaction occurred and hydrogen gas, H_2, was produced. Eventually, all the magnesium reacted and the reaction stopped.

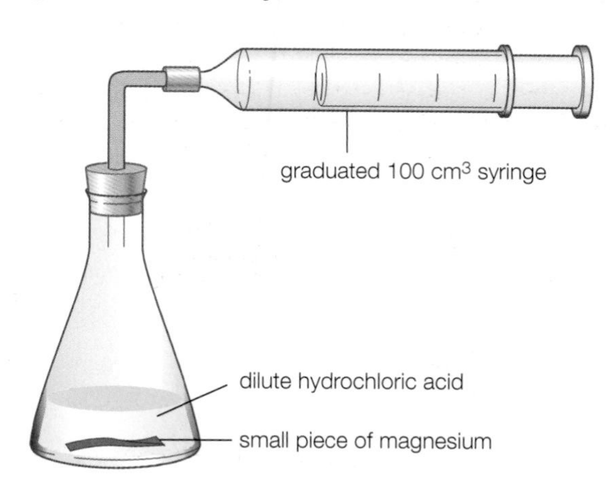

Figure 2.10 ▲ Measuring the volume of hydrogen produced when magnesium reacts with dilute hydrochloric acid.

Here are the results.

Mass of magnesium added = 0.061 g

Volume of hydrogen, H_2, produced = 60 cm^3

1 Why is it important to use excess hydrochloric acid?

2 What modifications could you make to prevent hydrogen escaping from the flask before the apparatus is re-connected after adding the magnesium?

3 How many moles of magnesium, Mg, reacted? (Mg = 24.3)

4 How many moles of hydrogen, H_2, were produced? (Assume that the molar volume of hydrogen is $24\,dm^3\,mol^{-1}$.)

5 Copy and complete the following statements.

____ mol magnesium, Mg, produced ____ mol hydrogen, H_2.

∴ 1 mol magnesium, Mg produces ____ mol hydrogen, H_2.

6 Use your result from Question 5 to write a balanced equation with state symbols for the reaction of magnesium with hydrochloric acid to form hydrogen. (Assume that magnesium chloride, $MgCl_2$, is also produced.)

Practical guidance

Test yourself

17 a) Copy and balance the equation below for the complete combustion of propane.

$C_3H_8(g)$ + _____ $O_2(g)$ → _____ $CO_2(g)$ + _____ $H_2O(g)$

b) What volume of oxygen reacts with $200\,cm^3$ of propane and what volume of carbon dioxide is produced? (Assume that all volumes are measured at room temperature and pressure.)

c) What is the mass of the carbon dioxide produced?

18 What volume of gas forms at room temperature and pressure when:

a) 0.654 g zinc reacts with excess dilute hydrochloric acid

b) 2.022 g potassium nitrate, KNO_3, decomposes on heating to potassium nitrite, KNO_2, and oxygen?

2.5 Calculating the concentration of solutions

The concentration of a solution tells us how much solute is dissolved in a certain volume of solution. It is usually measured in grams per cubic decimetre ($g\,dm^{-3}$) or moles per cubic decimetre ($mol\,dm^{-3}$). For example, a solution of sodium hydroxide that contains $1.0\,mol\,dm^{-3}$ comprises 1 mole of sodium hydroxide (40 g of NaOH) in $1\,dm^3$ ($1000\,cm^3$) of solution. Its concentration of $1.0\,mol\,dm^{-3}$ is sometimes shown as 1.0 M for short.

Concentrated solutions contain a lot of solute in a relatively small amount of solution, whereas dilute solutions have only a small amount of solute in a lot of solution. Dilute solutions usually have concentrations of $1\,mol\,dm^{-3}$ or less.

Figure 2.11 shows how to prepare a solution with a specified concentration.

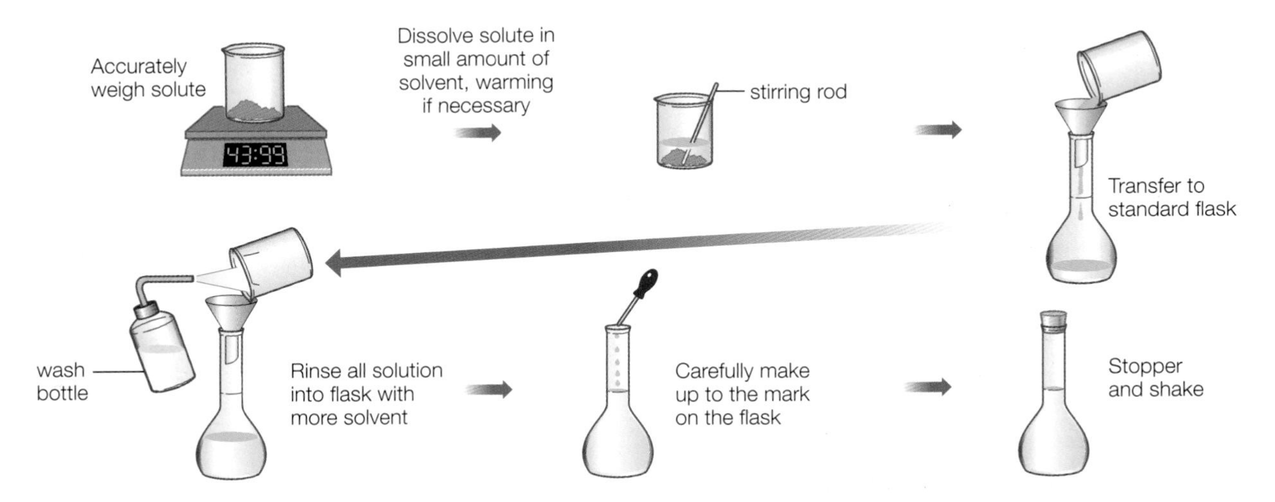

Figure 2.11 ▼ Using a standard flask to prepare a solution with a specified concentration.

Worked example

A car battery contains 2354.4 g of sulfuric acid, H_2SO_4, in $6\,dm^3$ of battery liquid. What is the concentration of sulfuric acid in:

a) $g\,dm^{-3}$

b) $mol\,dm^{-3}$?

a) $$\text{Concentration/g dm}^{-3} = \frac{\text{mass of solute/g}}{\text{volume of solution/dm}^3}$$

$$= \frac{2354.4\,g}{6\,dm^3}$$

$$= 392.4\,g\,dm^{-3}$$

b) $M(H_2SO_4) = 98.1\,g\,mol^{-1}$

$$\text{Amount of } H_2SO_4 \text{ in battery} = \frac{2354.4\,g}{98.1\,g\,mol^{-1}} = 24\,mol$$

$$\therefore \text{concentration} = \frac{\text{amount of solute/mol}}{\text{volume of solution/dm}^3}$$

$$= \frac{24\,mol}{6\,dm^3} = 4\,mol\,dm^{-3}$$

When ionic compounds dissolve, the ions separate in the solution. For example:

$$CaCl_2(s) \xrightarrow{(aq)} Ca^{2+}(aq) + 2Cl^-(aq)$$

So, if the concentration of $CaCl_2$ is $0.1\,mol\,dm^{-3}$, then the concentration of Ca^{2+} is also $0.1\,mol\,dm^{-3}$, but the concentration of Cl^- is $0.2\,mol\,dm^{-3}$.

Definition

Concentration/g dm^{-3} = $\frac{\text{mass of solute/g}}{\text{volume of solution/dm}^3}$

or

Concentration/mol dm^{-3} = $\frac{\text{amount of solute/mol}}{\text{volume of solution/dm}^3}$

Test yourself

19 Highland Spring mineral water contains 0.133 g of hydrogencarbonate, HCO_3^-, per cubic decimetre. What is this concentration in mol dm^{-3}?

20 What is the concentration in mol dm^{-3} of a solution that contains:

a) 4.25 g silver nitrate, $AgNO_3$, in 500 cm^3 solution

b) 20.75 g potassium iodide, KI, in 200 cm^3 of solution?

21 171 g of aluminium sulfate, $Al_2(SO_4)_3$, were added to 100 dm^3 of polluted river water to help with its purification.

a) What is the concentration of aluminium sulfate in the river water in mol dm^{-3}?

b) What is the concentration of the following ions in the river water in mol dm^{-3}:

i) aluminium ions, Al^{3+}

ii) sulfate ions, SO_4^{2-}?

22 What mass of solute is present in:

a) 50 cm^3 of 2.0 mol dm^{-3} sulfuric acid

b) 100 cm^3 of 0.01 mol dm^{-3} potassium manganate(VII), $KMnO_4$

c) 250 cm^3 of 0.2 mol dm^{-3} sodium carbonate, Na_2CO_3?

REVIEW QUESTIONS

1 Excess calcium reacted vigorously with 50 cm^3 of 0.1 mol dm^{-3} hydrochloric acid to produce calcium chloride solution and hydrogen.

a) Write a balanced equation with state symbols for the reaction. (3)

b) Draw a labelled diagram to show how the hydrogen gas could be collected during the reaction. (2)

c) Calculate:

i) the number of moles of hydrogen produced

ii) the volume of hydrogen produced at room temperature and pressure.

(Assume the molar volume is 24,000 $cm^3\ mol^{-1}$ at room temperature and pressure.) (3)

d) Calculate:

i) the number of moles of calcium chloride formed

ii) the mass of calcium chloride formed.

(Ca = 40, Cl = 35.5) (3)

2 a) How many molecules are present in 4.0 g of oxygen, O_2? (O = 16) (3)

b) How many ions are present in 9.4 g of potassium oxide, K_2O? (K = 39, O = 16) [Avogadro constant = $6.0 \times 10^{23}\ mol^{-1}$] (3)

3 Ammonium sulfate was prepared by adding ammonia solution to 25 cm^3 of 3.0 M sulfuric acid.

$2NH_3(aq) + H_2SO_4(aq) \rightarrow (NH_4)_2SO_4(aq)$

a) What volume of 3.0 mol dm^{-3} ammonia solution will just neutralise the sulfuric acid? (1)

b) How could you test to check that enough ammonia has been added to neutralise all of the acid? (2)

c) What is the concentration of:

i) $(NH_4)_2SO_4$

ii) NH_4^+ ions in the solution of ammonium sulfate produced? (4)

4 Balance the following equations.

a) $Cu_2S(s) + O_2(g) \rightarrow CuO(s) + SO_2(g)$ (1)

b) $FeS(s) + O_2(g) + SiO_2(s) \rightarrow FeSiO_3(s) + SO_2(g)$ (3)

c) $Fe(NO_3)_3(s) \rightarrow Fe_2O_3(s) + NO_2(g) + O_2(g)$ (3)

5 The concentration of cholesterol ($C_{27}H_{46}O$) in a patient's blood was found to be 6.0×10^{-3} mol dm^{-3}.

a) What is the molar mass of cholesterol? (2)

b) Calculate the concentration of cholesterol in the patient's blood in g dm^{-3}. (2)

c) 10 cm^3 of the patient's blood was taken so that various tests could be done. What was the mass of cholesterol in these 10 cm^3 of blood? (1)

6 Describe the experiment and the calculation you would carry out to find the percentage of water of crystallisation in hydrated iron(II) sulfate knowing that anhydrous iron(II) sulfate does not decompose on heating. (9)

3 Acids, bases and salts

There are millions upon millions of known chemicals and many of these can be classified as acids, bases or salts. Acids, bases and salts are part of everyday life: they are present in our food, in our bodies and in the soil. Acids and bases are chemical opposites. Acids donate H^+ ions, whereas bases accept them; and acids react with bases to form salts.

Figure 3.1 ▶
Acids and bases are everywhere, including the soil. Some plants, such as heathers, grow best in slightly acid soils, whereas other plants, such as beech trees, grow best in slightly basic (alkaline) soils.

3.1 Testing for acids and bases

The sharp taste of lemon juice and vinegar is caused by acids, whereas the sour flat taste of indigestion tablets, which contain magnesium hydroxide and calcium carbonate, is caused by bases. Some acids and bases can be identified by their taste, but it would be disastrous to rely on taste to identify acids and bases. Just imagine the danger of tasting an acid if it happened to be sulfuric acid or the danger of tasting a base if it happened to be sodium hydroxide.

Most acids are soluble in water, but this is not true of bases. The bases that do dissolve in water are called alkalis, and their solutions are described as alkaline.

The most convenient method of testing for acids and bases is to use indicators like litmus and universal indicator. Indicators are substances that change colour depending on the acidity or alkalinity of a solution:

- Acidic solutions turn litmus red and give an orange or red colour with universal indicator (Figure 3.2).
- Alkaline solutions turn litmus blue and give a green, blue or violet colour with universal indicator.

It would be awkward to use the colour of an indicator to describe how acidic or alkaline something is. Instead, chemists use a scale of numbers called the pH scale. The pH of a solution gives a measure of its acidity or alkalinity. On this scale:

- acids and acidic solutions have a pH lower than 7
- alkalis and alkaline solutions have a pH greater than 7
- neutral solutions, like pure water, have a pH of 7.

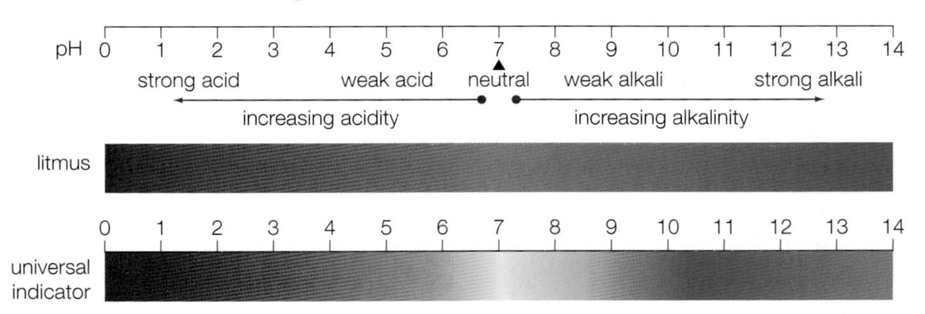

Note

Generally, alkalis are more dangerous to the eyes than acids of the same concentration. This means that you must always wear eye protection when using alkalis in the same way you would do when using acids.

Figure 3.2 ◀
Colours of litmus and universal indicator with solutions of different pH.

3.2 Bases and alkalis

Bases are substances that neutralise acids. They are the chemical opposite of acids. The largest group of common bases are metal oxides and metal hydroxides, such as sodium oxide, Na_2O, copper(II) oxide, CuO, sodium hydroxide, NaOH, and copper(II) hydroxide, $Cu(OH)_2$. Ammonia, NH_3, is another common base.

Alkalis are a special subset of bases. They are bases that are soluble in water and that form solutions with a pH greater than 7. The most common alkalis are sodium hydroxide, NaOH, calcium hydroxide, $Ca(OH)_2$, potassium hydroxide, KOH, and ammonia, NH_3. Calcium hydroxide is much less soluble than sodium hydroxide. A solution of calcium hydroxide in water is often called 'lime water'.

Sodium oxide, Na_2O, potassium oxide, K_2O, and calcium oxide, CaO, react with water to form hydroxides. The reactions of these three metal oxides with water produce alkalis. For example,

$$Na_2O(s) + H_2O(l) \rightarrow 2NaOH(aq)$$

$$CaO(s) + H_2O(l) \rightarrow Ca(OH)_2(aq)$$

Most other metal oxides and hydroxides are insoluble in water. These insoluble metal oxides and hydroxides are bases but not alkalis. The relationship between bases and alkalis is shown in the Venn diagram in Figure 3.3.

Definitions

Indicators are substances that change colour depending on how acidic or how alkaline a solution is.

The **pH scale** gives a measure of the acidity or alkalinity of a solution. The scale ranges in value from roughly 0 (very acidic) through 7 (neutral) to roughly 14 (very alkaline).

Universal indicator is the most widely used indicator. Its colour in acidic and alkaline solutions makes it possible to estimate their pH.

An **alkali** is a soluble base.

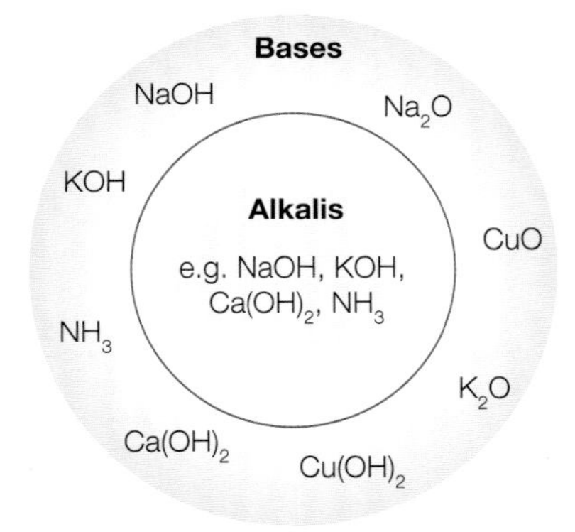

Figure 3.3 ▲
Alkalis are a subset of bases. They are soluble in water. This Venn diagram shows the relationship between bases and alkalis.

3.3 Properties of acids and bases

Solubility

Most acids and some bases are soluble in water, and many drinks contain acids and other substances dissolved in water. Fruity drinks often contain citrus fruits such as oranges, lemons and grapefruit. These fruits contain citric acid. Milk contains a small proportion of lactic acid and the proportion increases when milk goes sour. Some bases that dissolve in water to form alkalis are important in industry. Large quantities of sodium hydroxide solution are used in industry to make soaps, paper and ceramics and large amounts of ammonia are used to make fertilisers.

Figure 3.4 ▲
All fruits contain acids. These give some fruits, such as lemons, limes and oranges, a sharp taste. Other fruits, such as apples, peaches and strawberries, contain a lower concentration of acid, and the taste of the acid is masked by sugars and other constituents.

Electrolysis

Solutions of acids and bases (alkalis) conduct electricity and are decomposed by it. This shows that solutions of acids and bases contain ions. All acids produce hydrogen at the cathode when they are decomposed by electricity. This process is called electrolysis and the acid is said to be electrolysed. The formation of hydrogen at the cathode shows that acids release H^+ ions in aqueous solution. Table 3.1 shows the ions produced by some common acids and alkalis.

	Formula	Ions produced in solution
Acids		
Hydrochloric acid	HCl	$H^+(aq) + Cl^-(aq)$
Sulfuric acid	H_2SO_4	$2H^+(aq) + SO_4^{2-}(aq)$
Nitric acid	HNO_3	$H^+(aq) + NO_3^-(aq)$
Ethanoic acid	CH_3COOH	$H^+(aq) + CH_3COO^-(aq)$
Alkalis		
Sodium hydroxide	$NaOH$	$Na^+(aq) + OH^-(aq)$
Potassium hydroxide	KOH	$K^+(aq) + OH^-(aq)$
Calcium hydroxide	$Ca(OH)_2$	$Ca^{2+}(aq) + 2OH^-(aq)$

Table 3.1 ▲
The ions produced by some common acids and alkalis in aqueous solution.

In contrast to acids, the common alkalis – sodium hydroxide, potassium hydroxide and calcium hydroxide – release hydroxide ions, OH^-, in aqueous solution (Table 3.1).

Test yourself

1 a) Describe how you would check the pH of a soil sample.
  b) Why is it important for gardeners to know the pH of their soil?
2 Make a list of six different foods that contain acids and state what acid they each contain.
3 Write down the ions formed when the following dissolve in water:
  a) sulfuric acid
  b) hydrogen bromide
  c) lithium hydroxide.
4 Acids produce hydrogen at the cathode during electrolysis. Why does this suggest that acids contain H^+ ions?
5 Explain why all alkalis are bases but not all bases are alkalis.

3.4 The role of water in acidity and alkalinity

The properties of acids and alkalis described in Section 3.3 apply to their solutions in water. But, what happens when water is not present?

Dry hydrogen chloride and dry sulfuric acid have no effect on dry litmus paper or magnesium ribbon (Figure 3.5). This means that substances that we call acids do not behave like acids in the absence of water. When water is added, they become acidic straight away. Blue litmus paper turns red and bubbles of hydrogen are produced with magnesium.

Similar experiments with dry calcium hydroxide show that water must be present before it will behave like an alkali.

As a general rule, water must be present for substances to act as acids and alkalis. When acids dissolve in water, H^+ ions are formed. When the alkalis NaOH, KOH and $Ca(OH)_2$ dissolve in water, OH^- ions are formed.

In water, hydrogen ions, H^+, are attached to water molecules. Because of this, the H^+ ions are described as hydrated and are sometimes represented as $H_3O^+(aq)$ (that is, $H_2O + H^+ \rightarrow H_3O^+$) or more usually as $H^+(aq)$ for short.

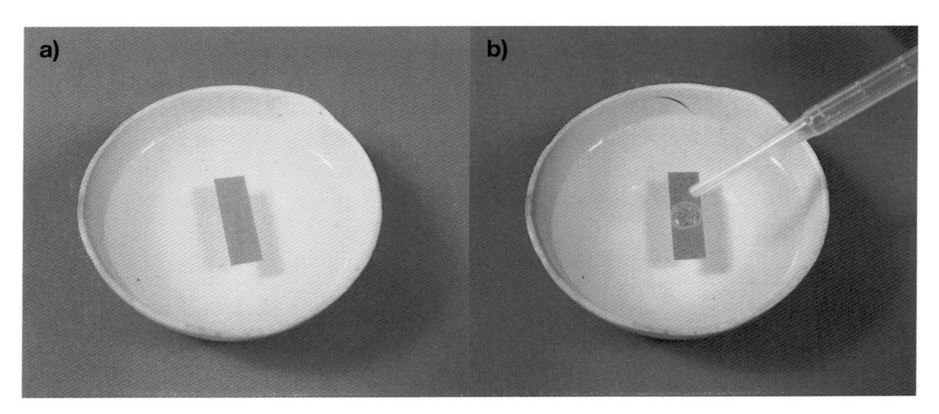

Figure 3.5 ◄
a) When dry blue litmus paper is placed on dry sulfuric acid, it remains blue.
b) When one drop of water is added, the litmus turns red.

Dry hydrogen chloride and dry sulfuric acid contain no hydrogen ions. If water is added, these substances react with water to produce hydrated H^+ ions, $H_3O^+(aq)$, which make them acidic.

$$HCl(g) + H_2O(l) \rightarrow H_3O^+(aq) + Cl^-(aq)$$

$$H_2SO_4(l) + 2H_2O(l) \rightarrow 2H_3O^+(aq) + SO_4^{2-}(aq)$$

In a similar way, alkalis like sodium hydroxide and calcium hydroxide cannot act as alkalis until hydroxide ions are present in aqueous solution.

$$NaOH(s) \xrightarrow{\text{water}} Na^+(aq) + OH^-(aq)$$

3.5 Reactions of acids with bases – neutralisation

Acids react with bases to form salts. The acidity of the acids is neutralised by the bases and this leads to the description of the reactions as neutralisations. One of the best known examples of neutralisation is the reaction between sodium hydroxide and hydrochloric acid to form sodium chloride, which is often called common salt.

$NaOH(aq)$	+	$HCl(aq)$	→	$NaCl(aq)$	+	$H_2O(l)$
sodium hydroxide		hydrochloric acid		sodium chloride (common salt)		water

Most neutralisation reactions take place between acids and the oxides and hydroxides of metals. All of these reactions produce salts, which are ionic compounds of metals with non-metals. For example, when black copper(II) oxide is added to warm sulfuric acid, the black solid disappears and a blue solution of the salt, copper sulfate, forms.

$$CuO(s) + H_2SO_4(aq) \rightarrow CuSO_4(aq) + H_2O(l)$$

When acids react with ammonia gas or aqueous ammonia, a similar reaction occurs. This results in the formation of ammonium salts containing the NH_4^+ ion. For example:

$NH_3(aq)$	+	$HCl(aq)$	→	$NH_4Cl(aq)$
ammonia		hydrochloric acid		ammonium chloride
$2NH_3(aq)$	+	$H_2SO_4(aq)$	→	$(NH_4)_2SO_4(aq)$
ammonia		sulfuric acid		ammonium sulfate

Acids also react with carbonates to form salts plus carbon dioxide and water. This explains why sulfuric acid in acid rain 'attacks' buildings made of limestone (calcium carbonate).

$CaCO_3(s)$	+	$H_2SO_4(aq)$	→	$CaSO_4(aq)$	+	$CO_2(g)$	+	$H_2O(l)$
calcium carbonate		sulfuric acid		calcium sulfate		carbon dioxide		water

Figure 3.6 ▲
These gravestones are made from limestone (calcium carbonate). Their worn and pitted appearance is caused by the action of sulfuric acid, sulfurous acid, nitric acid and carbonic acid in 'acid rain'.

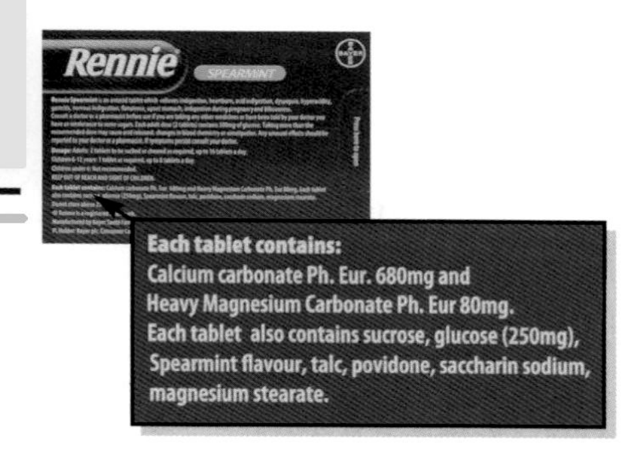

Figure 3.7 ▲
What substances in Rennie tablets help to neutralise acids in the stomach?

Neutralisation in practice

Indigestion cures

Adverts for indigestion cures usually talk of 'acid stomach and acid indigestion'. Medicines that ease stomach ache (such as Milk of Magnesia and Rennie) are called antacids, because they neutralise excess acid in the stomach. The bases in indigestion tablets include magnesium hydroxide and calcium carbonate.

Dental care

Toothpaste neutralises the acids produced when food is broken down in the mouth (Figure 3.8). Notice how the pH becomes lower and more acidic during and just after meals. This is because sugars in food are broken down into acids. These acids react with tooth enamel when the pH falls below 5.5.

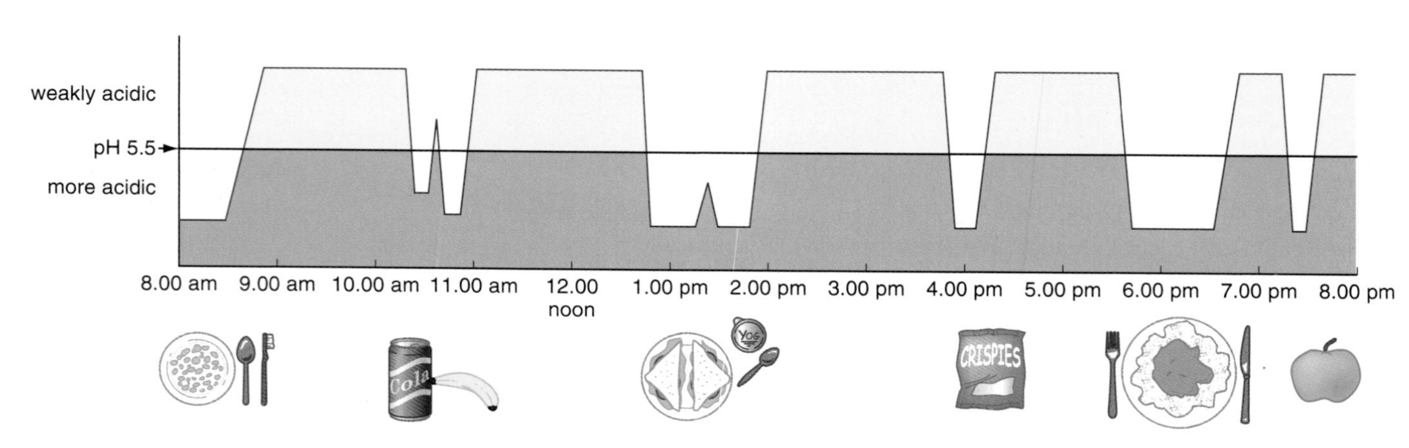

Figure 3.8 ▲
The pH in your mouth changes during the day.

Treating stings from plants and animals

The stings from certain plants and animals contain acids or bases, so they can be treated by neutralisation. For example, stings from nettles and wasps contain complex bases like histamine. They are treated with acidic substances called antihistamines. Unlike wasp stings, bee stings are acidic, so they are treated with bases.

Making fertilisers such as ammonium nitrate (Nitram)

This is a very important industrial application of neutralisation. Ammonium nitrate is manufactured by neutralising nitric acid, HNO_3, with ammonia, NH_3.

Test yourself

6 a) How does toothpaste work?
  b) Why is the pH of toothpaste important?

7 Acid rain contains sulfurous acid, $H_2SO_3(aq)$, and nitric acid, $HNO_3(aq)$, as well as sulfuric acid. Write equations, including state symbols, to show how these two acids react with limestone (calcium carbonate) to damage the stonework in some buildings.

8 a) Why can vinegar (which contains ethanoic acid, CH_3COOH) be used to remove the scale (calcium carbonate) in kettles?
  b) Why is sulfuric acid not used to descale kettles?
  c) The concentration of ethanoic acid in vinegar is about $1.0\ mol\ dm^{-3}$. What is this concentration in $g\ dm^{-3}$? (C = 12, O = 16, H = 1)

9 Complete the word equations and then write balanced chemical equations for the following reactions.
  a) ammonia + nitric acid →
  b) aluminium oxide + sulfuric acid →
  c) copper(II) carbonate + hydrochloric acid →

10 Ammonium sulfate is used as a lawn fertiliser. Describe how you would use ammonia solution to make some ammonium sulfate to use as a fertiliser.

3.6 Explaining neutralisation in terms of ions

All acids produce $H^+(aq)$ ions with water, and this is why they all react in a similar way. Hydrochloric acid, for example, contains equal numbers of $H^+(aq)$ and $Cl^-(aq)$ ions, so we can write $H^+(aq) + Cl^-(aq)$ in place of $HCl(aq)$ in equations (Table 3.1).

Similarly, sodium hydroxide and sodium chloride, as compounds of metals with non-metals, also consist of ions. This means that we can write $Na^+(aq) + OH^-(aq)$ in place of $NaOH(aq)$ and $Na^+(aq) + Cl^-(aq)$ in place of $NaCl(aq)$ in equations.

Water, carbon dioxide, ammonia and almost all compounds of non-metals (except acids and ammonium salts) consist of molecules and not ions. This means that water must be written as $H_2O(l)$, carbon dioxide as $CO_2(g)$ and ammonia as $NH_3(g)$ in equations. Our initial equation for the reaction of sodium hydroxide solution with hydrochloric acid was written as:

$$NaOH(aq) + HCl(aq) \rightarrow NaCl(aq) + H_2O(l)$$

We can now write an ionic equation for the reaction as:

> **Definition**
>
> An **ionic equation** is an equation in which ionic substances are written in terms of the ions they contain.

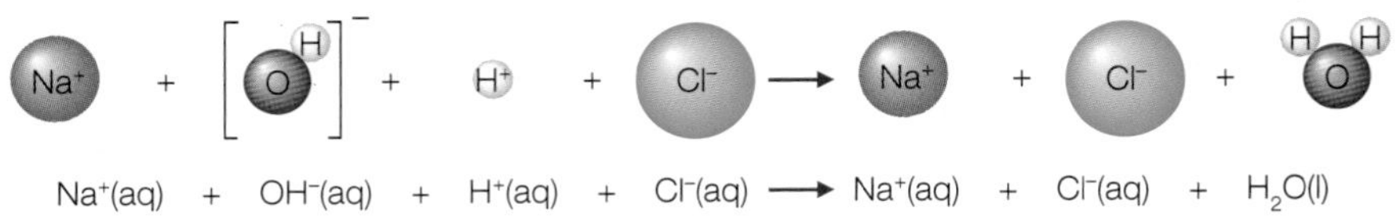

$$Na^+(aq) + OH^-(aq) + H^+(aq) + Cl^-(aq) \rightarrow Na^+(aq) + Cl^-(aq) + H_2O(l)$$

Figure 3.9 ▲
When sodium hydroxide solution reacts with hydrochloric acid, neutralisation occurs. H^+ and OH^- ions react to form water, which leaves Na^+ and Cl^- ions in the solution.

Notice that $Na^+(aq)$ and $Cl^-(aq)$ appear on both sides of the equation in Figure 3.9. These ions do not really react – they are just the same after the reaction as they were before. If we cancel $Na^+(aq)$ and $Cl^-(aq)$ on both sides of the equation, we get:

$$H^+(aq) + OH^-(aq) \rightarrow H_2O(l)$$

This final equation shows clearly what happens during the reaction. H^+ ions in the acid have reacted with OH^- ions in the base to form water. Na^+ and Cl^- ions have not changed. They have just 'stood by and watched' the reaction like spectators standing and watching a sports event. Ions that are present but take no part in the reaction, like Na^+ and Cl^- in this reaction, are called spectator ions.

Reactions of acid with metal oxides, metal carbonates and ammonia also involve ions, and we can re-write our previous equations in terms of ions.

Reactions with metal oxides

$$Cu^{2+}O^{2-}(s) + 2H^+(aq) + SO_4^{2-}(aq) \rightarrow Cu^{2+}(aq) + SO_4^{2-}(aq) + H_2O(l)$$

This time, the Cu^{2+} and SO_4^{2-} ions take no part in the reaction, and we can simplify the equation to:

$$2H^+(aq) + O^{2-}(s) \rightarrow H_2O(l)$$

Reactions with metal carbonates

$$Ca^{2+}CO_3^{2-}(s) + 2H^+(aq) + SO_4^{2-}(aq) \rightarrow Ca^{2+}(aq) + SO_4^{2-}(aq) + CO_2(g) + H_2O(l)$$

In this case, the Ca^{2+} and SO_4^{2-} are spectator ions, so the equation becomes:

$$2H^+(aq) + CO_3^{2-}(s) \rightarrow CO_2(g) + H_2O(l)$$

Reactions with ammonia

$$NH_3(aq) + H^+(aq) + Cl^-(aq) \rightarrow NH_4^+(aq) + Cl^-(aq)$$

Here, the Cl^- ions are spectators, and the net reaction is:

$$H^+(aq) + NH_3(aq) \rightarrow NH_4^+(aq)$$

Generalisations

Some important generalisations can be made from these reactions of acids.

- **Acids are H^+ ion donors, bases are H^+ ion acceptors**

In all of the reactions given above, acids donate H^+ ions to other substances. Metal oxides, metal hydroxides, metal carbonates and ammonia accept these H^+ ions. All these substances that accept H^+ ions are described and classified as bases.

Oxide ions, O^{2-}, in metal oxides and hydroxide ions, OH^-, in metal hydroxides readily accept H^+ ions from acids to form water.

$$O^{2-} + 2H^+ \rightarrow H_2O$$
$$OH^- + H^+ \rightarrow H_2O$$

Carbonate ions, CO_3^{2-}, in metal carbonates accept H^+ ions from acids to form carbon dioxide and water

$$CO_3^{2-} + 2H^+ \rightarrow CO_2 + H_2O$$

and ammonia accepts H^+ ions from acids to form ammonium salts

$$NH_3 + H^+ \rightarrow NH_4^+$$

- **Salts are ionic compounds containing metal ions or NH_4^+**

When acids react with bases, salts are produced. From the equations earlier in this section, you will see that salts are produced when the H^+ ions in acids are replaced by metal ions or NH_4^+ ions. For example, the salt copper(II) sulfate is produced when H^+ ions in H_2SO_4 are replaced by Cu^{2+} ions (Figure 3.10).

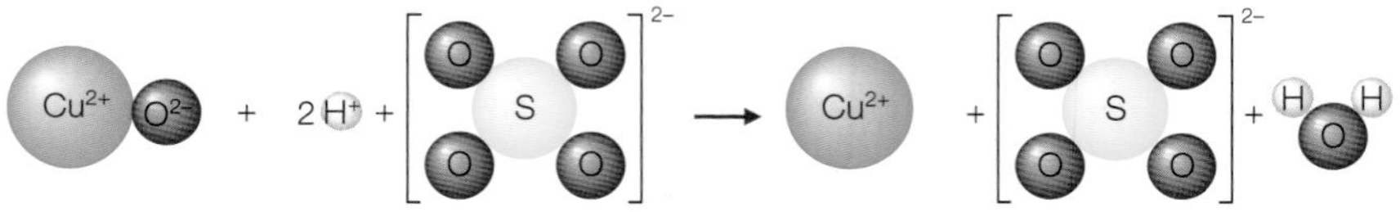

$$Cu^{2+}O^{2-}(s) + \underbrace{2H^+(aq) + SO_4^{2-}(aq)}_{\text{acid}} \rightarrow \underbrace{Cu^{2+}(aq) + SO_4^{2-}(aq)}_{\text{salt}} + H_2O(l)$$

Figure 3.10 ▲
When copper(II) oxide reacts with sulfuric acid, the salt copper(II) sulfate is formed as H^+ ions in the acid are replaced with Cu^{2+} ions.

In the same way, ammonium salts like ammonium chloride are produced when H^+ ions in HCl are replaced by NH_4^+ ions.

$$NH_3(aq) + \underbrace{H^+(aq) + Cl^-(aq)}_{\text{acid}} \rightarrow \underbrace{NH_4^+(aq) + Cl^-(aq)}_{\text{salt}}$$

Definitions

Acids are substances that donate H^+ ions.

Bases are substances that accept H^+ ions.

Spectator ions are ions that take no part in a reaction.

Definitions

Neutralisation occurs when H^+ ions from acids react with bases (such as O^{2-}, OH^-, CO_3^{2-} and NH_3) to form salts.

Salts are ionic compounds that are formed when an acid reacts with a base. H^+ ions in the acid are replaced by metal ions or NH_4^+ ions to form the salt.

Test yourself

11 Complete the following word equations and then write balanced ionic equations for the reactions:
  a) magnesium oxide + sulfuric acid → _________ + _________
  b) _________ carbonate + hydrochloric acid → zinc _________ + _________ + _________
12 a) How do indigestion tablets like Rennie work?
  b) An indigestion tablet contains magnesium hydroxide. Write an equation for the reaction of this with hydrochloric acid in someone's stomach.
  c) Re-write the equation in part b) as an ionic equation.
13 45 cm^3 of 0.2 mol dm^{-3} NaOH just react with 10 cm^3 of 0.3 mol dm^{-3} H_3PO_4 (phosphoric acid).
  a) How many moles of NaOH react?
  b) How many moles of H_3PO_4 react?
  c) How many moles of NaOH react with 1 mole of H_3PO_4?
  d) Write an equation for the reaction.
14 Suppose your best friend has missed the last few lessons and asks for help. How would you explain the differences between:
  a) acids and bases
  b) bases and alkalis
  c) bases and salts.
15 Limestone, $CaCO_3$, quicklime, CaO, and slaked lime, $Ca(OH)_2$, are all bases, but only slaked lime is an alkali. Explain why.

3.7 Salts

Salts are formed when acids react with bases and when acids react with metals (Section 4.3). Most salts contain a positive metal ion or NH_4^+ and a negative ion composed of one or two non-metals.

Salts:

- are ionic compounds
- have high melting points and boiling points
- conduct electricity when they are dissolved in water and when molten
- are often soluble in water.

The best known salt is sodium chloride (common salt). Many ores and minerals are composed of salts. These include chalk and limestone (calcium carbonate), gypsum (calcium sulfate) and copper pyrites (a mixture of copper(II) sulfide and iron(II) sulfide).

Figure 3.11 ◄ Stalagmites and stalactites in Black Spring Cave, South Wales. Stalagmites and stalactites are composed of the salt calcium carbonate. (Stalag**m**ites – **m**ount; stalac**t**ites – **t**ilt)

Definitions

A **hydrated salt** contains water as part of its structure.

Water of crystallisation is the water present in hydrated salts.

Anhydrous is the term used to describe a hydrated salt after it has lost its water of crystallisation.

Salt crystals, like those of sodium chloride, NaCl, are often formed by crystallisation from aqueous solution. When some salts, such as copper(II) sulfate, $CuSO_4.5H_2O$, crystallise from aqueous solution, water forms part of the crystal structure. This also occurs in washing soda, $Na_2CO_3.10H_2O$, and gypsum, $CaSO_4.2H_2O$. The water that forms part of the crystal structure is called the water of crystallisation, and salts that contain water of crystallisation are called hydrated salts or hydrates.

When hydrated salts are heated, they often lose their water of crystallisation, leaving a salt without water that does not decompose any further. The hydrated salt that has lost its water of crystallisation is described as anhydrous. For example:

$$\underset{\text{blue hydrated copper(II) sulfate}}{CuSO_4.5H_2O(s)} \xrightarrow{\text{heat}} \underset{\text{white anhydrous copper(II) sulfate}}{CuSO_4(s)} + 5H_2O(g)$$

Test yourself

16 Write the formulae of the ions present in the following minerals.
   a) common salt   b) chalk
   c) gypsum   d) copper pyrites.

17 When 10.80 g of hydrated iron(II) sulfate were heated, the water of crystallisation was driven off. This left 5.91 g of anhydrous iron(II) sulfate. (Fe = 56, H = 1, S = 32, O = 16)
   a) What are the masses of iron(II) sulfate, $FeSO_4$ and water, H_2O, in 10.80 g of the hydrated salt?
   b) How many moles of $FeSO_4$ and H_2O are present in 10.80 g of hydrated iron(II) sulfate?
   c) How many moles of H_2O are associated with 1 mole of $FeSO_4$ in the hydrated salt?
   d) What is the formula of hydrated iron(II) sulfate?

3.8 Finding the volumes of acids and alkalis that react

When an alkali neutralises an acid, the pH of the solution changes. We can follow this pH change by using an indicator. The indicator shows how much alkali just reacts with all the acid when there is no acid or alkali left over.

Activity

Comparing the acidity of different vinegars

Two students were asked to compare the acidity of different white wine vinegars. The acid in vinegar is ethanoic acid, CH_3COOH. The students decided to use 25 cm^3 of each vinegar and then measure how much 1.0 mol dm^{-3} sodium hydroxide just reacts with each vinegar using phenolphthalein indicator. Phenolphthalein is colourless in acid and red in alkali.

25 cm^3 of the first vinegar was measured into a conical flask using a pipette (Figure 3.12).

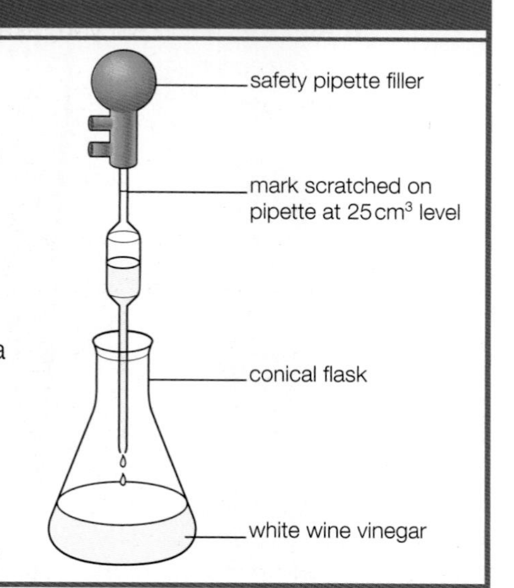

Figure 3.12 ▶ Measuring 25.0 cm^3 of vinegar into a conical flask using a pipette.

After adding 5 drops of phenolphthalein, the sodium hydroxide solution was slowly added from the burette (Figure 3.13) while the contents of the conical flask were kept well mixed. When the first permanent tinge of red appeared in the mixture, the students recorded the volume of sodium hydroxide added.

This method of adding one solution from a burette to a measured volume of another solution is called a titration. Titrations enable chemists to determine how much of the two solutions will just react.

The first titration is only a rough titration and the students then carried out three accurate titrations. To do this, they added 1 cm^3 less sodium hydroxide solution than the total amount required in their rough titration. They then added the sodium hydroxide solution one drop at a time until the first permanent tinge of pink/red appeared in the mixture.

The students' results are shown in Table 3.2, together with the results for the second white wine vinegar.

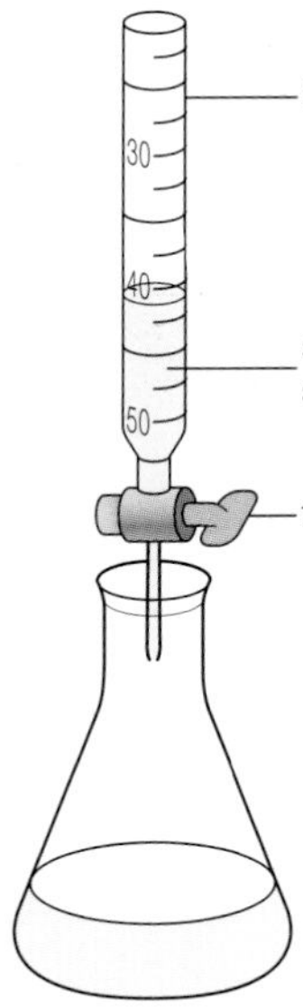

Figure 3.13 ▲ Adding sodium hydroxide solution from a burette to 25 cm^3 of white wine vinegar.

Experiment	Rough titration	Accurate titrations		
First wine vinegar				
Final burette reading/cm^3	20.50	40.15	19.60	30.30
Initial burette reading/cm^3	0.00	20.50	0.00	19.60
Second wine vinegar				
Final burette reading/cm^3	13.70	26.80	40.65	13.10
Initial burette reading/cm^3	0.00	13.70	26.80	0.00

Table 3.2 ◄

1 Why does the mixture in the conical flask suddenly change from colourless to pink/red during the addition of sodium hydroxide solution?

2 Why must the contents of the conical flask be well mixed during the titration?

3 Calculate the volumes of sodium hydroxide solution added in the rough and accurate titrations for both wine vinegars in Table 3.2.

4 Calculate an average value for the accurate titration with the first wine vinegar.

5 What did the students do to increase the reliability of their results?

6 a) What value will you take for the accurate titration with the second wine vinegar?

b) Explain why you have taken this value.

7 Compare the acidity of the two vinegars.

8 State three factors the students must control in their experiments in order to compare the two vinegars fairly.

3.9 Preparing for titrations

Titrations with glassware and other equipment – like that described in the last activity and like those used in school and college laboratories – are widely used in the food, pharmaceutical and other industries.

Many laboratories have automatic instruments for carrying out titrations, although the principles are exactly the same as in titrations in which the volumes are measured using traditional burettes and pipettes.

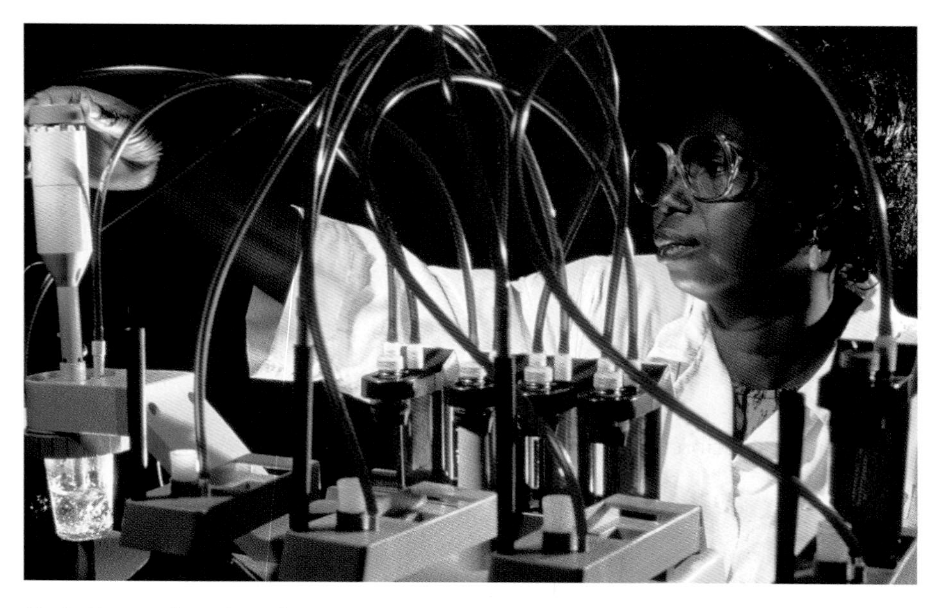

Figure 3.14 ▶
A scientist in Nigeria adjusting an automatic titration device. This is being used to check that a pharmaceutical product contains the correct amount of folic acid.

Solutions for titrations

All titrations involve two solutions. Usually, a measured volume of one solution is run into a flask from a pipette. The second solution is then added bit by bit from a burette until the colour change in an indicator, or a change in the signal from an instrument, shows that the reaction is complete. This procedure of measuring the volume of one solution that reacts with a given volume of another solution is called volumetric analysis.

Pipettes, burettes and graduated flasks allow chemists to measure out volumes of solutions very precisely during a titration and there are correct techniques for using this glassware.

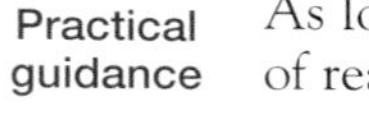

Practical guidance

Titrations will give accurate results only if the reaction between the two solutions is rapid and proceeds exactly as described in the chemical equation. As long as these two conditions apply, titrations can be used to study any type of reaction: acid–base, redox, etc.

Standard solutions

Standard solutions with accurately known concentrations are essential for titrations. A standard solution is prepared by dissolving a known mass of a chemical in water and then making the volume of solution up to a definite volume in a graduated flask (Figure 3.15).

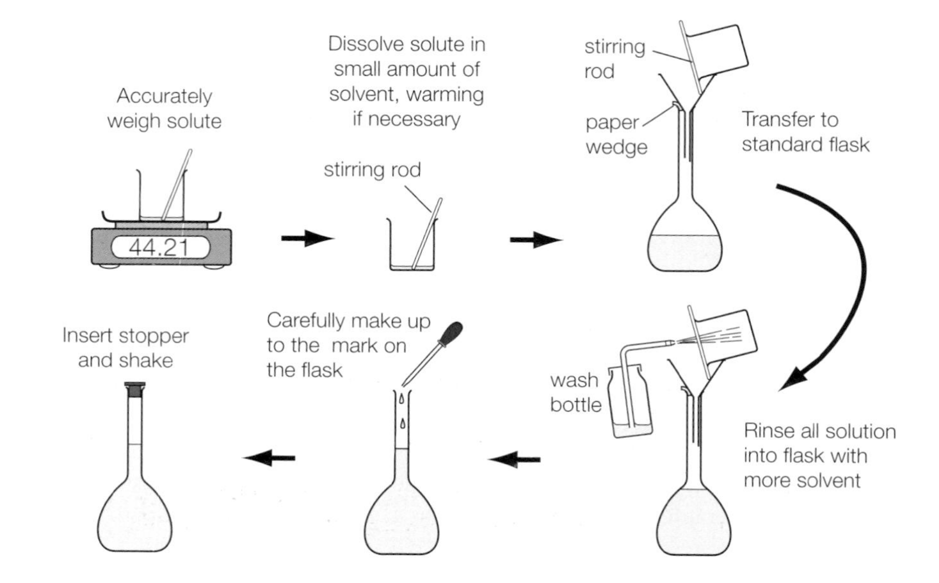

Figure 3.15 ▶
Preparing a standard solution.

Definitions

A **titration** involves adding one solution from a burette to a measured volume of another solution in order to determine how much of the two solutions just react.

A **standard solution** is a solution with an accurately known concentration.

A **primary standard** is a pure chemical that can be weighed out accurately to make up a standard solution.

This method of preparing a standard solution is only appropriate if the chemical:

- is very pure
- does not gain or lose mass when stored in air
- has a relatively high molar mass so that weighing errors are minimised.

Chemicals that meet these requirements are called primary standards. The use of primary standards means that it is possible to carry out titrations and measure the concentrations of other solutions very accurately.

Diluting solutions

Careful quantitative dilution is an important preparation before carrying out some titrations. Two common reasons for diluting solutions are:

- to make a solution of known concentration by diluting a standard solution
- to dilute a solution for analysis so that its concentration is suitable for titration.

In the procedure for dilution, a pipette (or burette) is used to run a measured volume of the more concentrated solution into a graduated flask. The flask is then filled carefully to the mark with pure water. The key to calculating the volume to use when diluting a solution is to remember that the amount, in moles, of the chemical dissolved in the diluted solution is the same as that in the volume taken from the concentrated solution.

If c is the concentration in mol dm^{-3} and V is the volume in dm^3:

- the amount, in moles, of the chemical taken from the concentrated solution is $c_c \times V_c$
- the amount, in moles, of the same chemical in the diluted solution is $c_d \times V_d$

These two amounts are the same, so $c_c \times V_c = c_d \times V_d$

Worked example

An analyst needs a 0.10 mol dm^{-3} solution of sodium hydroxide, NaOH(aq). She has a 250 cm^3 graduated flask and a supply of 0.50 mol dm^{-3} NaOH(aq). What volume of the concentrated solution should she pipette into the graduated flask?

c_c = 0.50 mol dm^{-3}
c_d = 0.10 mol dm^{-3}
V_c = to be calculated
V_d = 250 cm^3 = 0.25 dm^3

$$V_c = \frac{c_d \times V_d}{c_c}$$
$$= \frac{0.10 \text{ mol dm}^{-3} \times 0.25 \text{ dm}^3}{0.50 \text{ mol dm}^{-3}}$$
$$= 0.05 \text{ dm}^3$$
$$= 50.0 \text{ cm}^3$$

So, 50.0 cm^3 of the concentrated solution should be pipetted into the 250 cm^3 graduated flask and then made up to the mark with distilled water.

Test yourself

22 How would you prepare:

a) 0.05 mol dm^{-3} hydrochloric acid given a 1000 cm^3 graduated flask and a 1.00 mol dm^{-3} solution of the acid

b) 0.01 mol dm^{-3} sodium hydroxide solution given a 500 cm^3 graduated flask and a 0.50 mol dm^{-3} solution of the alkali?

23 What is the concentration of the solution after making up to the mark with pure water and then mixing:

a) 10.0 cm^3 of 0.01 mol dm^{-3} $AgNO_3$(aq) in a 100 cm^3 graduated flask

b) 50.0 cm^3 of 2.00 mol dm^{-3} HNO_3(aq) in a 250 cm^3 graduated flask?

Test yourself

18 Why might the following examples of analysis be important and why must the results be reliable?

a) The concentration of alcohol in blood

b) The concentration by mass of iron ore in a rock sample

c) The concentration of nitrogen oxides in the air

19 What are the key properties of a primary standard?

20 Why is sodium hydroxide unsuitable for use as a primary standard?

21 Why is anhydrous sodium carbonate a suitable primary standard but hydrated sodium carbonate unsuitable?

Figure 3.16 ▲
It is sensible to dilute antifreeze by the amount shown on the label in order to use it most efficiently.

Hint

Check dilution calculations by inspection. In the example shown, the solution has to be diluted by a factor of five, so the final volume of the diluted solution must be five times the volume of the concentrated solution pipetted into the flask.

Practical guidance

3.10 The principles of titrations

All titrations involve two solutions. A measured volume of one solution is pipetted into a conical flask. The second solution is then added bit by bit from a burette until the reaction is just complete.

Some titrations are used to investigate the stoichiometry of a reaction (that is, the relative molar amounts of the substances that react). In these experiments, the concentrations of both substances are known and the aim is to determine an equation for the reaction.

More often, titrations are used to measure the concentration of one solution when the concentration of the other solution and the equation for the reaction are known.

Figure 3.17 ▶ The apparatus used for a titration.

Figure 3.17 shows the apparatus used for a titration involving solution A and solution B. The apparatus is essentially the same for acid–base and redox titrations.

Practical guidance

Suppose that n_A moles of A react with n_B moles of B, according to the equation:

$$n_A A + n_B B \rightarrow \text{products}$$

The concentration of solution A in the burette is c_A mol dm^{-3} and that of solution B in the conical flask is c_B mol dm^{-3}.

The analyst uses a pipette to run a known volume of solution B, V_B, into the flask. Solution A is then added from the burette until an indicator shows that the reaction is just complete. This is called the end point of the titration, and the volume of solution A added, V_A, is called the titre. The analyst should now repeat the titration enough times to get at least two consistent results within 0.10 cm^3.

Definition

The **end point** of a titration is the point at which just sufficient solution has been added from the burette to react with the amount of solution pipetted into the conical flask.

Titration calculations

Although volumes are usually measured in cm^3, they should be converted to dm^3 in order to be consistent with the units of mol dm^{-3} used for concentrations.

The amount of B, in moles, pipetted into the flask = $c_B \times V_B$

The amount of A, in moles, added from the burette = $c_A \times V_A$

The ratio of these amounts must be the same as the ratio of the amounts of A and B, n_A and n_B, in the equation:

$$\therefore \quad \frac{c_A \times V_A}{c_B \times V_B} = \frac{n_A}{n_B}$$

This is sometimes called the titration formula.

Determining an equation

In titrations to determine equations, the objective is to find the ratio $\frac{n_A}{n_B}$. In this situation, the concentrations c_A and c_B are known and the volumes V_A and V_B are measured during the titration.

The ratio $\frac{n_A}{n_B}$ can be calculated using the titration formula given above.

Measuring the concentration of a solution

In these titrations, the equation for the reaction is given, so the ratio $\frac{n_A}{n_B}$ is known.

The concentration of one of the solutions is also known and the volumes V_A and V_B are measured during the titration. Substituting all the known quantities into the titration formula allows the unknown concentration to be calculated.

Worked example

41.0 g of the acid H_3PO_3 was dissolved in water and the solution was made up to 1 dm^3. 20 cm^3 of this solution just reacted with 25.0 cm^3 of 0.080 mol dm^{-3} sodium hydroxide. What is the equation for the reaction? (H = 1, O = 16, P = 31)

Concentration of NaOH(aq), c_A = 0.80 mol dm^{-3}

Volume of NaOH(aq), V_A = 25 cm^3 = 0.025 dm^3

Concentration of H_3PO_3(aq), c_B = 41 g dm^{-3}

Molar mass of H_3PO_3 = 82 g mol^{-1}

$$\Rightarrow \text{Concentration of } H_3PO_4,\ c_B = \frac{41\ \text{g dm}^{-3}}{82\ \text{g mol}^{-1}} = 0.5\ \text{mol dm}^{-3}$$

Volume of H_3PO_3(aq), V_B = 20 cm^3 = 0.02 dm^3

Using the titration formula: $\frac{c_A \times V_A}{c_B \times V_B} = \frac{n_A}{n_B}$

$$\Rightarrow \frac{0.80 \times 0.025}{0.50 \times 0.2} = 2 = \frac{n_A}{n_B}$$

So, the equation for the reaction is:

$$2NaOH(aq) + H_3PO_3(aq) \rightarrow Na_2HPO_3(aq) + 2H_2O(l)$$

Acid–base titrations

In acid–base titrations, indicators can be used to detect the end point of the reactions. Indicators are chemicals that change colour as the pH varies. Typically, an indicator completes its colour change over a range of about 2 pH units, as shown in Figure 3.18. Universal indicator is a mixture of indicators, however, and its colour changes continuously from pH 0 to pH 14.

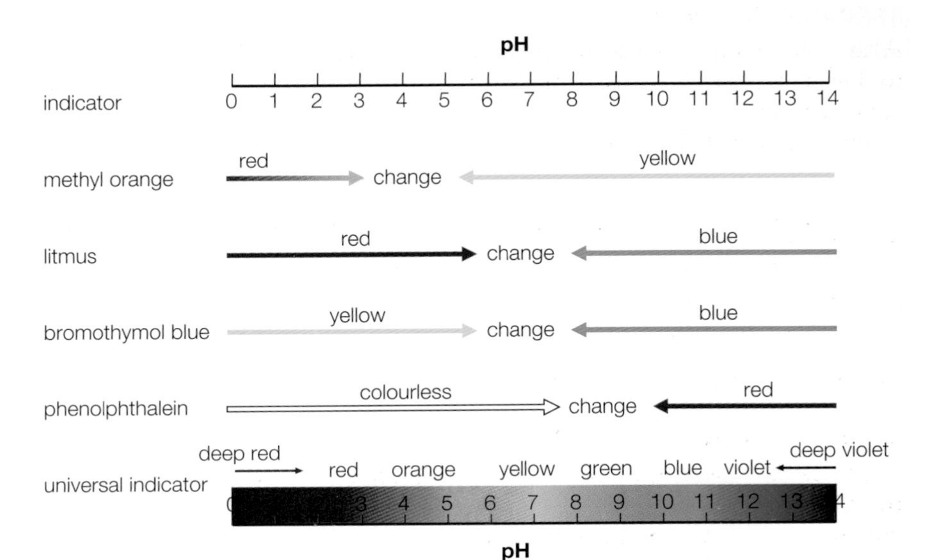

Figure 3.18 The pH ranges of some indicators.

In any acid–base titration, there is a sudden change of pH at the end point. The chosen indicator must therefore complete its colour change within the range of pH values spanned at the end point.

Test yourself

24 25 cm^3 of nitric acid was neutralised by 18.0 cm^3 of 0.15 mol dm^{-3} sodium hydroxide solution.
  a) Write an equation for the reaction.
  b) Use the titration equation to calculate the concentration of the nitric acid.

25 2.65 g of anhydrous sodium carbonate was dissolved in water and the solution was made up to 250 cm^3. In a titration, 25.0 cm^3 of this solution was pipetted into a flask and the end point was reached after adding 22.5 cm^3 of hydrochloric acid.
  a) Write an equation for the reaction.
  b) Calculate the concentration of the sodium carbonate solution in mol dm^{-3}.
  c) Calculate the concentration of the hydrochloric acid.

26 25 cm^3 of 0.050 mol dm^{-3} sodium hydroxide solution, NaOH(aq), just react with 27.5 cm^3 of 0.045 mol dm^{-3} ethanoic acid, CH_3COOH(aq).
  a) How many moles of sodium hydroxide react with one mole of ethanoic acid?
  b) Write an equation for the reaction.

Practical guidance

3.11 Evaluating titrations

There is some uncertainty in the measurements and results of every titration. It is important to assess this uncertainty because it shows how reliable the results are. Important decisions are based on the results of chemical analyses, which sometimes involve titrations in health care, in the food industry and in forensic science. It is crucial that the people making such decisions understand the extent to which they can rely on the data from analysis.

Poor accuracy but good precision

Poor accuracy and poor precision

Good accuracy and good precision

Figure 3.19 ▲ Using archery to illustrate the ideas of accuracy and precision.

Random errors in titrations

Every time an analyst carries out a titration, there are small differences in the results. This is not because the analyst has made mistakes but because of factors that are impossible to control. Unavoidable random errors arise in judging when the bottom of the meniscus is level with the graduation mark on a pipette, in judging the colour change at the end point and in taking the reading from a burette scale.

If these random errors are small, all the results will be close together. In other words, they will be precise. The precision of a set of titration results can be judged from the range in a number of repeated titrations. Precise results have a small random error.

Systematic errors in titrations

Systematic errors occur when the results differ from the true value by the same amount each time. The measurement is always too high or too low.

One source of systematic error is the tolerance allowed in the manufacture of graduated glassware. The tolerance for grade B 250 cm^3 graduated flasks is + or – 0.3 cm^3. When an analyst uses a particular flask filled correctly to the graduation mark, the volume of solution may be as little as 249.7 cm^3 or as much as 250.3 cm^3. The error is the same every time that particular flask is used. Similar tolerances are allowed for pipettes and burettes. For a class B 25 cm^{-3} pipette, the tolerance is + or – 0.05 cm^3, and for a class B 50 cm^3 burette, the tolerance is + or – 0.1 cm^3.

Systematic errors can be identified by calibrating the measuring instruments. For example, it is possible to calibrate a pipette by using it to measure out pure water and then weighing the water on an accurate balance.

Test yourself

27 Identify examples of random and systematic error when:
   a) using a pipette
   b) using a burette
   c) making up a standard solution in a graduated flask.

28 A 25 cm^3 pipette was filled to the mark with water, and the water was then run into a beaker. The beaker weighed 46.35 g empty and 71.33 g when it contained the water.
   a) What is the systematic error each time the pipette is used?
   b) Describe how you would calibrate a 250 cm^3 graduated flask for possible systematic error.

Activity

Determining the concentration of calcium hydroxide in lime water

Lime water is an aqueous solution of the alkali calcium hydroxide. It can be prepared by dissolving either calcium oxide or calcium hydroxide in water.

The concentration of calcium hydroxide in lime water can be determined by titrating a measured volume of lime water with a standard solution of hydrochloric acid using phenolphthalein as the indicator.

1 Copy and complete the results in Table 3.3.

In the flask: 25.0 cm^3 of lime water
In the burette: 0.050 mol dm^{-3} ________
The indicator was ________

Titration number	1 (rough)	2 (accurate)	3 (accurate)
Final burette reading/cm^3	22.00	42.70	22.15
Initial burette reading/cm^3	0.45	21.90	1.25
Titre/cm^3			

Table 3.3

2 Calculate the average accurate titre.

3 Before starting the titration, 3 drops of phenolphthalein were added to the lime water and then hydrochloric acid was added bit by bit.
   a) Use Figure 3.18 to identify the colour of phenolphthalein in the lime water.
   b) How would the end point of the titration be detected?

4 Write an equation for the reaction of calcium hydroxide with hydrochloric acid.

5 a) Complete the following statement.
   From the titration results, 25 cm^3 of lime water, $Ca(OH)_2(aq)$, react with ________ cm^3 of 0.050 mol dm^{-3} hydrochloric acid.
   b) Substitute values from the statement above into the following titration formula and calculate the concentration of calcium hydroxide in lime water in mol dm^{-3}.

$$\frac{c_A \times V_A}{c_B \times V_B} = \frac{n_A}{n_B}$$

6 Use the Practical guidance: Errors and uncertainty from your CD-ROM to calculate:

a) the uncertainty and percentage uncertainty in:
- i) the volume of calcium hydroxide solution pipetted
- ii) the volume of hydrochloric acid found by titration
- iii) the concentration of the standard hydrochloric acid

Practical guidance

b) the total percentage uncertainty in the concentration of calcium hydroxide in lime water.

7 Finally, write your result for the concentration of calcium hydroxide in lime water in the following forms:

a) $x \pm y$ mol dm^{-3}

b) $a \pm b$ g dm^{-3}.
(Ca = 40, O = 16, H = 1)

REVIEW QUESTIONS

Extension questions

1 Acids and bases are commonly used in our homes.

a) Baking powder contains sodium hydrogencarbonate, $NaHCO_3$, mixed with an acid.

i) When water is added, the baking powder releases carbon dioxide. Why does the reaction not occur until water is added? **(2)**

ii) The acid in baking powder can be written as H_2X. This produces two H^+ ions per molecule of H_2X. Write an equation to show the ions produced from H_2X when water is added to it. **(2)**

iii) Write an equation for the reaction between sodium hydrogencarbonate and H_2X when water is added to baking powder. **(2)**

b) Indigestion tablets contain bases that cure indigestion by neutralising excess hydrochloric acid in the stomach.

i) One type of indigestion tablet contains magnesium hydroxide. Write an equation, with state symbols, for the reaction of magnesium hydroxide with hydrochloric acid. **(2)**

ii) How does the pH in the stomach change after a person takes the tablets? **(1)**

iii) Explain why magnesium hydroxide acts as a base when it reacts with stomach acid. **(2)**

2 A solution of oven cleaner contains sodium hydroxide. 25 cm^3 of the oven cleaner was taken and titrated with hydrochloric acid containing 73 g dm^{-3} of hydrogen chloride, HCl.

a) Describe how the titration should be carried out. **(6)**

b) What safety precaution should be taken during the titration? **(1)**

c) What is the concentration of the hydrochloric acid in mol dm^{-3}? (H = 1.0, Cl = 35.5) **(2)**

d) 20 cm^3 of the hydrochloric acid just react with 25 cm^3 of the oven cleaner.

i) Write an equation with state symbols for the reaction of sodium hydroxide in the oven cleaner with hydrochloric acid. **(2)**

ii) What is the concentration of sodium hydroxide in the oven cleaner in mol dm^{-3}? **(2)**

3 The graph in Figure 3.20 shows how the pH of the solution changed when 0.1 mol dm^{-3} sodium hydroxide solution was slowly titrated with 50 cm^3 of 0.1 mol dm^{-3} ethanoic acid, CH_3COOH.

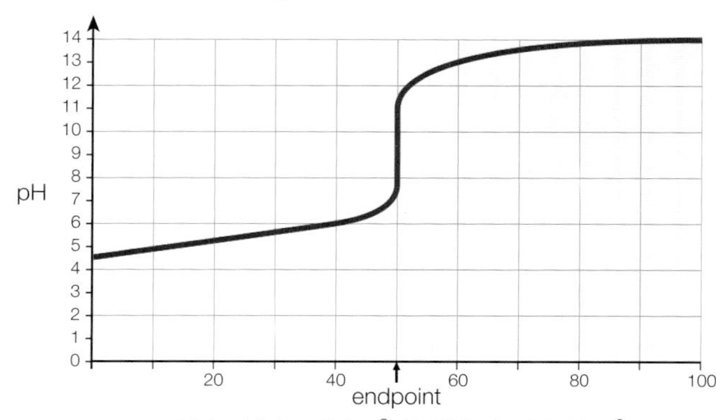

Figure 3.20 ▲
The change in pH when 0.1 mol dm^{-3} sodium hydroxide is added to 50 cm^3 of 0.1 mol dm^{-3} ethanoic acid.

a) What is the pH of the mixture when:

i) 25 cm^3 of 0.1 mol dm^{-3} sodium hydroxide have been added (1)

ii) 75 cm^3 of 0.1 mol dm^{-3} sodium hydroxide have been added? (1)

b) Imagine that the titration was carried out with universal indicator in the ethanoic acid. What colour would the mixture be when the following volumes of sodium hydroxide had been added:

i) 5.0 cm^3

ii) 60.0 cm^3? (2)

c) i) How does the pH change at the end point? (2)

ii) What would you observe at the end point? (2)

iii) What volume of sodium hydroxide is just sufficient to neutralise the ethanoic acid? (1)

d) Why would phenolphthalein, which changes colour between pH 8 and pH 10, be a suitable indicator for this titration? (2)

e) Would methyl orange, which changes colour between pH 3 and pH 5, be suitable as an indicator for this reaction? Explain your answer. (2)

4 a) Ammonium sulfate was prepared by adding ammonia solution to 25 cm^3 of 2.0 mol dm^{-3} sulfuric acid.

$2NH_3(aq) + H_2SO_4(aq) \rightarrow (NH_4)_2SO_4(aq)$

i) What volume of 2.0 mol dm^{-3} ammonia solution will just neutralise the sulfuric acid? (1)

ii) How could you test to check that enough ammonia had been added to neutralise all of the acid? (2)

b) Explain why hydrogen chloride, HCl, acts as an acid and ammonia, NH_3, acts as a base in the equation below.

$HCl(g) + NH_3(g) \rightarrow NH_4^+Cl^-(s)$ (4)

c) Which of the reactants acts as an acid in the following equation?

$HCO_3^-(aq) + H_2O(l) \rightarrow CO_3^{2-}(aq) + H_3O^+(aq)$ (1)

d) Which of the reactants acts as a base in the following equation?

$HCO_3^-(aq) + H_3O^+(aq) \rightarrow H_2CO_3(aq) + H_2O(l)$ (1)

5 A sample of limestone (impure calcium carbonate) weighing 0.20 g was reacted with 50 cm^3 of 0.10 mol dm^{-3} hydrochloric acid.

$CaCO_3(s) + 2HCl(aq) \rightarrow CaCl_2(aq) + CO_2(g) + H_2O(l)$

The hydrochloric acid was in excess, and this excess just reacted with 14 cm^3 of 0.10 mol dm^{-3} sodium hydroxide.

a) Write an equation, including state symbols, for the reaction of hydrochloric acid with sodium hydroxide. (2)

b) What excess volume of 0.10 mol dm^{-3} hydrochloric acid reacted with the 14 cm^3 of 0.10 mol dm^{-3} sodium hydroxide? (1)

c) i) What volume of 0.10 mol dm^{-3} hydrochloric acid reacted with the calcium carbonate in the limestone? (1)

ii) How many moles of hydrochloric acid reacted with the calcium carbonate? (1)

iii) How many moles of calcium carbonate reacted with the hydrochloric acid? (1)

iv) What mass of calcium carbonate reacted? (Ca = 40, C = 12, O = 16) (1)

d) What is the percentage by mass of calcium carbonate in the limestone? (1)

4 Redox

Classification of chemical reactions helps chemists make sense of the many different changes they study. An important class of reactions includes those that involve oxidation and reduction. Acid–base reactions involve the loss or gain of hydrogen ions. In the case of oxidation and reduction reactions, it is electrons that are lost or gained. Chemists have invented a system of oxidation numbers to keep track of the numbers of electrons involved.

4.1 Oxidation and reduction

Oxidation and reduction reactions are very common. Chemists have devised a number of ways of recognising and describing what happens during reactions of this kind.

Burning is perhaps the most common example of oxidation. Another example is rusting, which converts iron to a form of iron oxide. At its simplest, oxidation involves adding oxygen to an element or compound.

Reduction is the opposite of oxidation. Metal oxides are reduced during the extraction of metals from their ores. In a blast furnace, for example, carbon monoxide takes oxygen away from iron oxide to leave metallic iron (Figure 4.1).

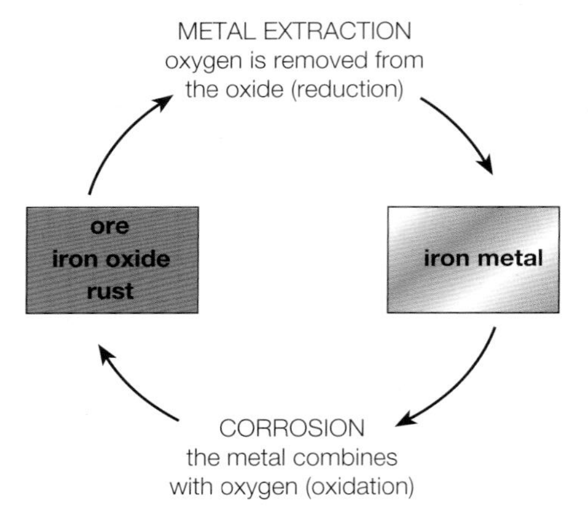

Figure 4.1 ▶
The cycle of extraction and corrosion for iron.

Test yourself

1 Which element or compound is oxidised and which is reduced in the reaction of:
  a) steam with hot magnesium
  b) copper(II) oxide with hydrogen
  c) aluminium with iron(III) oxide
  d) carbon dioxide with carbon to form carbon monoxide.

Electron transfer

Magnesium burns brightly in air. The product is a white solid: the ionic compound magnesium oxide, $Mg^{2+}O^{2-}$.

$$2Mg(s) + O_2(g) \rightarrow 2Mg^{2+}O^{2-}(s)$$

During the reaction, each magnesium atom gives up two electrons and turns into a magnesium ion:

$$2Mg \rightarrow 2Mg^{2+} + 4e^-$$

Oxygen takes up the electrons from the magnesium to produce oxide ions:

$$O_2 + 4e^- \rightarrow 2O^{2-}$$

In this way, electrons transfer from magnesium atoms to oxygen atoms and form ions from atoms.

Magnesium atoms also turn into ions when they react with other non-metals such as chlorine, bromine and sulfur.

In all of magnesium's reactions with non-metals, it loses electrons and its atoms become positive ions. The non-metals gain electrons and become negative ions. All of the reactions involve electron transfer. Magnesium is oxidised as it loses electrons, and the non-metal is reduced as it gains electrons.

Definition

Reduction and **ox**idation always go together, hence the term **redox** reaction.

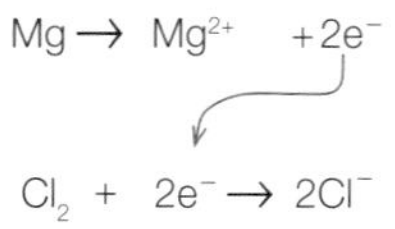

$$Mg \rightarrow Mg^{2+} + 2e^-$$

$$Cl_2 + 2e^- \rightarrow 2Cl^-$$

Figure 4.2 ▲
Electron transfer in the reaction of magnesium with chlorine.

Ionic half-equations

A half equation is an ionic equation used to describe either the gain or loss of electrons during a redox process. Half equations help to show what is happening during a reaction. Two half equations combine to give the overall balanced equation.

Zinc metal can reduce copper ions to copper. This can be shown as two half equations:

electron gain (reduction): $Cu^{2+}(aq) + 2e^- \rightarrow Cu(s)$

electron loss (oxidation): $Zn(s) \rightarrow Zn^{2+}(aq) + 2e^-$

When these two half equations are added together, they produce the full ionic equation. The electrons on each side must cancel out so. The electrons gained on one side of the equation equal the electrons lost on the other side.

$Cu^{2+}(aq) + 2e^- \rightarrow Cu(s)$

$Zn(s) \rightarrow Zn^{2+}(aq) + 2e^-$

$Cu^{2+}(aq) + Zn(s) \rightarrow Cu(s) + Zn^{2+}(s)$

Note

You may find the mnemonic **oil rig** helpful.

Oxidation	**R**eduction
Is	**I**s
Loss	**G**ain

Definitions

Ionic equations describe chemical changes by showing only the reacting ions and leaving out spectator ions.

A **half-equation** is an ionic equation used to describe the gain or loss of electrons during a redox reaction.

Test yourself

2 Write the separate ionic half equations for the reaction of:
 a) sodium with chlorine
 b) zinc with oxygen
 c) calcium with bromine.

3 Write the ionic half equations and the full ionic equation for the reaction of zinc with silver nitrate solution.

Data

4.2 Oxidation numbers

Chemists use oxidation numbers to keep track of the electrons transferred or shared during chemical changes. Oxidation numbers make it much easier to recognise redox reactions. Oxidation numbers also provide a useful way of organising the chemistry of elements such as chlorine, which can be oxidised or reduced to varying degrees. Chemists have also chosen to base the names of inorganic compounds on oxidation numbers.

Oxidation numbers and ions

Oxidation numbers show how many electrons are gained or lost by an element when atoms turn into ions and vice versa. In Figure 4.3 (on page 52), movement up the diagram involves the loss of electrons and a shift to more positive oxidation numbers – this is oxidation. Movement down the diagram involves the gain of electrons and a shift to less positive or more negative oxidation numbers – this is reduction.

The oxidation numbers of the elements are zero. In a simple ion, the oxidation number of the element is the charge on the ion. In calcium chloride, for example, the metal is present as the Ca^{2+} ion and the oxidation number of calcium is +2.

Oxidation numbers distinguish between the compounds of elements such as iron, which can exist in more than one oxidation state. In iron(II) chloride, the Roman number II shows that iron is in oxidation state +2. Iron atoms lose two electrons when they react with hydrogen chlorine to make iron(II) chloride.

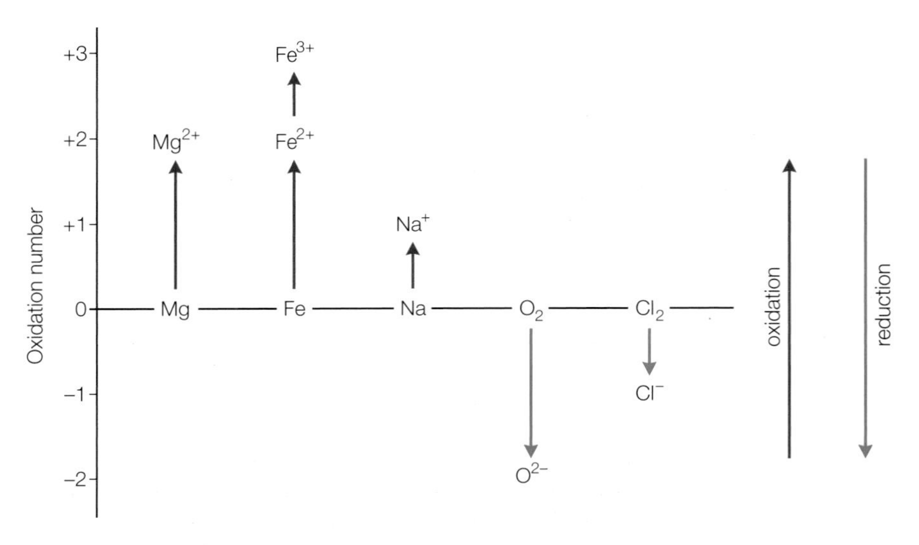

Figure 4.3 ▲
Oxidation numbers of atoms and ions.

Oxidation number rules

1. The oxidation number of uncombined elements is zero.
2. In simple ions the oxidation number of the element is the charge on the ion.
3. The sum of the oxidation numbers in a neutral compound is zero.
4. The sum of the oxidation numbers for an ion is the charge on the ion.
5. Some elements have fixed oxidation numbers in all their compounds.

Metals		**Non-metals**	
group 1 metals (e.g. Li, Na, K)	+1	hydrogen (except in metal hydrides, H^-)	+1
group 2 metals (e.g. Mg, Ca, Ba)	+2	fluorine	−1
aluminium	+3	oxygen (except in peroxides, O_2^{2-}, and compounds with fluorine)	−2
		chlorine (except in compounds with oxygen and fluorine)	−1

Figure 4.4 ▲
Oxidation number rules.

NH_4^+	MnO_4^-
−3 +1	+7 −2
SO_4^{2-}	$Cr_2O_7^{2-}$
+6 −2	+6 −2

Figure 4.5 ▲
Oxidation numbers in ions with more than one atom. Note that 2− is used for the electric charge on a sulfate ion (number first for ionic charges) but that −2 is used to refer to the oxidation state of oxygen in the ion (charge first for oxidation states in ions and molecules).

With the help of the rules in Figure 4.4, it is possible to extend the use of oxidation numbers to ions that consist of more than one atom. The charge on an ion, such as the sulfate ion, is the sum of the oxidation numbers of the atoms. The normal oxidation state of oxygen is −2. There are four oxygen atoms (four with an oxidation state of −2) in the sulfate ion, so the oxidation state of sulfur must be +6 to give an overall charge on the ion of −2 (Figure 4.5).

Test yourself

4 What is the oxidation number of:
a) aluminium in Al_2O_3
b) nitrogen in magnesium nitride, Mg_3N_2
c) nitrogen in barium nitrate, $Ba(NO_3)_2$
d) nitrogen in the ammonium ion, NH_4^+?

Definitions

Oxidation originally meant combination with oxygen, but the term now covers all reactions in which atoms, molecules or ions lose electrons. The definition is extended to cover molecules, as well as atoms and ions, by defining oxidation as a change that makes the oxidation number of an element more positive or less negative.

Reduction originally meant removal of oxygen or addition of hydrogen but the term now covers all reactions in which atoms, molecules or ions gain electrons. The definition is extended to cover molecules, as well as atoms and ions, by defining oxidation as a change that makes the oxidation number of an element more negative or less positive.

Oxidation numbers and molecules

The rules in Figure 4.4 make it possible to apply the definitions of oxidation and reduction to molecules. In most molecules, the oxidation state of an atom corresponds to the number of electrons from that atom that are shared in covalent bonds.

When two atoms are linked by covalent bonds, the more electronegative atom (Section 6.11) has the negative oxidation state. Fluorine always has a negative oxidation state of −1 because it is the most electronegative of all atoms. Oxygen normally has a negative oxidation state (−2), but it has a positive oxidation state (+2) when it is combined with fluorine.

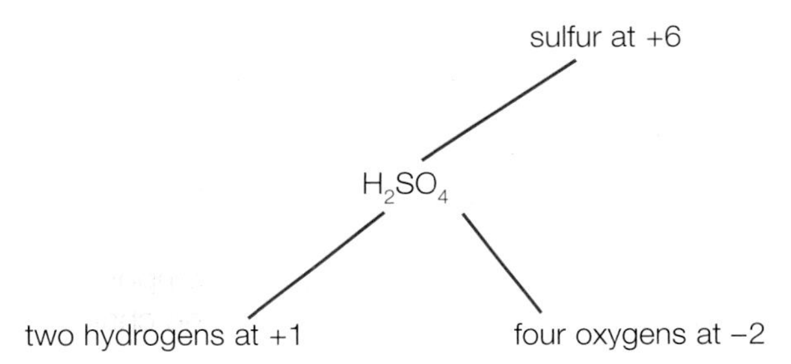

Figure 4.6 ◄
Oxidation numbers of the elements in sulfuric acid.

The reason oxidation numbers are written as +1, +2 and so on is to make quite clear that they do not refer to electric charges when dealing with molecules. Molecules are not charged, and the sum of the oxidation states for all of the atoms in a molecule is zero.

Test yourself

5 Are the following elements oxidised or reduced in these conversions?
a) calcium to calcium bromide
b) chlorine to lithium chloride
c) chlorine to chlorine dioxide
d) sulfur to hydrogen sulfide
e) sulfur to sulfuric acid

Oxidation numbers and the chemistry of elements

Oxidation numbers help to make sense of the chemistry of an element such as bromine (see Figure 4.7). A reaction that turns bromine into BrO^- ions involves oxidation of bromine. Conversion of BrO^- ions to BrO_3^- ions involves further oxidation of bromine.

The oxidation numbers of the elements lithium to chlorine in their oxides show a periodic pattern when plotted against atomic number (Figure 4.8). The most positive oxidation number for each element corresponds to the number of electrons in the outer shell of the atoms.

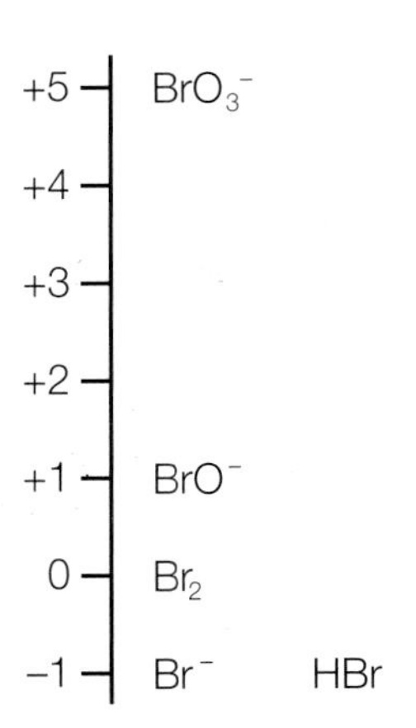

Figure 4.7 ▲
Oxidation numbers for bromine molecules and ions showing the four common oxidation states of the element.

Definition

The degrees of oxidation or reduction shown by an element are its **oxidation states**. The states are labelled with the oxidation number of the element in each state.

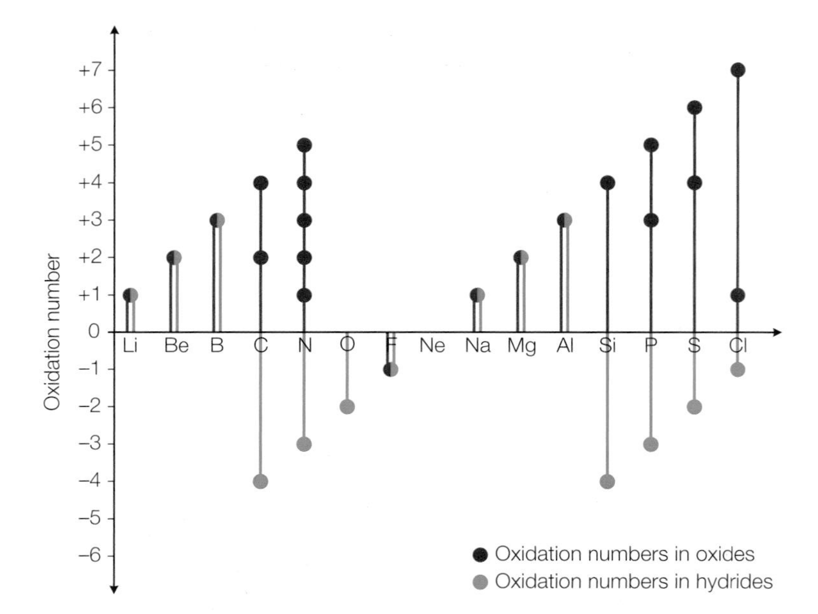

Figure 4.8 ▶
Oxidation numbers of elements in their oxides and hydrides. Note that some elements form oxides in a variety of oxidation states.

Test yourself

6 This question refers to Figure 4.8.
a) Give the formula of the oxide of lithium.
b) Give the formulae of the two oxides of carbon.
c) What are the oxidation states of nitrogen in these oxides: NO, N_2O, NO_2, N_2O_3 and N_2O_5?
d) Give the formulae of the hydrides of nitrogen and phosphorus.
e) Why is there only one element with a negative oxidation number in an oxide?

Oxidation numbers and the names of compounds

The names of inorganic compounds are becoming increasingly systematic, but chemists still use a mixture of names. Most chemists prefer the name 'copper sulfate' for the blue crystals with the formula $CuSO_4.5H_2O$. This is hydrated copper(II) sulfate. Its fully systematic name, tetraaquocopper(II) tetraoxosulfate(VI)-1-water, is used rarely. The systematic name has much more to say about the arrangement of atoms, molecules and ions in the blue crystals, but it is too cumbersome for normal use. The systematic name also shows the oxidation states of copper and sulfur in the compound.

The following list gives some of the basic rules for naming common inorganic compounds:

- The ending '-ide' shows that a compound contains just two elements mentioned in the name. The more electronegative element comes second – as in sodium sulfide, Na_2S, carbon dioxide, CO_2, and magnesium nitride, Mg_3N_2.
- The Roman numbers in names indicate the oxidation numbers of the elements – for example, iron(II) sulfate, $FeSO_4$, and iron(III) sulfate, $Fe_2(SO_4)_3$.
- The traditional names of oxoacids end in '-ic' or '-ous' as in sulfuric, H_2SO_4, sulfurous, H_2SO_3, nitric, HNO_3, and nitrous, HNO_2, acids. The '-ic' ending is for acids in which the central atom has the higher oxidation number.
- The corresponding traditional endings for the salts of oxoacids are '-ate' and '-ite', as in sulfate, SO_4^{2-}, sulfite, SO_3^{2-}, nitrate, NO_3^-, and nitrite, NO_2^-.
- The more systematic names for oxoacids and oxo salts use oxidation numbers, as in sulfate(VI) for sulfate, SO_4^{2-}, and sulfate(IV) for sulfite, SO_3^{2-}.

When in doubt, chemists give the name and the formula. In some cases, they may give two names: the systematic name and the traditional name.

Test yourself

7 Write the formulae of the following compounds:
   a) tin(II) oxide
   b) tin(IV) oxide
   c) iron(III) nitrate(V)
   d) potassium chromate(VI).

8 Give the systematic name for these ions:
   a) nitrate, NO_3^-
   b) nitrite, NO_2^-.

4.3 Recognising redox reactions

Oxidation numbers help to identify redox reactions. In the equation for any redox reaction, at least one element changes to a more positive oxidation state, while another changes to a less positive oxidation state. A reaction is not a redox reaction if no changes in oxidation states take place.

Oxidising and reducing agents

An agent is someone or something that gets things done. In spy stories, the main players are secret agents with a mission to make a change. In redox reactions, the chemicals with a mission are oxidising and reducing agents.

The term oxidising agent (or oxidant) describes chemical reagents that can oxidise other atoms, molecules or ions by taking electrons away from them. Common oxidising agents are oxygen, chlorine, nitric acid, potassium manganate(VII), potassium dichromate(VI) and hydrogen peroxide.

The term reducing agent (or reductant) describes chemical reagents that can reduce other atoms, molecules or ions by giving them electrons. Common reducing agents are hydrogen, sulfur dioxide and metals such as zinc or iron in acid.

It is easy to get into a mental tangle when using these terms. When an oxidising agent reacts, it is reduced. When a reducing agent reacts, it is oxidised. This is illustrated by the reaction of magnesium with chlorine (Figure 4.9).

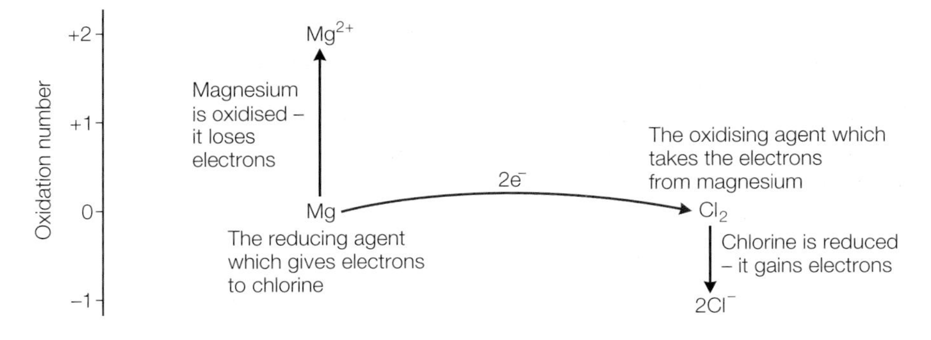

Figure 4.9 Magnesium is oxidised by loss of electrons. It is oxidised by chlorine, so chlorine is the oxidising agent. At the same time, chlorine gains electrons and is reduced by magnesium. Magnesium is the reducing agent.

Reactions of metals with dilute acids

The more reactive metals such as magnesium, zinc and iron react with dilute acids. These are redox reactions. The hydrogen ions in the solution of the acid react with the metal atoms.

The metals are oxidised as they dissolve to form aqueous positive ions. The hydrogen ions in the dilute acids are reduced to hydrogen gas.

$$\overset{0}{Zn}(s) + \underset{+1}{H_2}SO_4(aq) \rightarrow \overset{+2}{Zn}SO_4(aq) + \underset{0}{H_2}(g)$$

Test yourself

9 Write equations for the following reactions and use oxidation numbers to show that they are redox reactions.
a) Magnesium with dilute sulfuric acid
b) Zinc with dilute hydrochloric acid
c) Lithium with dilute hydrochloric acid

10 Classify these reactions of calcium and its compounds as redox, acid–base, precipitation or thermal decomposition.
a) $CaCl_2(aq) + K_2SO_4(aq) \rightarrow CaSO_4(s) + 2KCl(aq)$
b) $CaCO_3(s) \rightarrow CaO(s) + CO_2(g)$
c) $Ca(s) + 2H_2O(l) \rightarrow Ca(OH)_2(aq) + H_2(g)$
d) $Ca(OH)_2(s) + 2HCl(aq) \rightarrow CaCl_2(aq) + 2H_2O(l)$
e) $Ca(s) + 2HCl(aq) \rightarrow CaCl_2(aq) + H_2(g)$

Tutorial

4.4 Balancing redox equations

Oxidation numbers help in balancing redox equations, because the total decrease in oxidation number for the element reduced must equal the total increase in oxidation number for the element oxidised. This is illustrated below through the oxidation of hydrogen bromide by concentrated sulfuric acid. The main products are bromine, sulfur dioxide and water.

Step 1: Write down the formulae for the atoms, molecules and ions involved in the reaction

$$HBr + H_2SO_4 \rightarrow Br_2 + SO_2 + H_2O$$

Step 2: Identify the elements that change in oxidation number and the extent of change.

In this example, only bromine and sulfur show changes of oxidation state.

change of +1 (HBr → Br_2)

$$HBr + H_2SO_4 \rightarrow Br_2 + SO_2 + H_2O$$

change of −2 (H_2SO_4 → SO_2)

Step 3: Balance so that the total increase in oxidation number of one element equals the total decrease of the other element.

In this example, the increase of +1 in the oxidation number of two bromine atoms balances the −2 decrease of one sulfur atom.

$$2HBr + H_2SO_4 \rightarrow Br_2 + SO_2 + H_2O$$

Step 4: Balance for oxygen and hydrogen.

In this example, the four hydrogen atoms on the left of the equation join with the two remaining oxygen atoms to form two water molecules.

$$2HBr + H_2SO_4 \rightarrow Br_2 + SO_2 + 2H_2O$$

Step 5: Add state symbols.

$$2HBr(g) + H_2SO_4(l) \rightarrow Br_2(l) + SO_2(g) + 2H_2O(l)$$

Test yourself

11 Write balanced equations for these redox reactions. State which element is oxidised and which is reduced in each example.
a) Fe reacts with Br_2 to give iron(III) bromide
b) Fe reacts with $H_2SO_4(aq)$ to give iron(II) sulfate and hydrogen
c) F_2 reacts with H_2O to give HF and oxygen
d) Ba reacts with HCl(aq) to form $BaCl_2(aq)$ and hydrogen

Activity

Preparing a sample of an oxide of nitrogen

Lead(II) nitrate decomposes on heating and gives three products: lead(II) oxide, nitrogen dioxide and oxygen. Nitrogen dioxide, NO_2, is a brown gas. Some of the molecules of nitrogen dioxide pair up to form N_2O_4. Cooling the gas mixture condenses the N_2O_4 as a greenish liquid.

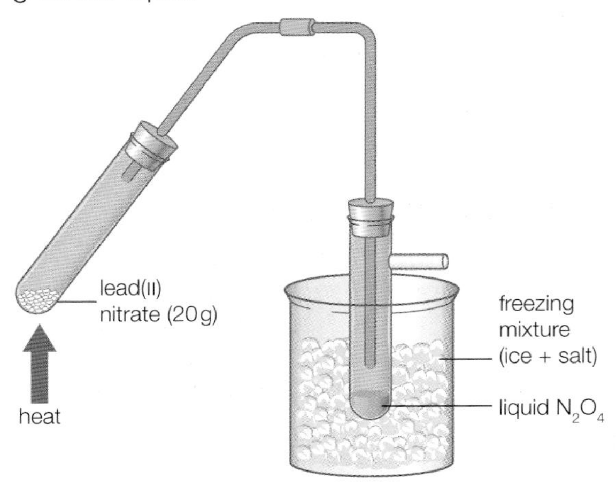

Figure 4.10 ◄
Heating lead(II) nitrate to collect a sample of N_2O_4.

1 Give the oxidation states of all the elements in:

a) lead(II) nitrate

b) lead(II) oxide

c) nitrogen dioxide

d) N_2O_4

e) oxygen gas.

2 Identify as oxidation, reduction or neither, the formation of:

a) lead(II) oxide from lead(II) nitrate

b) nitrogen dioxide from lead(II) nitrate

c) oxygen from lead(II) nitrate

d) N_2O_4 from NO_2.

3 Write balanced equations for:

a) the decomposition of lead(II) nitrate

b) the formation of N_2O_4 from NO_2.

4 Calculate the theoretical mass of N_2O_4 that could be collected by condensing the gases given off when 20 g lead(II) nitrate decomposes.

5 Suggest why the mass of N_2O_4 collected by heating 20 g lead(II) nitrate in the apparatus in Figure 4.10 would be less than your answer to Question 4.

REVIEW QUESTIONS

1 These are incomplete half equations for changes involving reduction. Complete and balance the half equations. You need to add the electrons.

a) $H^+(aq) \rightarrow H_2(g)$ (1)

b) $Fe^{3+}(aq) \rightarrow Fe^{2+}(aq)$ (1)

c) $Br_2(aq) \rightarrow 2Br^-(aq)$ (1)

2 These are incomplete half equations for changes involving oxidation. Complete and balance the half equations. You need to add the electrons.

a) $Na(s) \rightarrow Na^+(aq)$ (1)

b) $Zn(s) \rightarrow Zn^{2+}(aq)$ (1)

c) $I^-(aq) \rightarrow I_2(aq)$ (1)

3 Pick a reduction from Question 1 and an oxidation from Question 2 and combine them to give the full ionic equation for these reactions.

a) Bromine with iodide ions (1)

b) Iron(III) ions with iodide ions (2)

c) Hydrogen ions with zinc (2)

4 What are the oxidation numbers of chlorine in these ions: Cl^-, ClO^-, ClO_2^-, ClO_3^- and ClO_4^-? (5)

5 What are the oxidation numbers of nitrogen in these molecules: N_2, NH_3, N_2H_4, HNO_3, HNO_2, NH_2OH and NF_3? (7)

6 Give one example of a reaction to illustrate the statements in **a)** and **b)**. For each example, name the reactants and products, write an equation and show that it represents a redox reaction.

a) Metals are reducing agents that form positive ions when their atoms react. (3)

b) Non-metals are oxidising agents that generally form negative ions when their atoms react. (3)

c) Identify, with an example of a reaction, one exception to the general rule in part **b)**. (3)

5 Electronic structure

Since Rutherford proposed his nuclear model of atomic structure at the beginning of the twentieth century, scientists have been interested in the precise arrangement of electrons in different elements. Scientists have used measurements from mass spectrometers and emission spectra in the visible and ultraviolet regions of the electromagnetic spectrum to piece together an accurate description of the electronic structures of all elements. The patterns of electronic structure that have emerged are closely related to the positions of elements in the periodic table, and it is clear that the chemical properties of all elements are determined by their electronic structures.

5.1 Evidence for the electronic structure of atoms

When elements are investigated in a mass spectrometer, they must first be converted to positive ions in the gaseous or vapour state. Inside a mass spectrometer, a high vacuum allows the element being tested to vaporise more easily. In most cases, the element must also be heated to a high temperature in order to obtain individual gaseous atoms. Positive ions are then produced by bombarding the gaseous atoms with a beam of high-energy electrons. If a high-energy electron, e^-, hits an atom, X, it may knock an electron out of X leaving a positive ion, X^+.

e^-	+	$X(g)$	$\rightarrow$	$X^+(g)$	+	e^-	+	e^-
high-energy electron		gaseous atom		positive ion		electron knocked out of X		high-energy electron retreating

By varying the intensity of the beam of high-energy electrons, it is possible to measure the minimum amount of energy needed to remove electrons from the atoms of different elements. These measurements allow scientists to predict the electronic structures of atoms.

The energy needed to remove one electron from each atom in a mole of gaseous atoms is known as the first ionisation energy. It is given the symbol ΔH_{i1}. The product is one mole of gaseous ions with one positive charge.

The first ionisation energy of sodium is the energy required for the process:

$$Na(g) \rightarrow Na^+(g) + e^- \quad \Delta H_{i1} = +496\,kJ\,mol^{-1}$$

Figure 5.1 ◄ Neon lights in Las Vegas, United States.

In neon advertising signs, gaseous neon atoms are continually bombarded by electrons. This produces positive neon ions, which then reform as neon atoms. This is what happens when scientists measure ionisation energies.

Scientists can also determine ionisation energies from the emission spectra of atoms. Using data from spectra, it is possible to measure the energy needed to remove electrons from ions with increasing positive charges. A succession of ionisation energies, represented by the symbols ΔH_{i1}, ΔH_{i2}, ΔH_{i3} and so on, is obtained. For example:

$Na(g) \rightarrow Na^+(g) + e^-$ first ionisation energy, $\Delta H_{i1} = +496\,kJ\,mol^{-1}$

$Na^+(g) \rightarrow Na^{2+}(g) + e^-$ second ionisation energy, $\Delta H_{i2} = +4563\,kJ\,mol^{-1}$

$Na^{2+}(g) \rightarrow Na^{3+}(g) + e^-$ third ionisation energy, $\Delta H_{i3} = +6913\,kJ\,mol^{-1}$

There are 11 electrons in a sodium atom, so there are 11 successive ionisation energies for this element.

The successive ionisation energies of an element are all endothermic and they get bigger and bigger. This is not surprising because, having removed one electron, it is more difficult to remove a second electron from the positive ion formed.

The graph in Figure 5.2 shows a logarithmic plot of the successive ionisation energies of sodium against the number of electrons removed. This provides evidence to support the theory that electrons in an atom are arranged in a series of levels or shells around the nucleus. The logarithmic plot allows an extremely wide range of ionisation energies – from 496 kJ mol^{-1} to 159 080 kJ mol^{-1} – to be shown on the same graph.

Definitions

The **first ionisation energy** of an element is the energy needed to remove one electron from each atom in one mole of gaseous atoms.

A **successive ionisation energy** of an element measures the energy needed to remove a second, third or fourth electron and so on from one mole of gaseous ions of the element with the appropriate positive charge.

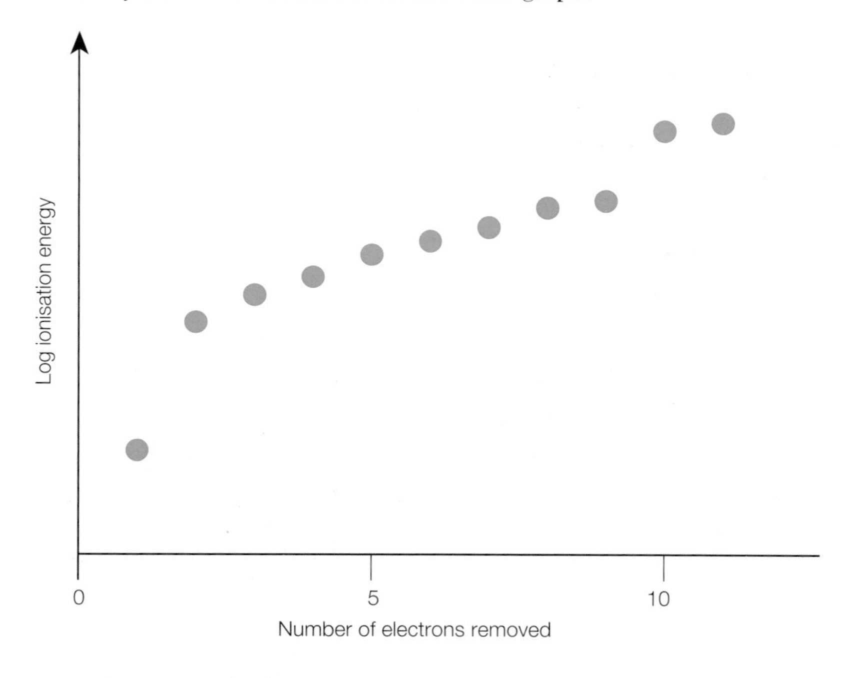

Figure 5.2 ▶
Log (ionisation energy) plotted against the number of electrons removed for sodium.

Note

The shells of electrons at fixed or specific levels are sometimes called **quantum shells**, because the word 'quantum' is used to describe something related to a fixed amount or fixed level.

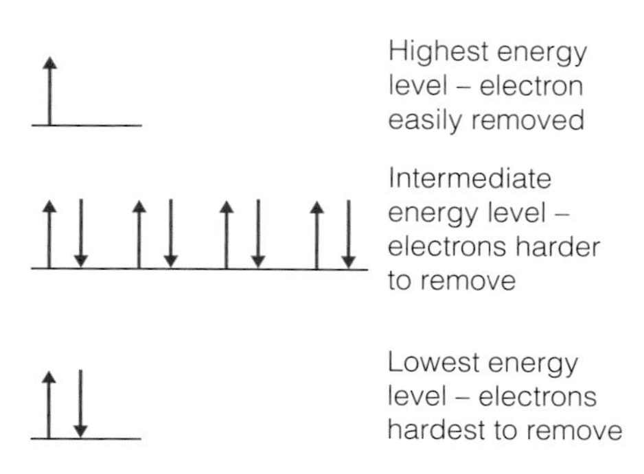

Figure 5.3 ▲
Energy levels of electrons in a sodium atom.

Notice in Figure 5.2 the big jumps in value between the first and second ionisation energies and again between the ninth and tenth ionisation energies. This suggests that sodium atoms have one electron in an outer shell or energy level furthest from the nucleus. This outer electron is easily removed because it is furthest from the nucleus and shielded from the full attraction of the positive nucleus by 10 inner electrons.

Below this outer single electron, sodium atoms seem to have eight electrons in a second shell – all at roughly the same energy level. These eight electrons are closer to the nucleus than the single outer electron and only have two inner electrons shielding them from the positive nucleus.

Finally, sodium atoms have two inner electrons in a shell or level closest to the nucleus. These two electrons feel the full attraction of the positive nucleus and are hardest to remove. They have the most endothermic ionisation energies.

As the successive ionisation energies of an element increase, there is a big jump in value each time electrons start to be removed from the next shell nearer the nucleus.

The electronic structure for a sodium atom is represented in the energy level diagram in Figure 5.3. The electron arrangement in sodium is usually written as 2,8,1.

In energy level diagrams such as that in Figure 5.3, the electrons are represented by arrows. When an energy level is filled, the electrons are paired up, and the electrons in each of these pairs spin in opposite directions.

Chemists believe that paired electrons can be stable only when they spin in opposite directions, because the magnetic attraction that results from their opposite spins can then counteract the electrical repulsion from their negative charges. In energy level diagrams, such as Figure 5.3, the opposite spins of the paired electrons are shown by drawing the arrows in opposite directions.

As soon as the successive ionisation energies of a few elements had been measured, scientists realised that the quantum shells of electrons correspond to the periods of elements in the periodic table. By noting where the first big jump comes in the successive ionisation energies, it is possible to predict the group to which an element belongs. For example, the first big jump in the successive ionisation energies for sodium comes after the first electron is removed. This suggests that sodium has just one electron in its outermost shell, so it must be in Group 1.

Test yourself

1 The successive ionisation energies of beryllium are 900, 1757, 14 849 and 21 007 kJ mol^{-1}.
   a) What is the atomic number of beryllium?
   b) Why do the successive ionisation energies of beryllium always get more endothermic?
   c) Draw an energy level diagram for the electrons in beryllium and predict its electronic structure.
   d) To which group in the periodic table does beryllium belong?

2 The first six ionisation energies of an element in kJ mol^{-1} are 578, 1817, 2745, 11 578, 14 831 and 18 378.
   a) How many electrons are there in the outer shell of the atoms of this element?
   b) To which group of the periodic table does this element belong?

3 Sketch a graph of log(ionisation energy) against number of electrons removed when all the electrons are successively removed from a phosphorus atom.

5.2 Evidence for sub-shells of electrons

By studying the first ionisation energies of successive elements in the periodic table, we can compare how easy it is to remove an electron from the highest energy level in different atoms. This provides us with evidence for the arrangement of electrons in sub-shells.

Factors affecting ionisation energies

The ionisation energy of an atom is influenced by three atomic properties:

- **Distance of the outermost electron from the nucleus** As the distance from the nucleus increases, the attraction of the positive nucleus for the negative electron decreases, and this tends to reduce the ionisation energy.
- **Size of the positive nuclear charge** As the positive nuclear charge increases with atomic number, its attraction for outermost electrons increases, and this tends to increase the ionisation energy.
- **Shielding effect of inner shells of electrons** Electrons in inner shells exert a repelling effect on electrons in the outermost shell of an atom. This effect shields the outermost electrons from the attractive force of the nucleus and reduces its pull on them. This shielding means that the 'effective nuclear charge' attracting electrons in the outer shell is much less than the full positive charge of the nucleus. As expected, the shielding effect increases significantly as the number of inner shells increases.

5.3 Understanding the pattern in ionisation energies

From the study of ionisation energies and spectra, we know that the electrons in atoms are grouped together in quantum shells or energy levels. Principal quantum numbers 1, 2, 3 and so on are used to denote these shells working out from the nucleus.

Activity

Studying the first ionisation energies of successive elements in the periodic table

1. Use the data on your CD-ROM headed 'First ionisation energies of successive elements in the periodic table' to plot a graph of the first ionisation energy for the first 20 elements in the periodic table. Put the first ionisation energy on the vertical axis and the atomic number of the element on the horizontal axis.

2. When you have plotted the points, draw lines from one point to the next to show a pattern of peaks and troughs. Label each point with the symbol of its corresponding element.
3. **a)** Where do the alkali metals in Group 1 appear in the pattern?
   b) Where do the noble gases in Group O appear in the pattern?
4. What similarities do you notice in the patterns for elements in period 2 (lithium to neon) and elements in period 3 (sodium to argon)?
5. Identify three sub-groups of points in both period 2 and period 3. How many elements are there in each sub-group?
6. What is the general trend in ionisation energies across periods 2 and 3?
7. Suggest an explanation for the general trend in ionisation energies across periods 2 and 3.
8. What evidence in your graph suggests that the values of ionisation energies decrease down groups in the periodic table?

Each quantum shell can hold only a limited number of electrons. If all the shells in the atoms of an element are full, it will be very stable with a highly endothermic first ionisation energy.

The first quantum shell nearest the nucleus ($n = 1$) is full and stable when it contains just two electrons. This is the case in helium atoms, which are very stable and unreactive with a higher first ionisation energy than neighbouring elements in the periodic table (Figure 5.4).

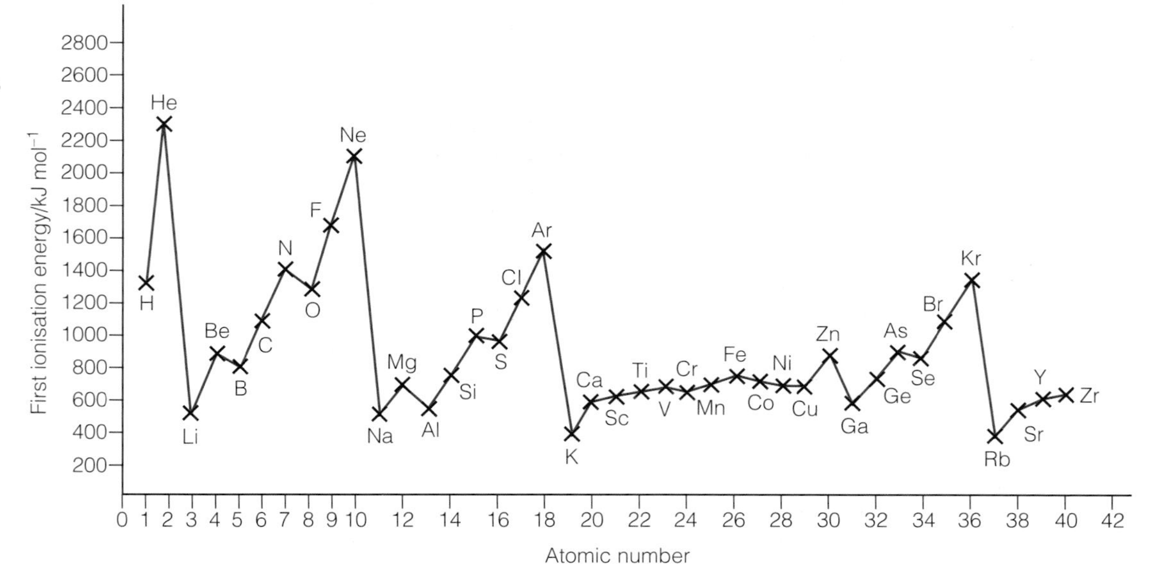

Figure 5.4 ▶ A graph showing the first ionisation energies of the elements plotted against their atomic numbers.

The second quantum shell (energy level, $n = 2$) is full and stable when it contains eight electrons. This is the situation in neon atoms. Neon has a filled first shell with 2 electrons and a filled second shell with 8 electrons. Its electronic structure is 2,8 and neon has a higher first ionisation energy than its neighbours in the periodic table like helium. The high first ionisation energies of helium and neon show that their electronic structures are very stable and explain why the elements are so unreactive.

The points between one noble gas and the next in Figure 5.4 can be divided into sub-sections. These sub-sections provide evidence for sub-shells of electrons.

After both He and Ne in Figure 5.4, there are deep troughs followed by small intermediate peaks at Be and Mg. These are subsections with just two points.

Immediately after Be and Mg there are similar sub-sections of six points (B to Ne and Al to Ar) or two sub-sections of three points. Further along in Figure 5.4, there is another sub-section of 6 points (Ga to Kr) and just before this is a sub-section containing 10 points from Sc to Zn.

The points in Figure 5.4 between one noble gas and the next correspond to one shell of electrons and the sub-sections of points correspond to sub-shells of electrons.

By studying ionisation energies in this way, with data like that in Figure 5.4, chemists have deduced the following:

- The $n = 1$ shell can hold 2 electrons in the same sub-shell.
- The $n = 2$ shell can hold 8 electrons: 2 in one sub-shell and 6 in a slightly higher sub-shell.
- The $n = 3$ shell can hold 18 electrons: 2 in one sub-shell, 6 in a slightly higher sub-shell and 10 electrons in a sub-shell that is slightly higher still.
- The $n = 4$ shell can hold 32 electrons, with sub-shells containing 2, 6, 10 and 14 electrons.

Electron structures

The sub-shells that make up the main shells are given names:

- the sub-shells that can hold up to 2 electrons are called s sub-shells
- those that can hold up to 6 electrons are called p sub-shells
- those that can hold up to 10 electrons are called d sub-shells and
- those that can hold up to 14 electrons are called f sub-shells.

The labels s, p, d and f are left over from early studies of the spectra of different elements, which used the words sharp, principal, diffuse and fundamental to describe different lines. These terms now have no special significance.

The electron structure of an atom can be described in terms of the shells occupied by electrons. In terms of shells, the electron structure of lithium is 2,1 and that of sodium is 2,8,1. It is also possible to describe the electron structure of an atom more precisely in terms of sub-shells.

When sub-shells (energy sub-levels) are being filled, electrons always occupy the lowest available energy level first. Figure 5.5 shows the relative energy levels of the various sub-shells in the first four quantum shells. From Figure 5.5 we can deduce that the order in which the sub-shells are filled in the first four principal quantum shells is:

- 1s in the $n = 1$ shell
- 2s then 2p in the $n = 2$ shell
- 3s then 3p in the $n = 3$ shell
- 4s in the $n = 4$ shell
- 3d in the $n = 3$ shell
- 4p in the $n = 4$ shell.

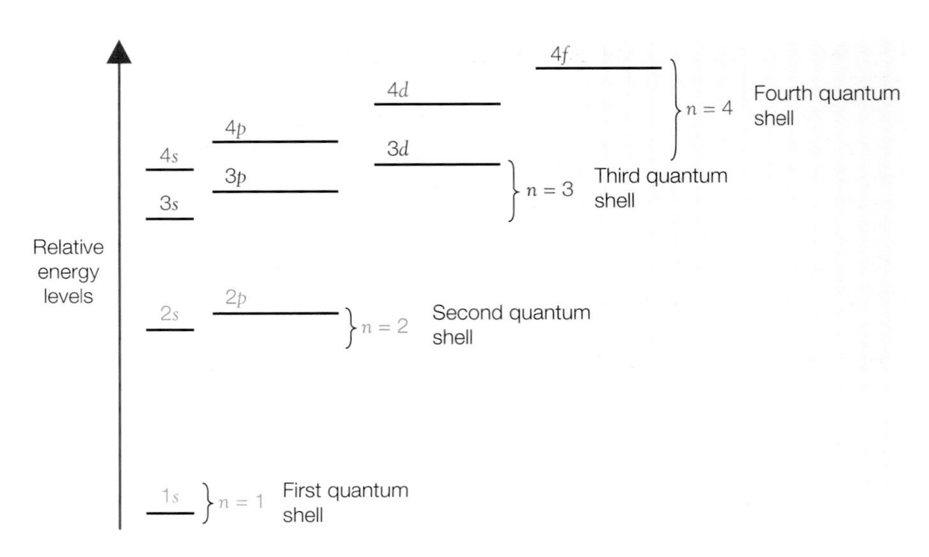

Figure 5.5 ▶ Relative energy levels of sub-shells in the first four quantum shells.

So, the single electron in a hydrogen atom goes in the 1s sub-shell, and the electronic structure of hydrogen can be written in sub-shell notation as $1s^1$. Following on, the electronic structure of helium is $1s^2$, then lithium is $1s^2 2s^1$ and so on.

The electron shell and sub-shell structures of the first 20 elements in the periodic table are shown in Figure 5.6.

Figure 5.6 ▶ The electron shell and sub-shell structure of the first 20 elements in the periodic table.

Period 1			H					He
Atomic no.			1					2
Electron shell structure			1					2
Electron sub-shell structure			$1s^1$					$1s^2$
Period 2	Li	Be	B	C	N	O	F	Ne
Atomic no.	3	4	5	6	7	8	9	10
Electron shell structure	2, 1	2, 2	2, 3	2, 4	2, 5	2, 6	2, 7	2, 8
Electron sub-shell structure	$1s^2$ $2s^1$	$1s^2$ $2s^2$	$1s^2$ $2s^2 2p^1$	$1s^2$ $2s^2 2p^2$	$1s^2$ $2s^2 2p^3$	$1s^2$ $2s^2 2p^4$	$1s^2$ $2s^2 2p^5$	$1s^2$ $2s^2 2p^6$
Period 3	Na	Mg	Al	Si	P	S	Cl	Ar
Atomic no.	11	12	13	14	15	16	17	18
Electron shell structure	2, 8, 1	2, 8, 2	2, 8, 3	2, 8, 4	2, 8, 5	2, 8, 6	2, 8, 7	2, 8, 8
Electron sub-shell structure	$1s^2$ $2s^2 2p^6$ $3s^1$	$1s^2$ $2s^2 2p^6$ $3s^2$	$1s^2$ $2s^2 2p^6$ $3s^2 3p^1$	$1s^2$ $2s^2 2p^6$ $3s^2 3p^2$	$1s^2$ $2s^2 2p^6$ $3s^2 3p^3$	$1s^2$ $2s^2 2p^6$ $3s^2 3p^4$	$1s^2$ $2s^2 2p^6$ $3s^2 3p^5$	$1s^2$ $2s^2 2p^6$ $3s^2 3p^6$
Period 4	K	Ca						
Atomic no.	19	20						
Electron shell structure	2, 8, 8, 1	2, 8, 8, 2						
Electron sub-shell structure	$1s^2$ $2s^2 2p^6$ $3s^2 3p^6$ $4s^1$	$1s^2$ $2s^2 2p^6$ $3s^2 3p^6$ $4s^2$						

In Figure 5.6, notice that the electron structure of calcium is 2,8,8,2. In sub-shell notation, this becomes $1s^2 2s^2 2p^6 3s^2 3p^6 4s^2$.

For calcium (atomic number $Z = 20$), all sub-shells up to and including 4s are filled. The next 10 electrons go into the 3d sub-shell, which covers the elements from scandium ($Z = 21$) to zinc ($Z = 30$) in the periodic table.

The electron structure of iron, which is six places after calcium, is therefore 2,8,14,2, and its sub-shell structure is $1s^2 2s^2 2p^6 3s^2 3p^6 3d^6 4s^2$ (Figure 5.7).

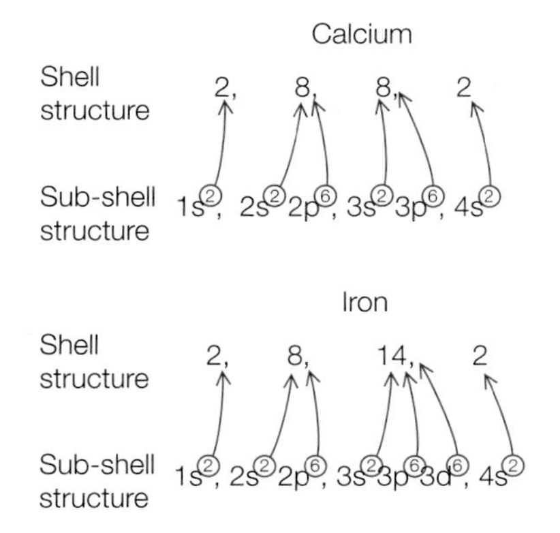

Figure 5.7 ▲
The relationship between the shell and sub-shell structures of calcium and iron.

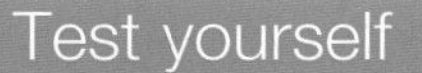

Test yourself

4 Write the electron structures of the following atoms. (Hint: for sodium, this would be 2,8,1.)
 a) Oxygen
 b) Neon
 c) Silicon
 d) Potassium
 e) Titanium ($Z = 22$)

5 Write the electronic sub-shell structures for the atoms in Question 4. (Hint: for sodium, this would be $1s^2 2s^2 2p^6 3s^1$.)

6 Write the electron shell structures of the following ions.
 a) Be^{2+}
 b) Na^+
 c) Cl^-
 d) Ca^{2+}
 e) Br^-

7 Identify the elements with the following electron structures in their outermost shells.
 a) $1s^2$
 b) $2s^2 2p^2$
 c) $3s^2$
 d) $3s^2 3p^4$
 e) $3d^3 4s^2$
 f) $4p^3$

5.4 Electrons and orbitals

Chemists have used complex mathematics to develop their ideas about the arrangement of electrons in sub-shells. As the masses and charges of protons and electrons are known, it is possible to calculate the probability of finding an electron at any point in an atom. These calculations have led chemists to believe that there is a high probability of finding an electron or a pair of electrons in certain regions around the nucleus of an atom. These regions are called orbitals.

Definition

An **orbital** is a region around the nucleus of an atom that can hold up to two electrons with opposite spins.

By pinpointing the likely position of an electron at millions of nanosecond intervals, it is possible to build up a picture showing the electron 'smeared out' over its orbital as a negatively charged cloud. These smeared-out pictures are sometimes described as electron density plots or electron density maps. The plots are darkest where the electrons are more likely to be and lightest where the electrons are less likely to be.

The overall shapes of orbitals are derived from electron density plots by showing the boundary of the region in which there is a 95% chance of finding an electron or a pair of electrons with opposite spins.

The s sub-shells contain one orbital that is best described as a spherical ring – like extra thick peel on an orange (Figure 5.8). The p sub-shells contain three orbitals labelled p_x, p_y and p_z. These are shaped like dumbbells and arranged at right angles to each other (Figure 5.8). The d sub-shells contain five orbitals that can hold a maximum of ten electrons.

So, the single orbital in the first quantum shell is an s orbital (1s). The second shell has four orbitals consisting of an s orbital (2s) and three p orbitals ($2p_x$, $2p_y$ and $2p_z$).

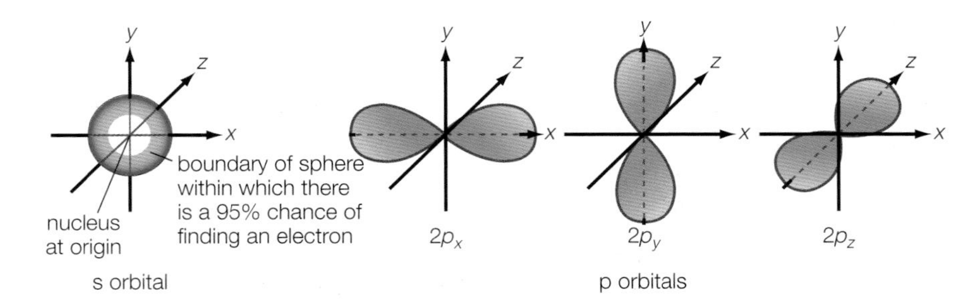

Figure 5.8 ▲
The shapes of s and p orbitals.

From these ideas about orbitals, chemists have developed an 'electrons in boxes' representation for the electronic structures of atoms. Using this method, each box represents an orbital.

The electrons in an atom fill the orbitals according to a set of three rules. The three rules are:

- Electrons go into the orbital with the lowest energy level first.
- Each orbital can hold only one electron or, at the most, two electrons with opposite spins.
- When there are two or more orbitals at the same energy level, electrons occupy these orbitals singly before they pair up.

Chemists sometimes use the term 'Aufbau principle' to describe these rules. This comes from the German word 'Aufbau', which means 'build up'. This is a reminder that electronic structures build up from the bottom, starting with the lowest energy level. From the 'Aufbau principle', it is possible to deduce the electron structures of all atoms.

Figure 5.9 shows the 'electrons in boxes' (orbitals) representations and the sub-shell notations for the electronic structures of beryllium, nitrogen and sodium. These descriptions of the arrangement of electrons in the atoms of elements are called electron configurations.

Figure 5.9 ►
'The electrons in boxes' representations and the sub-shell (s, p, d, f) notations for the electronic structures of beryllium, nitrogen and sodium.

Element	Electrons-in-boxes notation of electronic structure: 1s	2s	2p			3s	s, p, d, f electron notation
Beryllium	↑↓	↑↓					$1s^2 2s^2$
Nitrogen	↑↓	↑↓	↑	↑	↑		$1s^2 2s^2 2p^3$
Sodium	↑↓	↑↓	↑↓	↑↓	↑↓	↑	$1s^2 2s^2 2p^6 3s^1$

Definition

Electron configurations describe the number and arrangement of electrons in an atom. A shortened form of electron configuration uses the symbol of the previous noble gas in square brackets to stand for the inner shells. Using this convention, the electron configuration of sodium is $[Ne]3s^1$.

The development of knowledge and understanding about electronic structures illustrates how chemists use the results of their experiments, such as the measurements of ionisation energies, to explain the structure and properties of materials. It also illustrates the important distinction between evidence and experimental data on the one hand and ideas, theories and explanations on the other.

Ionisation energies and spectra have provided chemists with evidence and information that has caused them to develop and modify their models and theories about electron structure. Early ideas about electrons arranged in shells have been developed to take in the evidence for sub-shells and then modified to include ideas about orbitals.

Test yourself

8 Copy and complete the following information for the quantum shell with principal quantum number 3.
 a) Total number of sub-shells = ________
 b) Total number of orbitals = ________
 c) Number of different types of orbital = ________
 d) Maximum number of electrons in the shell = ________

9 Draw the 'electrons in boxes' structure for the following elements:
 a) boron
 b) fluorine
 c) phosphorus
 d) potassium.

5.5 Electron structures and the periodic table

The periodic table helps chemists to see order and patterns in the vast amount of information they have discovered about all the elements and their compounds.

In the modern periodic table, elements are arranged in order of atomic number. The horizontal rows in the table are called periods and each of these ends with a noble gas.

The vertical columns in the table are called groups and these can be divided into four blocks: the s-block, p-block, d-block and f-block on the basis of the electron structures of the elements (Figure 5.10). Our modern arrangement of elements in the periodic table reflects the pattern in electronic structures of the atoms, while the more sophisticated model of electron structure in terms of orbitals allows chemists to explain the properties of elements more effectively. The four blocks in the periodic table are shown in different colours in Figure 5.10.

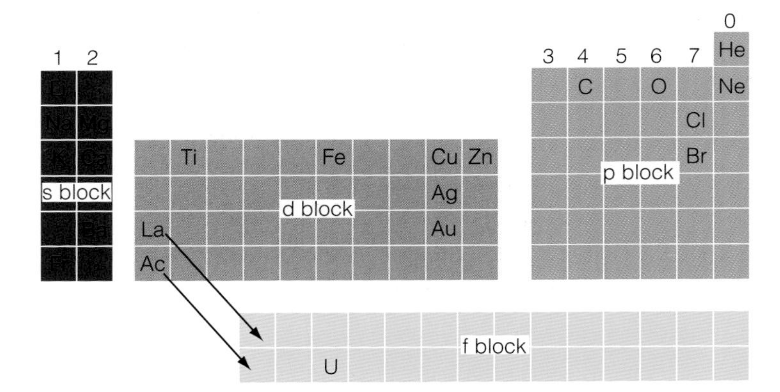

Figure 5.10 s, p, d and f blocks in the periodic table.

Definitions

In the periodic table:

- **periods** are the horizontal rows of elements
- **groups** are the vertical columns of similar elements (that is, the chemical families).

- The **s-block** comprises the reactive metals in Groups 1 and 2, such as potassium, sodium, calcium and magnesium. In these metals, the last electron added goes into an s orbital in the outer shell.
- The **p-block** comprises elements in Groups 3, 4, 5, 6, 7 and 0 on the right of the periodic table. These elements include relatively poor metals, such as tin and lead, as well as all the non-metals. In these elements, the last electron added goes into a p orbital in the outer shell.
- The **d-block** elements occupy a rectangle across Periods 4, 5, 6 and 7 between Groups 2 and 3. The d-block elements are all metals. They include titanium, iron, copper and silver in which the last electron added goes into a d orbital. These metals are much less reactive than the s-block metals in Groups 1 and 2. Within the d-block, there are marked similarities across the periods as well as the usual vertical similarities. The d-block elements are sometimes loosely called transition metals.

- The **f-block** elements occupy a low rectangle across Periods 6 and 7 within the d-block, although they are usually placed below the main table to prevent it becoming too wide to fit the page. Like the d-block elements, those in the f-block are all metals. In these, the last added electron is in an f orbital. The f-block elements are often called the lanthanides and actinides because they comprise the 14 elements immediately following lanthanum, La, and the 14 elements immediately following actinium, Ac, in the periodic table. Another name for the f-block elements is the inner transition elements.

In the periodic table, the elements in each group have similar properties because they have similar electron structures. This important point is well illustrated by the alkali metals in Group 1.

Group 1: the alkali metals

Look at Figure 5.11 and notice that each alkali metal has one s electron in its outer shell. This similarity in their electron structures explains why they have similar properties.

Alkali metals:

- are very reactive because they lose their single outer electron so easily
- form ions with a charge of 1+ (Li^+, Na^+, K^+, etc), so the formulae of their compounds are similar
- form very stable ions with an electron structure like a noble gas.

Group 1

The alkali metals
Lithium Li 2, 1 ($1s^2 2s^1$)
Sodium Na 2, 8, 1 ($1s^2 2s^2 2p^6 3s^1$)
Potassium K 2, 8, 8, 1 ($1s^2 2s^2 2p^6 3s^2 3p^6 4s^1$)

Figure 5.11 ▲
Electron structures of the first three alkali metals.

As the atomic numbers of the alkali metals increase, the outer electron is further from the positive nucleus. As a result, the outer electron is held less strongly by the nucleus and lost more readily. This explains why the alkali metals become more reactive as their atomic number increases.

The chemical properties of all other elements are also determined by their electronic structures. In fact, chemistry is mostly about electrons in the outer shells of atoms. The reactivity of an element depends largely on the number of electrons in its outer shell and how strongly they are held by the nuclear charge. This is a fundamental feature of chemistry that we will return to again and again. It is also an essential principle that governs the way in which chemists think and work.

Test yourself

10 Why are sodium and potassium so alike?

11 Why are the noble gases so unreactive?

12 **a)** Write down the electron shell structures and sub-shell structures of fluorine and chlorine in group 7.

b) Why do you think fluorine and chlorine are so reactive with metals?

c) Why do the compounds of fluorine and chlorine with metals have similar formulae?

13 Elements in the d-block are sometimes called transition metals. State four properties that are common among d-block elements.

REVIEW QUESTIONS

Extension questions

1 The table below shows the electron structures and first ionisation energies of the first five elements in Period 2 of the periodic table.

Element	Li	Be	B	C	N
Electron structure	$1s^22s^1$	$1s^22s^2$	$1s^22s^22p^1$	$1s^22s^22p^2$	$1s^22s^22p^3$
First ionisation energy /kJ mol^{-1}	520	900	801	1086	1402

a) Describe the **general** trend in first ionisation energies from Li to N. (1)

b) Explain this general trend. (3)

c) Why do you think boron, B, has a lower first ionisation energy than beryllium, Be? (3)

2 The table below shows the first and second ionisation energies of sodium and potassium in Group 1 of the periodic table.

	First ionisation energy/kJ mol^{-1}	Second ionisation energy/kJ mol^{-1}
Sodium	496	4563
Potassium	419	3051

a) Write an equation with state symbols for the second ionisation energy of sodium. (2)

b) Why are the second ionisation energies of sodium and potassium greater than their first ionisation energies? (3)

c) Why are the first and second ionisation energies of potassium less than those of sodium? (4)

d) The first five successive ionisation energies in kJ mol^{-1} of an element X in Period 3 of the periodic table are 578, 1817, 2745, 11 578 and 14 831.

i) Identify element X. (1)

ii) Explain how you obtained your answer. (2)

3 a) The first ionisation energies of three consecutive elements in the periodic table are 1251, 1521 and 419 kJ mol^{-1}, respectively. Sketch a graph of these ionisation energies against atomic number. Continue the sketch to show the pattern of the first ionisation energies of the next three elements in the periodic table, assuming no transition metals are involved. (2)

b) The first five ionisation energies of an element are 738, 1451, 7733, 10 541 and 13 629 kJ mol^{-1}, respectively. Explain why the element cannot have an atomic number less than 12. (4)

4 Describe the positions relative to the nucleus of the sub-shells with principal quantum numbers 1 and 2 and the relative energies, number and shape of the orbitals within these sub-shells. (10)

6 Bonding and structure

One of the central aims of chemistry is to explain the properties of elements and compounds in terms of their bonding and structure. In this topic, we will be studying the arrangement and structure of atoms, molecules and ions in different materials and the theories of bonding that account for the forces that hold atoms or ions together. There are three types of strong bonding – ionic, covalent and metallic – all of which depend on electrostatic attractions to hold their ions and atoms together. Covalent bonding differs from ionic and metallic bonding, however, in that it is directional. This leads to some interesting shapes for various molecules and for ions in which the atoms are held together by covalent bonds.

In some compounds, such as sodium chloride and caesium fluoride, the bonding can be regarded as entirely ionic, while the bonding in methane is purely covalent. In most compounds, however, the bonding is neither purely ionic nor purely covalent.

Ionic, covalent and metallic bonds account for the properties of materials like salt, diamond and metals. Equally important are the much weaker attractive forces between molecules. Without these intermolecular forces, there would be no rivers or oceans and our bodies would fall apart.

6.1 Investigating structure and bonding

The word structure has different levels of meaning in science. On a grand scale, engineers design the structures of buildings and bridges. On the smallest scale, chemists and physicists explore the inner structure of atoms. Not surprisingly, scientists use different models and different theories to explain the structure and properties of materials at these different levels. In the same way tourists who visit London must use different types of maps to negotiate its road and rail systems.

Scientists have developed increasingly sophisticated models to account for the structure, bonding and properties of materials as their knowledge has increased. No one model can be used to explain the properties of elements and compounds at all levels. Similarly, no one map can be used by all visitors to London. We must live with these limitations and appreciate which map and which model are the most appropriate in different contexts.

The regular shapes of crystals suggest an underlying pattern in the arrangement of atoms, ions or molecules in their structure. Until the early part of the twentieth century, scientists could only guess at the arrangement of invisible atoms in crystals. Sir Lawrence Bragg (1890–1971) then realised that X-rays could be used to investigate crystal structures, because their wavelengths are about the same as the distances between atoms in a crystal.

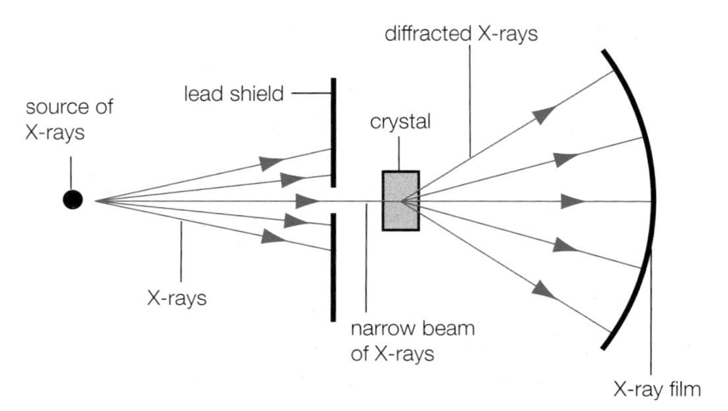

Figure 6.1 ▶ Using X-rays to study the structure of atoms or ions in a crystal.

A narrow beam of X-rays is directed at a crystal of the substance being studied (Figure 6.2). Atoms or ions in the crystal scatter the X-rays producing a pattern of diffracted rays. The diffracted X-rays were originally photographed using X-ray film, but they can now be recorded electronically. It is possible to deduce the three-dimensional structure of crystals from the pattern of diffracted X-rays.

If we know how the atoms or ions in a substance are arranged (the structure) and how they are held together (the bonding), we can explain its properties. The clear links between structure, bonding and properties help us to understand the uses of different substances and materials. For example, they explain why metals are used as conductors, why graphite is used in pencils and why clay can be used to make pots (Figure 6.3).

Figure 6.2 ▲
An X-ray diffraction pattern of lysozyme, a protein found in egg white.

Figure 6.3 ▲
Wet clay is easily moulded by the potter because water molecules can get between its flat two-dimensional particles and allow them to slide over one another. When the clay is fired, the water is driven out and atoms in one layer bond to those in the layers above and below. This gives the clay a three-dimensional structure making it hard and rigid for use as pots and ceramics.

6.2 Two types of structure

Broadly speaking, there are two types of structure: giant structures and simple molecular structures.

Materials with giant structures form crystals in which all the atoms or ions are linked by a network of strong bonds extending throughout the crystal. These strong bonds result in giant structures with high melting points and boiling points.

Substances with simple molecular structures consist of small groups of atoms. The bonds linking the atoms in the molecules (intramolecular forces) are relatively strong, but the forces between molecules (intermolecular forces) are weak. These weak intermolecular forces allow the molecules to be separated easily. This means that molecular substances have low melting and boiling points.

The main types of giant structures are ionic solids, giant covalent solids and metals. All of these materials are solids that depend for their properties on three types of strong bonding: ionic bonding, covalent bonding and metallic bonding.

These three types of strong bonding will be the main focus of interest in the next sections of this topic. For each type of bonding, the strength of the bond depends on electrostatic attractions between positive and negative charges.

Figure 6.4 ▲
Crystals of rock salt (sodium chloride, NaCl).

Test yourself

1 Look up the melting and boiling points of the following elements and decide whether they have giant or simple molecular structures: boron, fluorine, silicon, sulfur, manganese, iodine.

2 Look at the crystals of rock salt in Figure 6.4.

a) What shape are most of the crystals of rock salt?

b) How do you think the ions are arranged in rock salt?

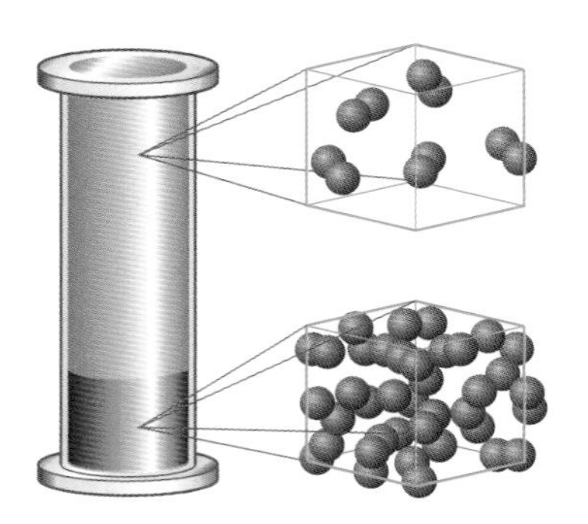

Figure 6.5 ▲
Molecules in bromine liquid and vapour. Many molecular elements and compounds are liquids or gases at room temperature because little energy is needed to overcome the weak forces between their molecules.

Figure 6.6 ▲
Metal cables in the electricity grid supported by steel pylons are a reminder that metals are strong, bendable and good conductors of electricity. Ceramic insulators between the conducting cables and the pylons prevent the electric current leaking away to Earth.

6.3 Ionic bonding and structures

Atoms into ions

Compounds of metals with non-metals, such as sodium chloride and calcium oxide, are composed of ions. When compounds form between metals and non-metals, the metal atoms lose electrons and become positive ions (cations). At the same time, the non-metal atoms gain electrons and become negative ions (anions). For example, when sodium reacts with chlorine (Figure 6.7), each sodium atom loses its one outer electron forming a sodium ion, Na^+. At the same time, chlorine atoms gain the electrons lost from the sodium atoms and form chloride ions, Cl^- (Figure 6.8). This transfer of electrons involves both oxidation and reduction (Topic 4).

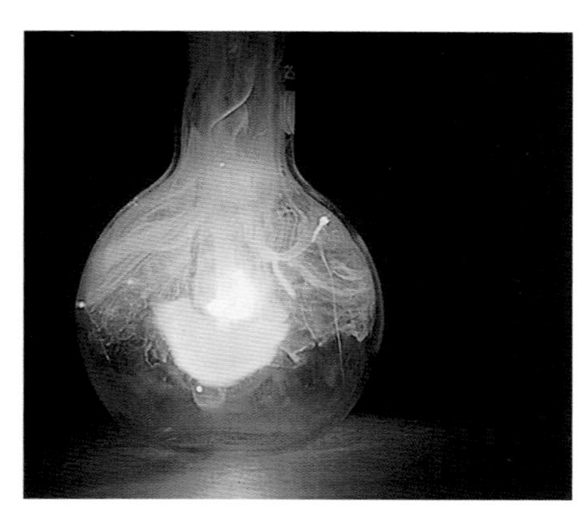

Figure 6.7 ▲
Hot sodium reacting with chlorine gas.

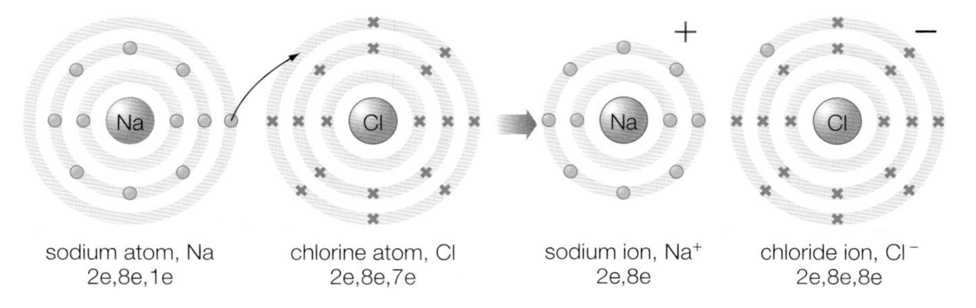

Figure 6.8 ►
Formation of ions in sodium chloride when sodium reacts with chlorine.

Chemists describe diagrams like that in Figure 6.8 as '*dot-and-cross*' diagrams. The electrons belonging to one reactant are shown as dots and those belonging to the other reactant are shown as crosses. Remember, though, that all the electrons are the same. The dots and crosses are simply used to show which electrons come from the metal and which come from the non-metal. The '*dot-and-cross*' diagrams are useful because they provide a balance sheet that keeps track of the electrons when ionic compounds form. Figure 6.9 shows simplified '*dot-and-cross*' diagrams for the formation of sodium chloride and calcium fluoride, in which only the outer shell electrons are drawn.

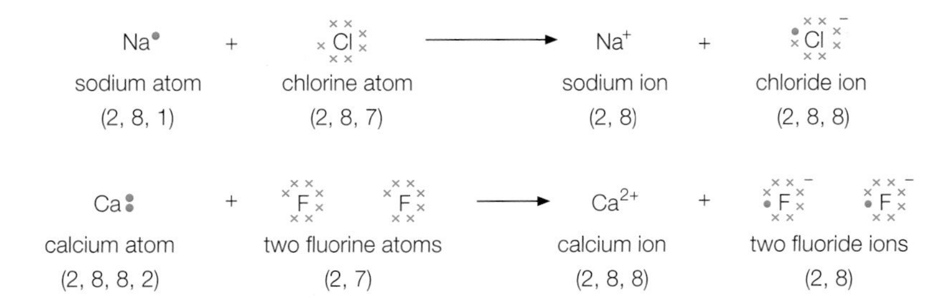

Figure 6.9 ▲
'*Dot-and-cross*' diagrams for the formation of sodium chloride and calcium fluoride showing only the electrons in the outer shells of the reacting atoms.

Ionic bonding

When metals react with non-metals, the ions produced form ionic crystals. These crystals are giant ionic lattices containing billions of positive and negative ions packed together in a regular pattern. Figure 6.10 shows how the ions are arranged in one layer of sodium chloride, NaCl, and Figure 6.11 shows a three-dimensional model of its structure.

Notice that each Na^+ ion is surrounded by Cl^- ions and that each Cl^- ion is surrounded by Na^+ ions. This means that, overall, there are strong electrostatic attractions between ions in all directions throughout the lattice. Many other compounds have the same cubic structure as sodium chloride, including the chlorides, bromides and iodides of Li, Na and K and the oxides and sulfides of Mg, Ca, Sr and Ba.

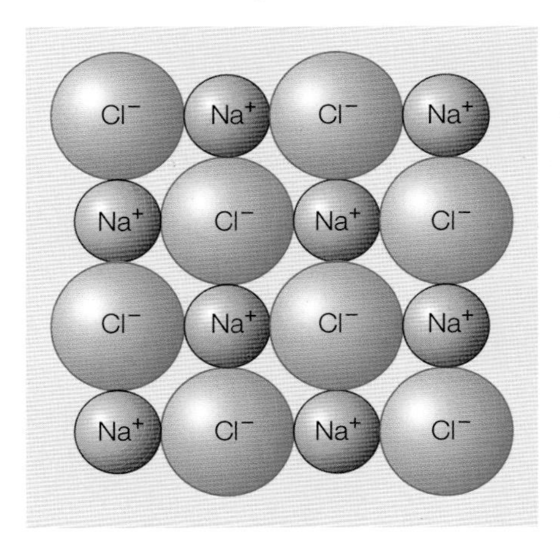

Figure 6.10 ▲
The arrangement of ions in one layer of a sodium chloride crystal.

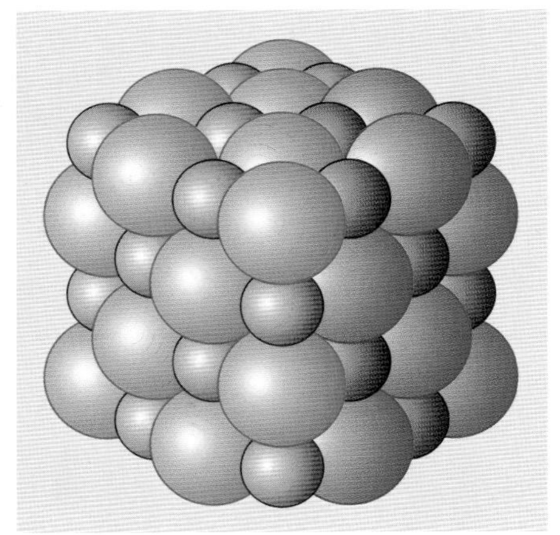

Figure 6.11 ▲
A three-dimensional model of the structure of sodium chloride. The smaller red balls represent Na^+ ions (A_r(Na) = 23.0) and the larger green balls represent Cl^- ions (A_r(Cl) = 35.5).

Activity

Atoms, ions and the periodic table

Look carefully at Table 6.1. This shows the electron structures of the atoms and ions of the elements in Period 3 of the periodic table.

Elements in Period 3	Na	Mg	Al	Si	P	S	Cl	Ar
Group	1	2	3	4	5	6	7	0
Electron structure	2,8,1	2,8,2	2,8,3	2,8,4	2,8,5	2,8,6	2,8,7	2,8,8
Number of electrons in outer shell	1	2	3	4	5	6	7	8
Common ion	Na^+	Mg^{2+}	Al^{3+}	No ion	P^{3-}	S^{2-}	Cl^-	No ion
Electron structure of ion	2,8	2,8	2,8	–	2,8,8	2,8,8	2,8,8	–

Table 6.1 ▲
Electron structures of the atoms and ions of elements in Period 3.

1 What pattern can you see in the ions formed by elements in Groups 1, 2 and 3?

2 Predict the ions formed by the following atoms:

a) Li **b)** Be **c)** K **d)** Ca.

3 We can write an equation for the formation of a sodium ion, Na^+, when a sodium atom, Na, loses an electron as:

$$\underset{(2,8,1)}{Na} \rightarrow \underset{(2,8)^+}{Na^+} + \underset{\text{electron}}{e^-}$$

Write a similar equation for the formation of a magnesium ion, Mg^{2+}, from a magnesium atom, Mg.

4 What pattern can you see in the ions formed by elements in Groups 5, 6 and 7?

5 Predict the ions formed by the following atoms:

a) N **b)** O **c)** F.

6 Why does argon not form an ion?

7 Use your answer to Question 6 to explain why argon is used to fill electric light bulbs.

8 Suggest why silicon does not form Si^{4+} or Si^{4-} ions.

Test yourself

3 Draw '*dot-and-cross*' diagrams for:
  a) lithium fluoride
  b) magnesium chloride
  c) lithium oxide
  d) calcium oxide.

4 With the help of a periodic table, predict the charges on ions of the following elements:
  a) caesium
  b) strontium
  c) gallium
  d) selenium
  e) astatine.

5 Why do metals form positive ions and non-metals form negative ions?

Properties of ionic compounds

Strong ionic bonds hold the ions firmly together in ionic compounds. This explains why ionic compounds:

- are hard crystalline substances
- have high melting points and boiling points
- are often soluble in water and other polar solvents but insoluble in non-polar solvents (Section 6.15)
- do not conduct electricity when solid because their ions cannot move away from fixed positions in the giant lattice
- conduct electricity when they are melted or dissolved in water because the charged ions are then free to move.

Definitions

Ionic bonding is the strong electrostatic attraction between oppositely charged ions in giant ionic lattices such as sodium chloride.

Chemists use the word **lattice** to describe the regular arrangement of atoms or ions in a crystal.

Test yourself

6 Look carefully at Figures 6.10 and 6.11.
   a) How many Cl^- ions surround one Na^+ ion in a layer of the NaCl crystal?
   b) How many Cl^- ions surround one Na^+ ion in the three-dimensional crystal?
   c) How many Na^+ ions surround one Cl^- ion in the three-dimensional crystal?
   d) The structure of crystalline sodium chloride is described as 6:6 co-ordination. Why is this?

7 The melting point of sodium fluoride is 993 °C but that of magnesium oxide is 2852 °C.
   a) Write the formulae of these two compounds showing charges on the ions.
   b) Suggest why the melting point of magnesium oxide is so much higher than that of sodium fluoride.

6.4 Covalent bonding and structures

Living things are composed mainly of water and a huge variety of molecular compounds of carbon with other non-metals such as hydrogen, oxygen and nitrogen. These compounds differ greatly from minerals in the rocks of the Earth's crust even though some of the minerals consist of giant structures of the non-metals silicon and oxygen with other elements. So, chemists must understand the structure and bonding of non-metals and their compounds if they are to explain the properties of the living (organic) and non-living (inorganic) worlds.

Simple molecular structures

In most non-metals, atoms are joined together in small molecules such as hydrogen, H_2, nitrogen, N_2, phosphorus, P_4, sulfur, S_8, and chlorine, Cl_2. Most of the compounds of non-metals with other non-metals also have simple molecular structures. This is true of simple compounds such as water, carbon dioxide, ammonia, methane and hydrogen chloride. It is also true of the many thousands of carbon compounds (Topic 10).

Figure 6.12 ◀ Clouds and the atmosphere contain simple molecular substances such as oxygen, nitrogen and water vapour. The oceans and hydrosphere contain ionic substances, such as sodium chloride, dissolved in water. Rocks in the Earth (the lithosphere) contain giant structures of atoms and ions.

The covalent bonds that hold atoms together within simple molecular structures are strong, so the molecules do not easily break up into atoms. The forces between separate molecules (intermolecular forces), however, are weak, so it is quite easy to separate the molecules. This means that molecular substances are often liquids or gases at room temperature and molecular solids are usually easy to melt and evaporate (Figure 6.13).

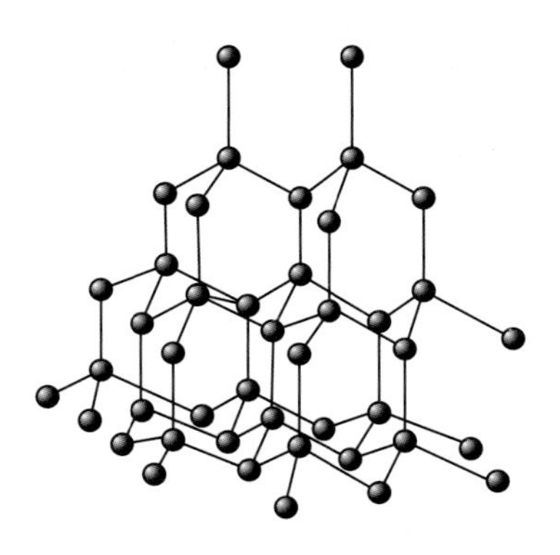

Figure 6.14 ▲
Part of the giant covalent structure in diamond. Each carbon atom is linked to four other atoms in a lattice extending throughout the giant structure.

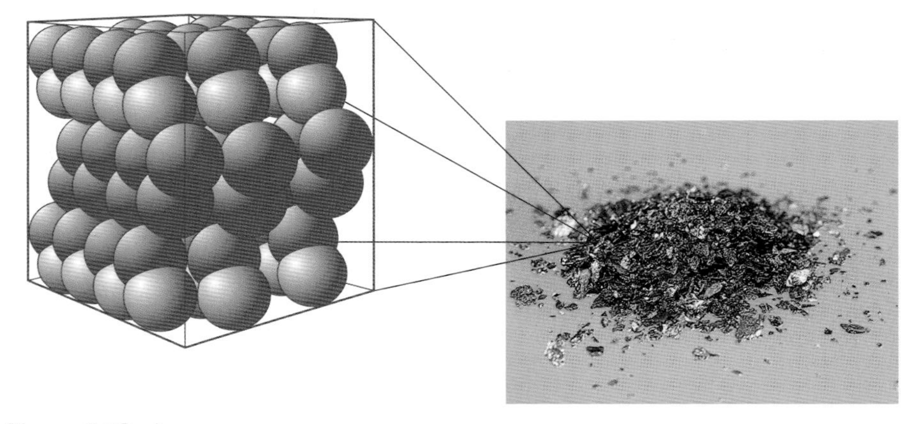

Figure 6.13 ▲
The structure of iodine showing the arrangement of I_2 molecules. The forces between I_2 molecules are so weak that iodine sublimes, changing directly from solid to vapour at low temperatures.

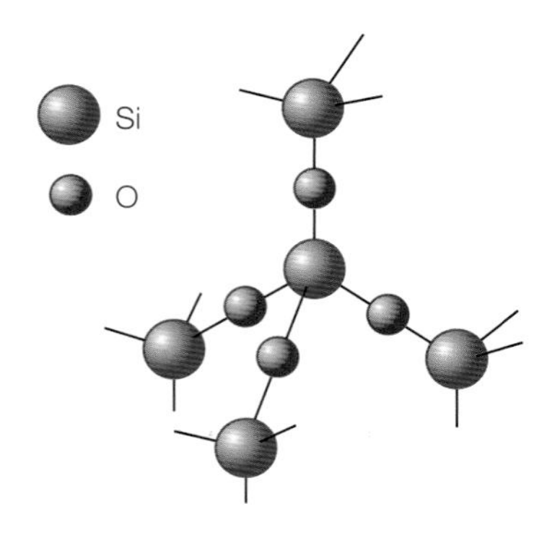

Figure 6.15 ▲
Part of the giant structure of silicon dioxide in the mineral quartz. Silicon atoms are arranged in the same way as carbon atoms in diamond but with an oxygen atom between every pair of silicon atoms. Sandstone and sand consist mainly of silicon dioxide.

Figure 6.16 ▲
An amethyst gemstone. Amethyst is crystalline quartz (silicon dioxide, SiO_2) coloured purple because of the presence of iron(III) ions.

Properties of simple molecular substances

In general, simple molecular elements and compounds:

- are usually gases, liquids or soft solids at room temperature
- have relatively low melting points and boiling points
- do not conduct electricity as solids, liquids or in solution because they contain neither ions nor free electrons to carry the electric charge
- are usually more soluble in non-polar solvents such as hexane than in water.

Giant covalent lattices

A few non-metal elements, including the two forms of carbon (diamond and graphite) and silicon, consist of giant lattices of atoms held together by covalent bonding. The covalent bonds in diamond are strong and point in a definite direction, so diamond is very hard with a very high melting point (Figure 6.14). Diamond does not conduct electricity because the electrons in covalent bonds are fixed (localised) between pairs of atoms.

Some compounds of non-metals, such as silicon dioxide and boron nitride, also form giant covalent lattices (Figure 6.15). These compounds are also hard non-conductors with very high melting points.

Test yourself

8 Compare the melting and boiling points of the non-metals boron, fluorine, white phosphorus, germanium and bromine. Which of these elements contain simple molecules and which have giant covalent lattices?

9 a) Compare the formulae and physical properties of carbon dioxide and silicon dioxide.

b) Explain the differences in terms of the structures of the two compounds.

6.5 Covalent bonding

Strong covalent bonds hold the atoms of non-metals together in simple molecules and giant lattices. Covalent bonds form when atoms share electron pairs. A single covalent bond consists of a shared pair of electrons. The atoms are held together by the electrostatic attraction between the positive charges on their nuclei and the negative charge on the shared electrons. The electron shell structure of fluorine is 2,7, with seven electrons in the outer shell. When two fluorine atoms combine to form a molecule, they share two electrons. The electron configuration of each atom is then like that of neon – the nearest noble gas (Figure 6.17).

Chemists draw a line between symbols to represent a covalent bond (Section 1.2), so they write a fluorine molecule as F–F. This is the structural formula, which shows the atoms and bonding. The molecular formula of fluorine is F_2.

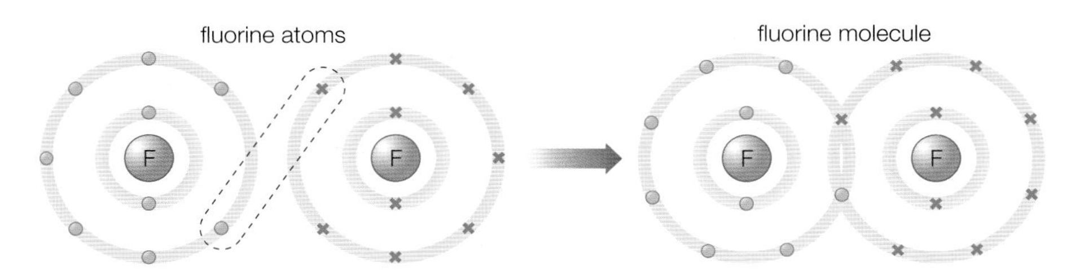

Figure 6.17 ◄
Covalent bonding in a fluorine molecule.

Definition

A **covalent bond** involves a shared pair of electrons between two atoms. In a normal covalent bond, each atom contributes one electron to the shared pair.

Covalent bonds also link the atoms in non-metal compounds and Figure 6.18 shows the covalent bonding in methane.

'*Dot-and-cross*' diagrams showing only the electrons in outer shells provide a simple way of representing covalent bonding. Three of these '*dot-and-cross*' diagrams are shown in Figure 6.19. Remember from Section 1.2 that each non-metal usually forms the same number of covalent bonds in all its compounds. This should help you predict the structures of different molecules (Table 1.1).

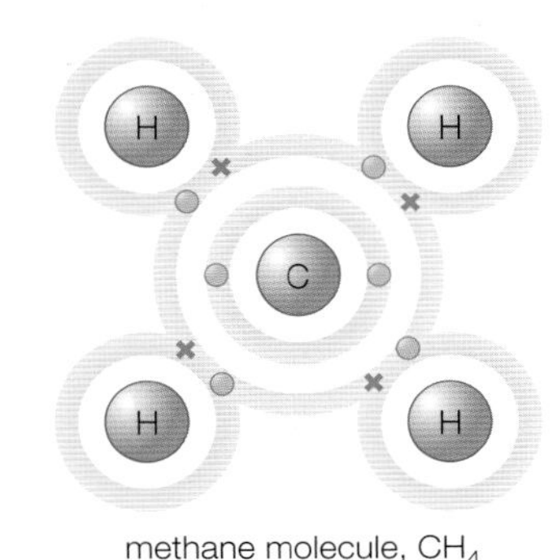

methane molecule, CH_4

Figure 6.18 ▲
Covalent bonding in methane.

Cl—Cl

chlorine

H—O—H (water)

water

H—N—H with H (ammonia)

ammonia

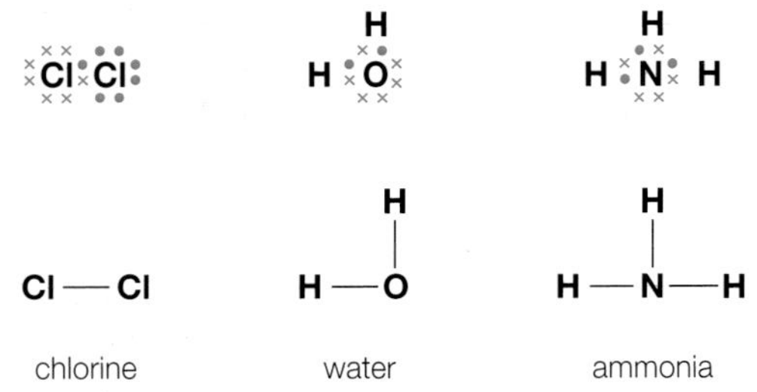

Figure 6.19 ▲
'*Dot-and-cross*' diagrams show the single covalent bonds in molecules. A simpler way of showing the bonding in molecules is also included. This shows each covalent bond as a line between two symbols.

Multiple bonds

One shared pair of electrons makes a single bond. Double bonds and triple bonds with two or three shared pairs are also possible.

There are two covalent bonds between the two oxygen atoms in an oxygen molecule and two covalent bonds between both the oxygen atoms and the carbon atom in carbon dioxide (Figure 6.20). When two electron pairs are involved in the bonding, there is a region of high electron density between the two atoms joined by a double bond.

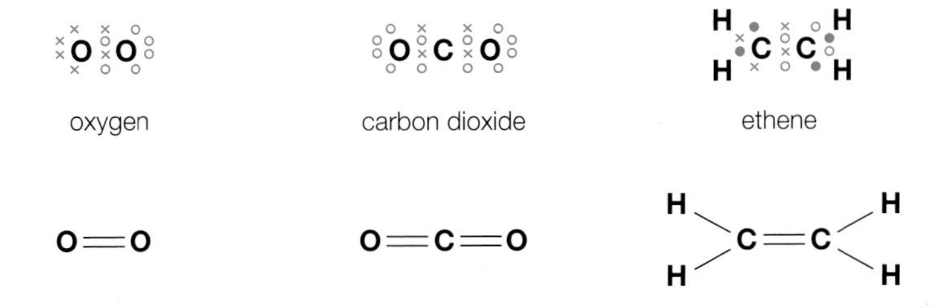

Figure 6.20 ▲
Three molecules with double covalent bonds.

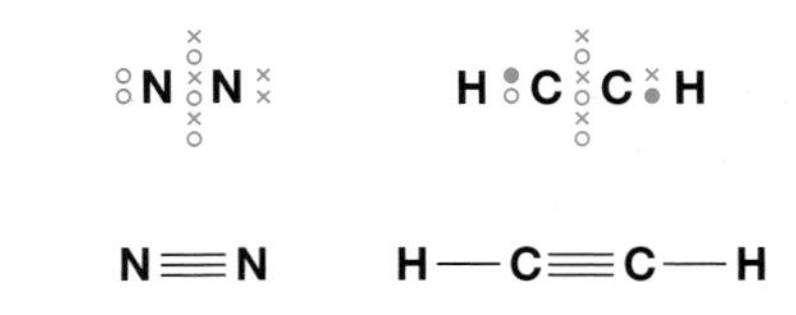

Figure 6.21 ▲
Two molecules with triple covalent bonds.

Lone pairs of electrons

In many molecules and ions, there are atoms with pairs of electrons in their outer shells which are not involved in the bonding between atoms. Chemists call these 'lone pairs' of electrons. Lone pairs of electrons:

- affect the shapes of molecules (Section 6.8)
- form dative covalent (co-ordinate) bonds
- are important in the chemical reactions of some compounds, including water and ammonia.

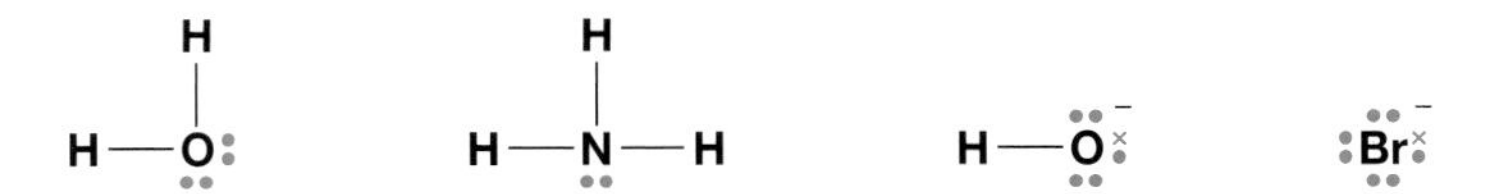

Figure 6.22 ▲
Molecules and ions with lone pairs of electrons.

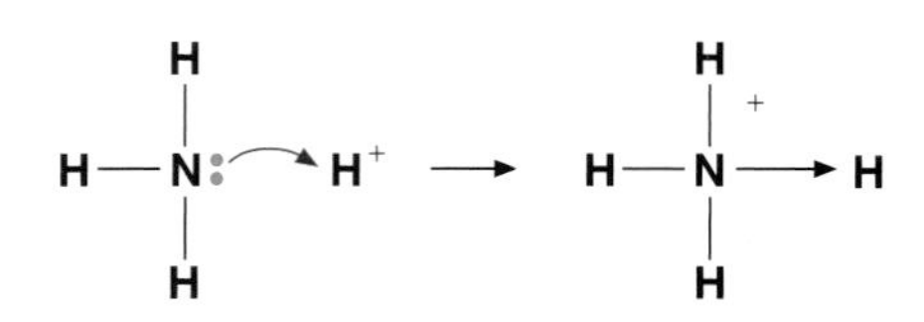

Figure 6.23 ▲
Formation of an ammonium ion.

Dative covalent bonds

In a covalent bond, two atoms share a pair of electrons. Usually, each atom supplies one electron to make up the pair. Sometimes, however, one atom provides both electrons and chemists call this a dative covalent bond. The word 'dative' means 'giving', and one atom gives both the electrons to make the covalent bond. An alternative name for a dative covalent bond is co-ordinate bond. Once formed, there is no difference between dative bonds and normal covalent bonds.

Ammonia forms a dative covalent bond when it reacts with a hydrogen ion to make an ammonium ion, NH_4^+ (Figure 6.23). Dative bonds are represented by an arrow in displayed formulae like NH_4^+ in Figure 6.23. The arrow points from the atom that donates the electron pair to the atom receiving the electrons.

Dative covalent (co-ordinate) bonding also accounts for the structures of the oxonium ion, H_3O^+ in which water molecules combine with H^+ ions, nitric acid and carbon monoxide (Figure 6.24).

Figure 6.24 ►
Dative covalent (co-ordinate) bonds in the oxonium ion, nitric acid and carbon monoxide.

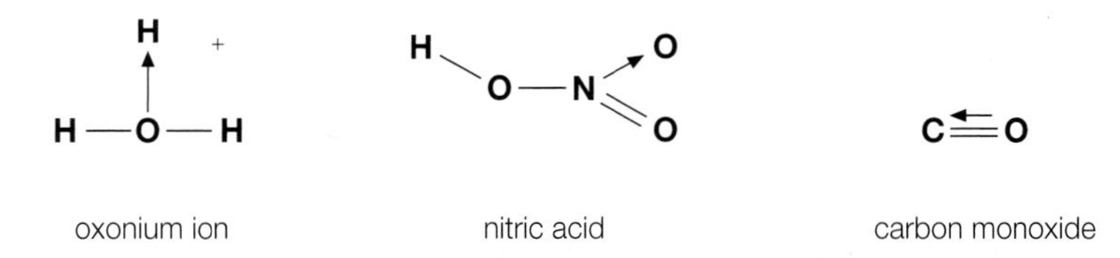

Definition

A **dative covalent bond** (co-ordinate bond) is a bond in which two atoms share a pair of electrons, with both electrons donated by one atom.

Test yourself

10 How do the physical properties of giant covalent lattices, such as diamond and silicon dioxide, provide evidence for strong covalent bonds between non-metal atoms?

11 Draw diagrams showing all the electrons in the shells and in covalent bonds in:
a) hydrogen, H_2 b) hydrogen chloride, HCl c) ammonia, NH_3.

12 Draw '*dot-and-cross*' diagrams to show the covalent bonding in:
a) hydrogen sulfide, H_2S b) ethane, C_2H_6 c) carbon disulfide, CS_2
d) nitrogen trifluoride, NF_3 e) phosphine, PH_3.

13 Identify the atoms with lone pairs of electrons in the following molecules and state the number of lone pairs:
a) ammonia b) water c) hydrogen fluoride d) carbon dioxide.

14 Draw '*dot-and-cross*' diagrams to show the outer shell electrons in the atoms of:
a) carbon monoxide b) nitric acid.

15 a) In aqueous solution, acids donate H^+ ions to water molecules forming H_3O^+ ions. Draw a '*dot-and-cross*' diagram to show the formation of an H_3O^+ ion.
b) Boron fluoride forms molecules with the formula BF_3. Draw a '*dot-and-cross*' diagram for BF_3 and explain why BF_3 molecules readily react with ammonia molecules, NH_3, to form the compound NH_3BF_3.

6.6 Metallic bonding and structures

Metals are important and useful materials. Look around and notice the uses of different metals – in vehicles, bridges, pipes, taps, radiators, cutlery, pans, jewellery and ornaments. X-ray studies show that the atoms in most metals are packed together as close as possible. This arrangement is called close packing.

Figure 6.25 shows a model of a few atoms in one layer of a metal crystal. Notice that each atom in the middle of the layer 'touches' six other atoms in the same layer. When a second layer is placed on top of the first, atoms in the second layer sink into the dips between atoms in the first layer (Figure 6.26).

Definitions

Delocalised electrons are bonding electrons that are not fixed between two atoms in a bond. They are free to move and are shared by several or many more atoms.

Metallic bonding is the strong attraction between a lattice of positive metal ions and a 'sea' of delocalised electrons.

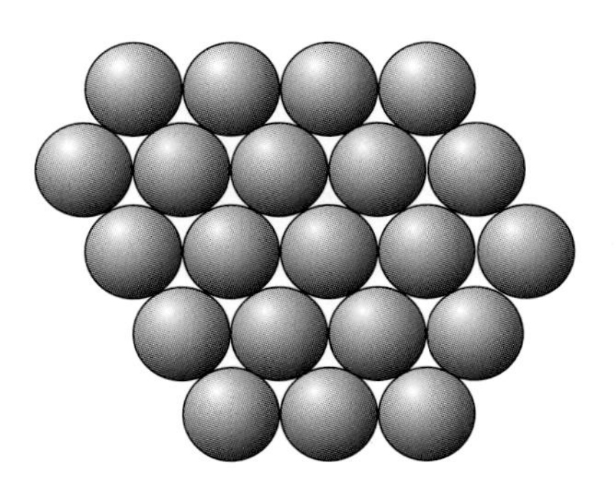

Figure 6.25 ▲
The close packing of atoms in one layer of a metal.

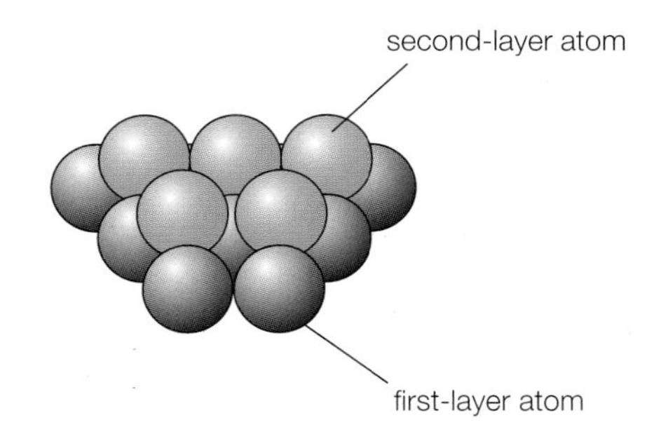

Figure 6.26 ▲
Atoms in two layers of a metal crystal.

This close packing allows atoms in one layer to get as close as possible to those in the next layer forming a giant lattice of closely packed atoms in a regular pattern. In this giant lattice, electrons in the outer shell of each metal atom are free to drift through the whole structure. These electrons do not have fixed positions. They are described as 'delocalised'. Metals therefore consist of giant lattices of positive ions, with electrons moving around and between them in a 'sea' of delocalised negative charge (Figure 6.27). The strong electrostatic attractions between the positive metal ions and the 'sea' of delocalised electrons result in strong forces between the metal atoms.

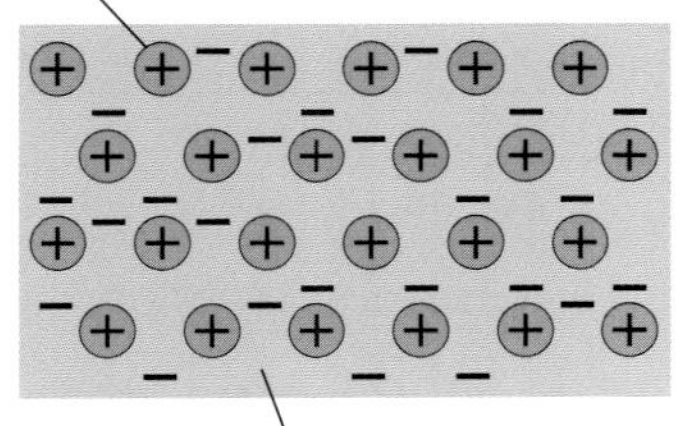

Figure 6.27 ▲
Metallic bonding results from the strong attractions between metal ions and the 'sea' of delocalised electrons.

Test yourself

16 Why are the electrons in the outermost shell of metal atoms described as 'delocalised'?

17 All of the properties of metals can be explained and interpreted in terms of their close-packed structure and delocalised electrons. Use these ideas and models to explain why most metals:
- a) have high densities
- b) have high melting points
- c) are good conductors of electricity.

18 Name a metal or alloy and its particular use to illustrate each of the following typical properties of metals:
- a) shiny
- b) conduct electricity
- c) bend without breaking
- d) high tensile strength.

19 Consider the patterns of metal properties in the periodic table.
- a) Which metals have:
  - i) relatively high densities
  - ii) relatively low densities?
- b) Which metals have:
  - i) relatively high melting points
  - ii) relatively low melting points?

Figure 6.28 ▲
Blacksmiths rely on the malleability of metals to hammer and bend them into useful shapes. Metals are malleable because their bonding is not directional and the layers of atoms can slide over each other.

Properties of metals

In general, metals:

- have high melting and boiling points
- have high densities
- are good conductors of heat and electricity
- are malleable (can be bent or hammered into different shapes).

Activity

Choosing metals for different uses

Various properties of six metals are shown in Table 6.2.

Metal	Density/ $g\ cm^{-3}$	Tensile strength/ $10^7\ Nm^{-2}$	Melting point/°C	Electrical resistivity $/10^{-8}$ ohm m	Thermal conductivity $/Js^{-1}\ cm^{-1}\ K^{-1}$	Cost per tonne/£
Aluminium	2.7	8	660	2.5	2.4	960
Copper	8.9	33	1 083	1.6	3.9	1,200
Iron	7.9	21	1 535	8.9	0.8	130
Silver	10.5	25	962	1.5	4.2	250 000
Titanium	4.5	23	1 660	43.0	0.2	27 000
Zinc	7.1	14	420	5.5	1.1	750

Table 6.2 ▲
Various properties of six metals.

1 Use the information in Table 6.2 to explain the following statements.

a) Copper is used in most electrical wires and cables, but high-tension cables in the National Grid are made of aluminium.
b) Bridges are built from steel, which is mainly iron, even though the tensile strength of iron is lower than that of some other metals.
c) Metal gates and dustbins are made from steel, coated (galvanised) with zinc.
d) Silver is no longer used to make coins in the UK.
e) Aircraft are now constructed from an aluminium/titanium alloy rather than pure aluminium.
f) The base of high-quality saucepans is copper rather than steel (iron).

2 If the atoms in a metal pack closer, the density should be higher, the bonds between atoms should be stronger and so the melting point should be higher. This suggests there should be a relationship between the density and melting point of a metal.

a) Use the data in the table to check whether there is a relationship between density and melting point.
b) Is there a relationship between the densities and melting points of the metals? Say 'yes' or 'no' and explain your answer.

3 The explanation of both electrical and thermal conductivity in metals uses the concept of delocalised electrons. This suggests there should be a relationship between the electrical and thermal conductivities of metals.

a) Use the data in the table to check whether there is a relationship. (Hint: Electrical resistivity is the reciprocal of electrical conductivity.)
b) Is there a relationship? Say 'yes' or 'no' and explain your answer.

6.7 Directional bonds

In covalent bonds, two atoms share a pair of electrons between their nuclei, and these two atoms may share pairs of electrons with other atoms (Section 6.5). This means that atoms always stay in the same positions relative to each other with bonds in the same direction, and this gives molecules a fixed shape. This constant three-dimensional spatial relationship between the atoms in a molecule is important, because it can affect the physical and chemical properties of covalently bonded compounds.

Figure 6.29 ◀
The three-dimensional shape of molecules is crucial in the structure of enzymes that control the rate of chemical reactions in living things. This photograph shows the use of an insecticide that has just the right shape to interfere with an enzyme in the insects and prevent their growth. Fortunately, the insecticide has little or no effect on the adjacent plants, which thrive once the insects have been destroyed.

In contrast to covalent bonds, ionic bonding and metallic bonding involve the attraction of oppositely charged particles around each other in every direction. Because of this difference, covalent bonding is described as directional, whereas ionic and metallic bonding are described as non-directional.

> **Definition**
>
> **Bond length** is the distance between the nuclei of two atoms linked by one or more covalent bonds.
>
> **Bond angle** is the angle between two covalent bonds in a molecule or giant covalent structure.

6.8 The shapes of molecules and ions

X-ray diffraction studies and other instrumental methods allow chemists to measure bond lengths and bond angles very accurately in molecules and ions that have covalent bonds, like NH_4^+. The results show that covalent bonds have a definite direction and a definite length. For example, X-ray diffraction studies show that all the C–H bond lengths in methane, CH_4, are 0.109 nm and all the H–C–H bond angles are 109.5° (Figure 6.30).

Chemists have developed a very simple theory to explain and predict the shapes and bond angles of simple molecules and ions containing covalently bonded atoms. The theory is based on repulsion between electron pairs in the outermost shell of the central atom. The theory says that electron pairs in the outer shell of atoms and ions will repel each other and get as far apart as possible. This is sometimes called the electron pair repulsion theory.

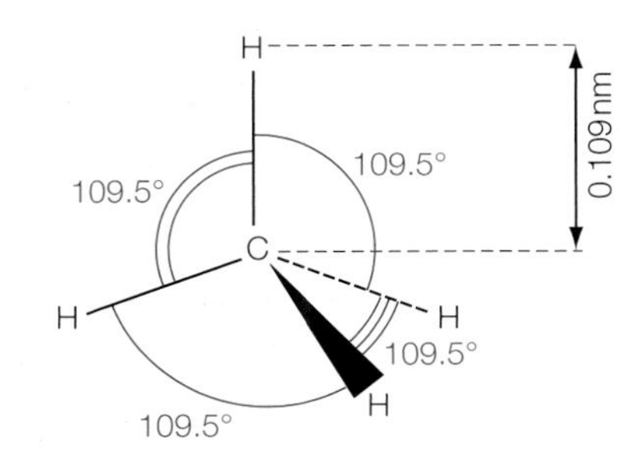

Figure 6.30 ▲
All the bond angles in methane are 109.5° and all the C–H bond lengths are 0.109 nm.

Look at the molecules of beryllium chloride and boron trifluoride in Figure 6.31. Beryllium chloride, $BeCl_2$, provides the simplest example of the electron pair repulsion theory. In the $BeCl_2$ molecule, beryllium has only two pairs of electrons in its outer shell. In order to get as far apart as possible, these two pairs of electrons must be on opposite sides of the beryllium atom. The shape of the molecule is described as linear, and the Cl–Be–Cl bond angle is 180°.

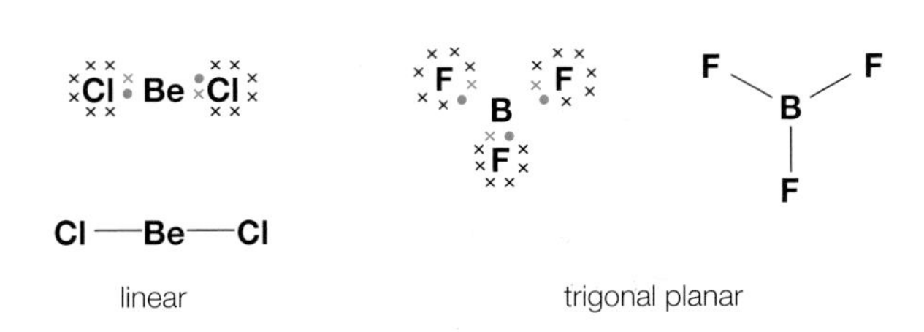

Figure 6.31 ◀
The shapes of molecules with two and three electron pairs around the central atom.

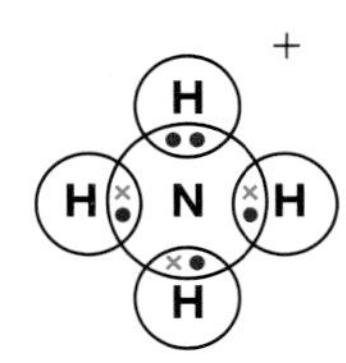

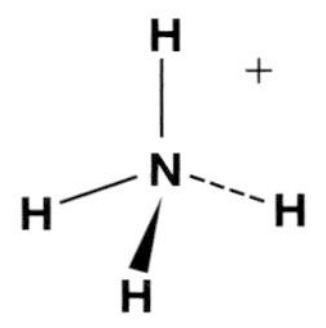

tetrahedral

Figure 6.32 ▲
The shape of the NH_4^+ ion.

The next simplest example of the electron pair repulsion theory is shown by boron trifluoride, BF_3. In the BF_3 molecule, boron has three electron pairs in its outer shell. This time, to get as far apart as possible, the three pairs must occupy the corners of a triangle around the boron atom. The shape of this molecule is described as trigonal planar and the F–B–F bond angles are 120°.

Now look at the shape of the ammonium ion, NH_4^+, in Figure 6.32. In this ion, nitrogen has four electron pairs in its outer shell and each is bonded to a hydrogen atom. To get as far apart as possible, the four pairs are in positions that would be at the corners of a tetrahedron. The four hydrogen atoms are in similar positions further away from the nitrogen atom. The shape of the NH_4^+ ion is described as tetrahedral and all of the H–N–H bond angles are 109.5°.

Table 6.3 summarises the shapes of molecules with two, three, four, five and six pairs of electrons based on the electron pair repulsion theory. In each case, the electron pairs are repelled as far apart as possible. The table also shows the predicted bond angles for each molecule.

The electron pair repulsion theory shows how chemists can make generalisations from their results and then use these generalisations to make predictions.

Number of electron pairs	Shape	Bond angle ∠XMX	Example
2	X–M–X Linear	180°	Cl—Be—Cl $BeCl_2$
3	Trigonal planar	120°	BF_3
4	Tetrahedral	109.5°	CH_4
5	Trigonal bipyramidal (Two triangle-based pyramids joined at the bases.)	90° and 120°	gaseous PCl_5
6	Octahedral (Two square-based pyramids joined at the bases.)	90°	SF_6

Table 6.3 ▶
The shapes of molecules with two to six pairs of bonded electrons around the central atom.

6.9 Molecules and ions with lone pairs and multiple bonds

Tutorial
Part 2

Lone pairs

Some molecules, such as ammonia and water, contain non-bonded or lone pairs of electrons, as well as bonded pairs (Figure 6.33).

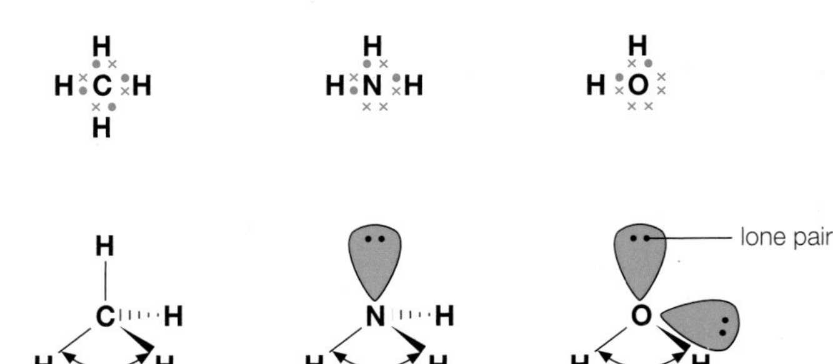

Figure 6.33 ▲
The shapes and bond angles in molecules with bonded pairs and lone pairs of electrons.

Ammonia and water have exactly the same electron structure as methane (Figure 6.33) with four pairs of electrons in the outer shell of the central atom. In methane, all four pairs of electrons are bonded pairs between the central carbon atom and each hydrogen atom. In ammonia, three of the four pairs make up N–H bonds as bonded pairs, but the fourth is a lone pair. Each of these four electron pairs repels the others, so they form a tetrahedral shape around the nitrogen atom. But, the positions of the atoms in the NH_3 molecule make a shape that is pyramidal – a triangle-based pyramid with a nitrogen atom at the top and hydrogen atoms at the three corners of its base. Water also has four pairs of electrons around the central atom – two bonded pairs and two lone pairs. The shape formed by these electron pairs is tetrahedral again, but the shape of the water molecule, H–O–H, is described as V-shaped or non-linear.

Test yourself

20 Why are covalent bonds described as directional bonds?

21 a) Draw a '*dot-and-cross*' diagram for a molecule of carbon dioxide showing only electrons in the outer shell.
   b) Predict the shape of CO_2 molecules.

22 Aluminium chloride sublimes at quite low temperatures forming simple molecules of $AlCl_3$ in which the Al atom is bonded to each Cl atom by a single covalent bond.
   a) Draw a '*dot-and-cross*' diagram for an $AlCl_3$ molecule showing only electrons in the outer shell of all four atoms.
   b) Describe the shape of $AlCl_3$ molecules and predict the Cl–Al–Cl bond angle.

23 Draw 'dot-and-cross' diagrams of the following simple molecules showing only electrons in the outer shell of all atoms:
   a) PF_5 b) SiH_4 c) BCl_3.

24 Predict the shape and bond angle in:
   a) PF_5 b) SiH_4 c) BCl_3.

Lone pairs of electrons are held closer to the central atom than bonded pairs. This means that lone pairs repel more strongly than bonded pairs and, therefore, the strength of repulsion between electron pairs is:

lone pair-lone pair > lone pair-bonded pair > bonded pair-bonded pair.

This explains why the bond angle in ammonia with one lone pair is less than the bond angle in methane and why the bond angle in water with two lone pairs is less than that in ammonia. Similar predictions about shapes and bond angles can be made for ions such as H_3O^+, BF_4^- and NH_2^- (Figure 6.34).

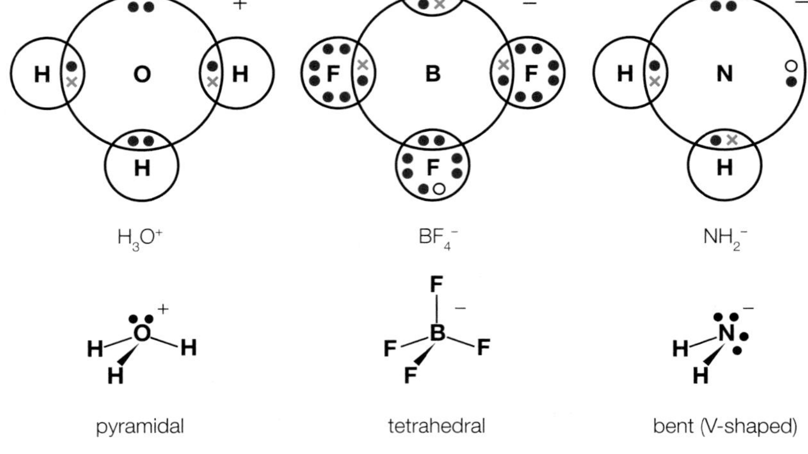

Figure 6.34 ▶
'*Dot-and-cross*' diagrams and shapes of H_3O^+, BF_4^- and NH_2^-.

Test yourself

25 Draw '*dot-and-cross*' diagrams for the following simple molecules and predict their shapes:
 a) PH_3
 b) HCN
 c) N_2
 d) CCl_4.

26 a) Draw a '*dot-and-cross*' diagram for CH_3Cl.
 b) What is the shape of the CH_3Cl molecule?
 c) The H–C–H bond angles in CH_3Cl are 111°, whereas the Cl–C–H bond angles are 108°. Why do you think the Cl–C–H bond angles are smaller than the H–C–H bond angles? (Hint: The C–H bond is much shorter than the C–Cl bond.)

27 a) Draw '*dot-and-cross*' diagrams for CH_4 and H_2O.
 b) Each of these molecules has four pairs of electrons around the central atoms (carbon and oxygen, respectively). In CH_4, the H-C-H bond angle is 109.5°. In H_2O, the H-O-H bond angle is 104.5°.
  i) Explain why the bond angle in CH_4 is 109.5°.
  ii) Both CH_4 and H_2O have four pairs of electrons around their central atoms. Why do you think the bond angle in H_2O is less than 109.5°.
  iii) Suggest a name for the shape of the H_2O molecule.

Multiple bonds

The electrons in double bonds and triple bonds are more or less between the nuclei of the two atoms they join. Double bonds and triple bonds are therefore equivalent to one centre of negative charge when predicting molecular shapes (Figure 6.35).

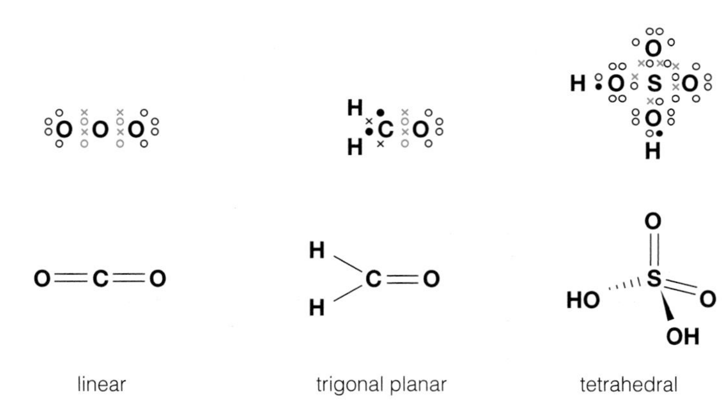

Figure 6.35 ▲
The shapes of some molecules with multiple bonds.

From our knowledge of lone pairs of electrons and multiple bonds, it is clear that lone pairs, double bonds and triple bonds all affect the shapes of molecules and ions in a similar way to electrons in single bonds. All of these (single bonds, lone pairs, double bonds and triple bonds) can be regarded as separate centres of negative charge when predicting the overall shapes of molecules and ions.

6.10 A spectrum of bonding

The electron pair in a covalent bond is not shared equally if the two atoms joined by the bond are different. The nucleus of one atom will attract the electrons more strongly than the nucleus of the other. This means that one end of the bond has a slight excess of negative charge (δ–). The other end of the bond has a slight deficit of negative charge and the charge cloud of electrons does not cancel the positive charge on the nucleus. This end of the bond has a partial positive charge (δ+) and the bond is described as polar.

This existence of partial charges at the ends of some covalent bonds results in a spectrum of bonding from pure covalent bonding in molecules such as chlorine and oxygen through increasing separation of partial charges in molecules such as hydrogen chloride and hydrogen fluoride to pure ionic bonding in compounds such as potassium fluoride and sodium chloride.

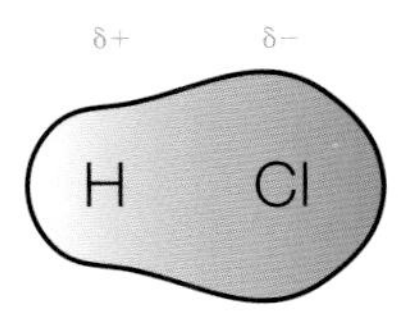

Figure 6.36 ▲
A polar covalent bond in hydrogen chloride. The chlorine atom, with 17 protons in its nucleus, attracts the shared electrons in the covalent bond more strongly than the hydrogen atom with only one proton in its nucleus. The uneven distribution of electrons leads to partial charges at the ends of the covalent bond but overall, the molecule is uncharged.

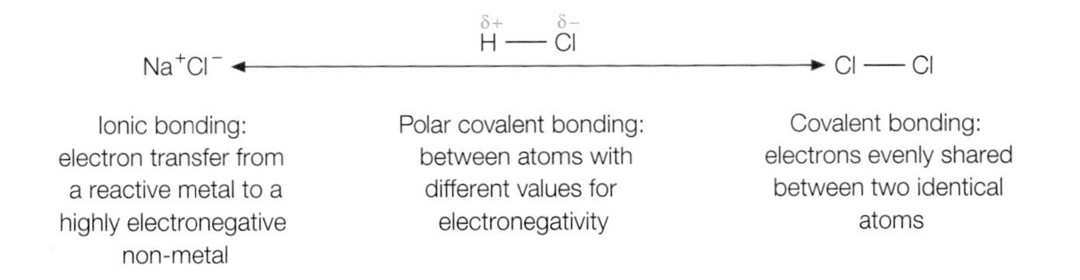

Figure 6.37 ▲
A spectrum from purely ionic to purely covalent bonding.

Definition

Polar covalent bonds are bonds between atoms of different elements. The shared electrons are drawn towards the atom with the stronger pull on the electrons. The bonds have a positive pole at one end and a negative pole at the other.

6.11 Electronegativity

Chemists use the term electronegativity to describe the ability of an atom to attract the bonded electrons in a covalent bond. They have also developed electronegativity values to predict the extent to which the bonds between different atoms are polar. The stronger the pull of an atom on the electrons it shares with other atoms, the higher its electronegativity. Oxygen is more electronegative than hydrogen, so an O–H bond is polar with a slight negative charge on the oxygen atom and a slight positive charge on the hydrogen atom.

Two quantitative scales of electronegativity are in common use: one developed by Linus Pauling (1901–1994) and the other by Robert Mulliken (1896–1986). In each of these scales, the electronegativity values compare one element with another qualitatively, so it is enough to know the trends in electronegativity across and down the periodic table.

Highly electronegative elements, such as fluorine and oxygen, are at the top right of the periodic table. The least electronegative elements, such as caesium, are at the bottom left.

Note

Scientists use the Greek letter 'delta', δ, for a small difference or change in something. So δ+ and δ– are used for the small charges at the ends of a polar bond. The capital Greek 'delta', Δ, is used for larger changes or differences (Section 16.2).

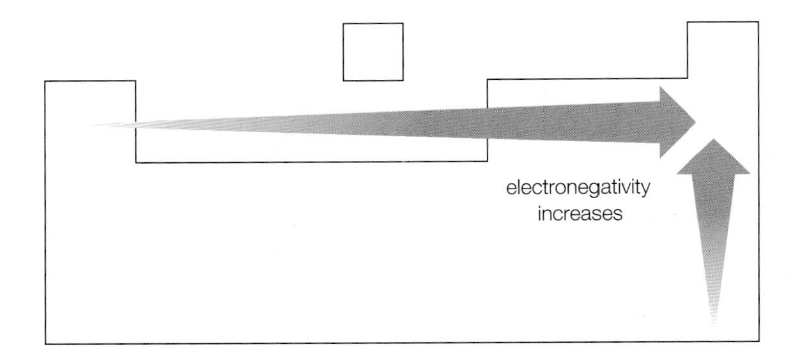

Figure 6.38 ◄
Trends in electronegativity in the periodic table.

Definition

Electronegativity is the ability of an atom of an element to attract the shared electrons in a covalent bond. In a polar bond, the shared electrons are drawn towards the more electronegative atom.

The bigger the difference in electronegativity of the elements forming a bond, the more polar and possibly more ionic the bond. The bonding in a compound becomes ionic if the difference in electronegativity is large enough for the more electronegative element to remove electrons completely from the other element. This happens in compounds such as sodium chloride, magnesium oxide and calcium fluoride.

Activity

Interpreting electronegativity values

Table 6.4 shows electronegativity values for selected elements in the periodic table. The values are those from the Pauling scale.

		H 2.1				
Li 1.0	Be 1.5	B 2.0	C 2.5	N 3.0	O 3.5	F 4.0
Na 0.9	Mg 1.2	Al 1.5	Si 1.8	P 2.1	S 2.5	Cl 3.0
K 0.8	Ca 1.0					Br 2.8
Rb 0.8	Sr 1.0					I 2.5

Table 6.4 ▲
Electronegativity values for selected elements.

1 Why are electronegativity values for He and Ne not included in the table?

2 Identify three pairs of elements in the table that would combine to form molecules with the least polar covalent bonds.

3 Identify the two elements in the table that would form the most purely ionic compound.

4 What is the trend in electronegativity from left to right across a period?

5 a) Draw diagrams showing the electrons in the main shells for lithium and fluorine.

b) Use your diagrams and the concept of shielding to explain why fluorine is much more electronegative than lithium.

6 What is the trend in electronegativity down a group?

7 a) Draw diagrams showing the electrons in the main shells for fluorine and chlorine.

b) Use your diagrams and the concept of shielding to explain why fluorine is more electronegative than chlorine.

Test yourself

28 Use Figure 6.38 and Table 6.4 to predict the polarity of the bonds in:
a) H_2S b) NO c) CCl_4 d) ICl.

29 Put these bonds in order of polarity with the most polar first: C–I, C–H, C–Cl, C–O, C–F and C–Br.

6.12 Polar molecules

The covalent bonds in hydrogen chloride are polar and the molecules are also polar because there is only one bond in each molecule. There are, however, molecules with polar bonds that are not polar. One example is tetrachloromethane (Figure 6.39). The four polar bonds in CCl_4 are arranged symmetrically around the central carbon atom, so that, overall, they cancel each other out.

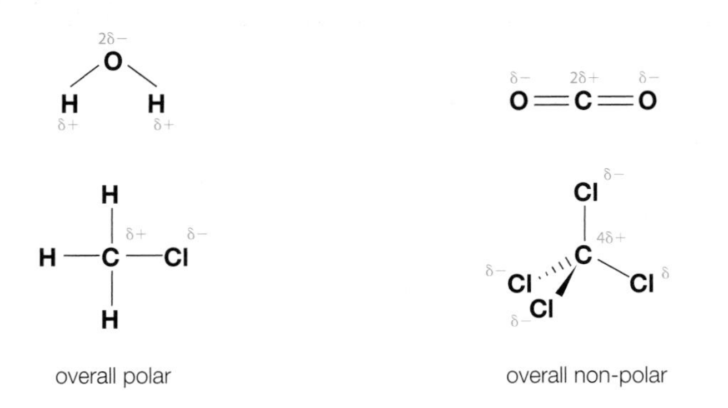

Figure 6.39 ▲
Molecules with polar bonds. In the examples on the left, the net result of all the polar bonds is a polar molecule. In the examples on the right, the overall effect of symmetrical polar bonds is a non-polar molecule.

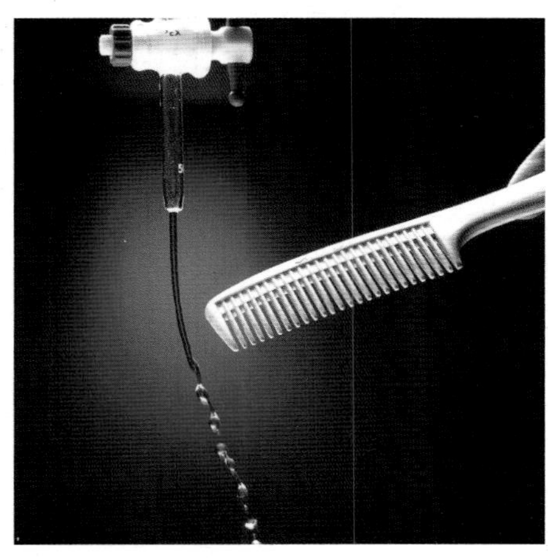

Figure 6.40 ▲
A thin stream of water is attracted to a nearby comb that is carrying an electrostatic charge. This is because the polar water molecules tend to rotate and move as the charge on one side of the molecules is attracted to the opposite charge on the comb.

Polar molecules have a positive electric pole and a negative electric pole. These two poles of opposite charge in a polar molecule form a permanent dipole. Molecules with permanent dipoles are affected by electric fields (Figure 6.40).

Definition

Polar molecules have polar bonds that do not cancel each other out so the whole molecule has a permanent dipole.

Test yourself

30 Put these sets of compounds in order of the character of the bonding, with the most ionic on the left and the most covalent on the right:
a) Al_2O_3, Na_2O, MgO and SiO_2
b) LiI, NaI, KI and CsI.

31 Consider the shape of the following molecules and the polarity of their bonds. Then, divide the molecules into two groups – polar and non-polar: HBr, $CHBr_3$, CBr_4, CO_2, SO_2 and C_2H_6.

Tutorial

6.13 Intermolecular forces

Van der Waals' forces

It is not obvious why there are weak attractions between uncharged non-polar molecules, such as those of iodine, oxygen and the noble gases. If there were no attractions between these non-polar molecules, it would be impossible to turn them into liquids or solids. During the 1870s, the Dutch physicist Johannes van der Waals (1837–1923) developed a theory of intermolecular forces to explain the properties of non-polar gases like oxygen, so the forces between uncharged non-polar molecules are sometimes called van der Waals' forces.

When non-polar molecules meet, there are fleeting repulsions and attractions between the nuclei of their atoms and the surrounding clouds of electrons. Temporary displacements of the electrons lead to temporary dipoles. These temporary instantaneous dipoles can induce dipoles in neighbouring molecules. Positive dipoles induce negative dipoles and vice versa resulting in attractions between these instantaneous and induced dipoles. These attractions give rise to the weakest intermolecular forces, which are roughly 100 times weaker than covalent bonds.

Figure 6.41 ▶
Geckos can climb up glass and hang from ceilings thanks to the intricate design of their feet and intermolecular forces. A gecko's toes are lined with microscopic hairs, and each hair is branched into finer structures. Weak intermolecular forces over the large surface area of the hairs are strong enough to grip on any surface, but weak enough to break as the gecko moves by peeling its feet away.

As the number of electrons in a molecule increases, the more likely it is to form instantaneous dipoles and therefore the greater the possibility for induced dipole–dipole attractions. This explains why the boiling points increase down Group 7 (the halogens) and Group 0 (the noble gases). For the same reason, the boiling points of alkanes increase as the number of carbon atoms increases (Section 11.2).

Definition

Intermolecular forces are weak attractive forces between molecules.

Permanent dipoles create the forces between polar molecules such as hydrogen chloride.

Van der Waals' forces are the forces that result from temporary instantaneous dipoles in non-polar molecules such as oxygen and the noble gases.

Note

Chemists disagree about the definition of van der Waals' forces. In this specification, the term is used to mean only the weakest attractions between temporary induced dipoles.

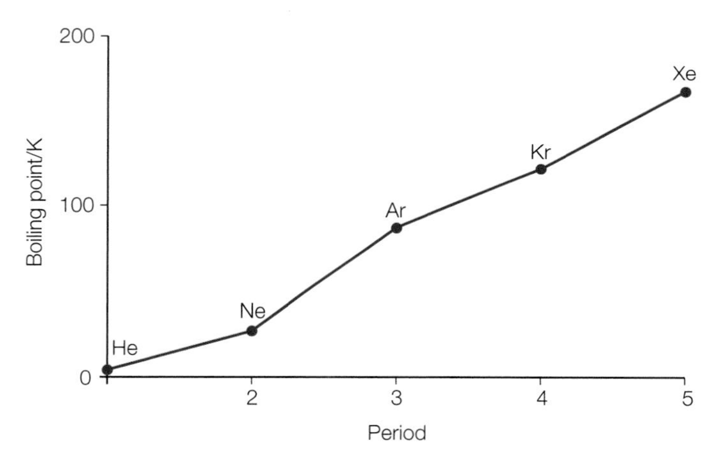

Figure 6.42 ▲
Boiling points of the noble gases plotted against the period they belong to.

The shapes of molecules can also affect the overall size of intermolecular forces. The attractions between long thin molecules are stronger than those between short fat molecules. This is because long thin molecules can lie close to each other and the attractions can take effect over a larger surface area.

Permanent dipole–dipole attractions

Molecules with permanent dipoles attract each other a little more strongly than non-polar molecules. The positive dipole of one molecule tends to attract the negative dipoles of others and vice versa (Section 6.12).

This contribution to intermolecular forces from permanent dipoles occurs in addition to the instantaneous induced dipoles of van der Waals' forces that act between all molecules.

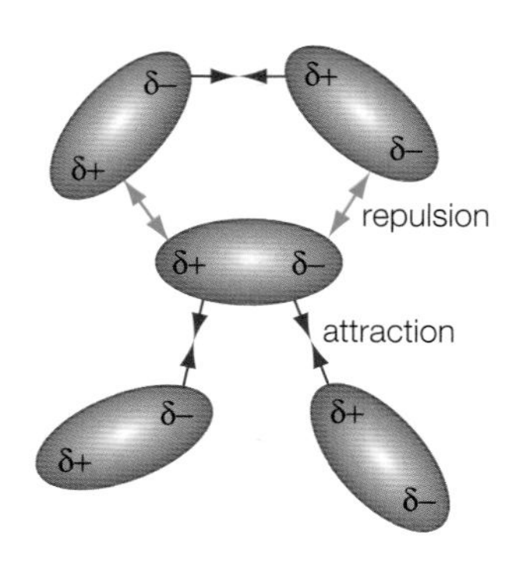

Figure 6.43 ▲
Attractions between molecules with permanent dipoles.

Test yourself

32 Explain how Figure 6.42 illustrates that intermolecular forces increase with the number of electrons in the molecules of monatomic gases.

33 Account for the states of the halogens at room temperatures: chlorine is a gas, bromine is a liquid and iodine is a solid.

34 Account for the difference in boiling points between the following pairs of molecules by considering both van der Waals' forces and permanent dipole–dipole attractions:

a) ethane, which boils at –88 °C and fluoromethane, which boils at –78 °C

b) butane, which boils at –0.5 °C and propanone, CH_3COCH_3, which boils at 56 °C.

6.14 Hydrogen bonding

Tutorial Part 1 Tutorial Part 2

Hydrogen bonding is a special form of permanent dipole–dipole attraction between molecules. It is much stronger than the other two kinds of intermolecular force but still at least 10 times weaker than covalent bonds. Hydrogen bonding affects molecules in which hydrogen is covalently bonded to one of the three highly electronegative elements – fluorine, oxygen and nitrogen. Like permanent dipole attractions, hydrogen bonding acts in addition to van der Waals' forces.

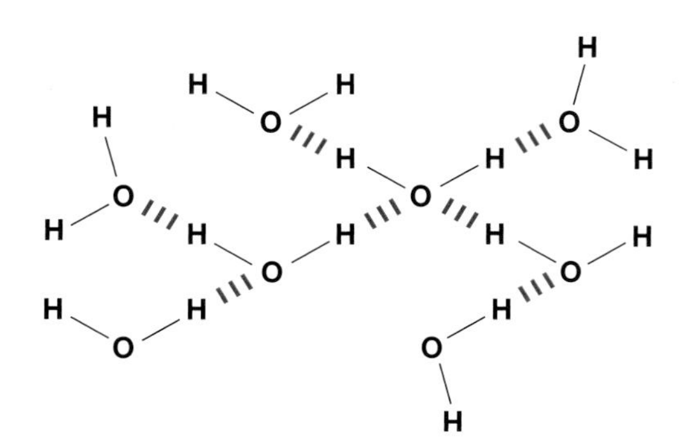

Figure 6.44 ◀ Hydrogen bonding in water.

In a hydrogen bond, the hydrogen atom lies between two highly electronegative atoms. It is hydrogen bonded to one of them and covalently bonded to the other. The covalent bond is highly polar. The small hydrogen atom, with its δ+ dipole, has no inner shells of electrons to shield its nucleus. Because of this, it is attracted strongly and closely to a lone pair of electrons on another electronegative atom (δ–) in a neighbouring molecule. The three atoms associated with a hydrogen bond are always in a straight line.

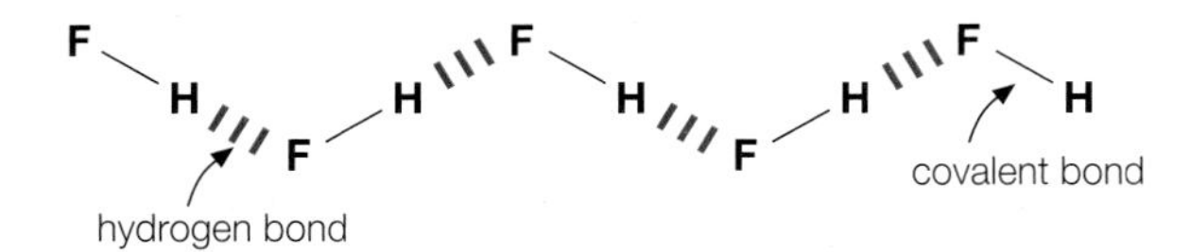

Figure 6.45 ◀ Hydrogen bonding in hydrogen fluoride.

The essential requirements for a hydrogen bond are:

- a hydrogen atom covalently bonded to a highly electronegative atom, and
- an unshared pair of electrons on a second electronegative atom.

A water molecule contains two O–H bonds and two unshared electron pairs on the oxygen atom. This means that each oxygen atom in a water molecule can form two hydrogen bonds, which helps to explain the three-dimensional structure of ice and the anomalous properties of water.

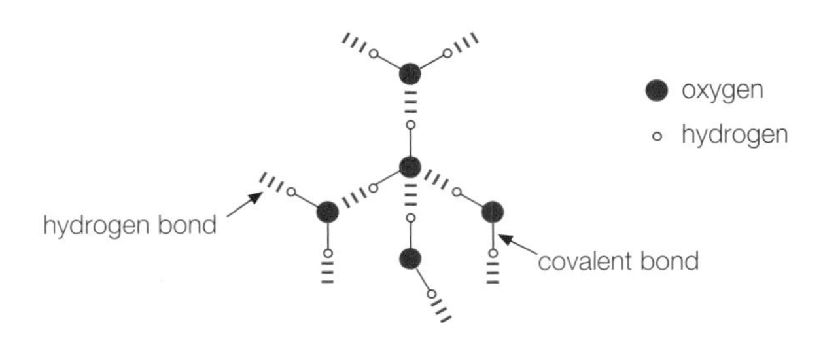

Figure 6.46 ▶
Molecules in ice are held together by hydrogen bonding. The molecules form a giant lattice in which each oxygen atom is bonded to two hydrogen atoms by covalent bonds and two others by hydrogen bonds.

Hydrogen bonding accounts for:

- the open structure of ice in which H_2O molecules are, on average, further apart than H_2O molecules in water. This leads to the unusual situation in which ice is less dense than water
- the relatively high freezing points and boiling points of water, ammonia and hydrogen fluoride, which are out of line with those of the other hydrides in Groups 5, 6 and 7.

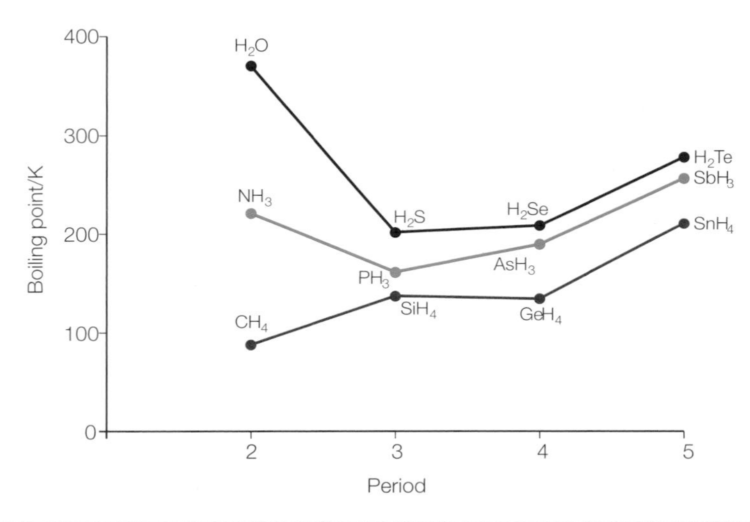

Figure 6.47 ▶
Boiling points of the hydrides of the elements in Groups 4, 5 and 6.

Figure 6.48 ▶
An iceberg in Antarctica. About 10% of the ice is above the surface of the sea because ice is less dense than water at 0 °C.

Definition

Hydrogen bonds are extra-strong permanent dipole attractions that are much stronger than other intermolecular forces, but much weaker than covalent bonds.

Test yourself

35 a) The three atoms associated with a hydrogen bond are always in a straight line. Why is this?
b) State and explain the maximum number of hydrogen bonds per molecule in:
i) hydrogen fluoride, HF
ii) ammonia, NH_3.

36 Draw diagrams to show hydrogen bonding between water molecules and:
a) ammonia molecules in a solution of ammonia, NH_3
b) ethanol molecules in a solution of ethanol, CH_3CH_2OH.

37 a) Use the data sheet for Group 7 to plot a graph that shows how the boiling points of the hydrogen halides vary with the atomic number of the halogen.
b) Describe and explain the pattern shown in the graph by referring to the types of intermolecular forces that act between the molecules.

Data

38 Explain the differences between the plot for the hydrides of Group 6 and the plot for the hydrides of Group 4 in Figure 6.47.

39 Which types of intermolecular force hold the molecules together in:
a) hydrogen bromide, HBr
b) methane, CH_4
c) methanol, CH_3OH?

6.15 Solutions and solubility

Patterns of solubility

As a rough guide to solubility, we can say, 'like dissolves like'. Water, which is highly polar, dissolves many ionic compounds and compounds with –OH groups such as alcohols and sugars. Non-polar solvents, such as cyclohexane, dissolve hydrocarbons, molecular elements and molecular compounds.

But, there is a limit to the quantity of a chemical that can dissolve in a solvent. A solution is saturated when it contains as much of the dissolved solute as possible at a particular temperature.

Definitions

A **solute** is a chemical that dissolves in a **solvent** to make a **solution**.

A **saturated solution** contains as much of the solute as possible at a particular temperature.

Solubility is a measure of the concentration of a saturated solution of a solute at a specified temperature. Solubilities are usually recorded as 'mol per 100 g water' or 'g per 100 g water' at 25 °C (298 K).

Soluble or insoluble?

No chemicals are completely soluble or completely insoluble in water. Even so, chemists use a rough classification of solubility based on what they see when a little solid is shaken with water in a test tube.

- With very soluble chemicals, like potassium nitrate, plenty of the solid dissolves quickly.
- With soluble chemicals, like copper(II) sulfate, crystals visibly dissolve to a significant extent.
- With sparingly or slightly soluble chemicals, like calcium hydroxide, a little solid seems to dissolve.
- With insoluble chemicals, like iron(III) oxide, there is no sign that any of the material dissolves.

A similar rough classification applies to gases dissolving in water. Ammonia and hydrogen chloride are very soluble. Sulfur dioxide and chlorine are soluble. Carbon dioxide and oxygen are slightly soluble. Helium is insoluble.

Solubility and intermolecular forces

Patterns of solubility for molecular solids are determined by intermolecular forces. Three interactions are involved:

- intermolecular forces between solute molecules
- intermolecular forces between solvent molecules and
- intermolecular forces between solute and solvent molecules.

When all three interactions are about the same strength, the solute dissolves freely in the solvent. This means that non-polar molecules, such as those of a

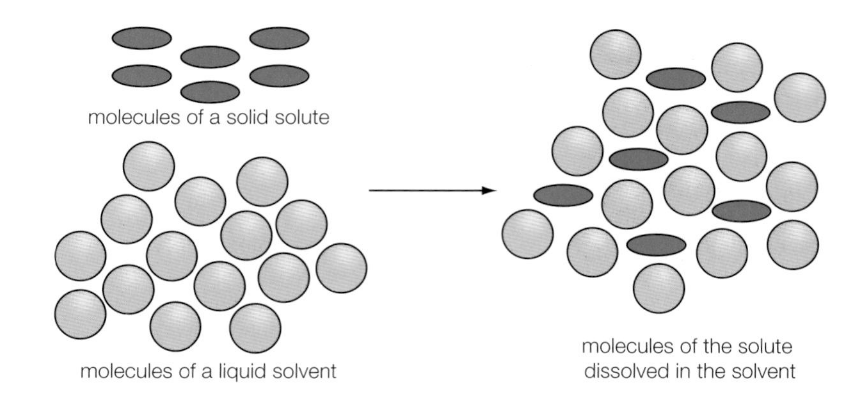

Figure 6.49 ▶ Molecular chemicals dissolve if the energy needed to break intermolecular forces and separate the molecules in the solute and in the solvent is about the same as the energy released when the solute forms new intermolecular forces with the solvent.

Test yourself

40 Explain why methane is insoluble in water, but ammonia is freely soluble.

41 Explain why iodine is soluble in cyclohexane, but almost insoluble in water.

42 Explain why methanol is very soluble in water.

43 With the help of data tables, classify the following salts as very soluble, soluble, slightly soluble or insoluble according to their solubility in water: barium sulfate, caesium fluoride, calcium hydroxide, calcium sulfate, lithium fluoride, lithium chloride, magnesium chloride, magnesium sulfate, potassium iodide.

Data

hydrocarbon wax, dissolve and mix freely with non-polar liquids such as cyclohexane. All of the intermolecular forces involved are van der Waals' forces.

However, non-polar molecules, such as hydrocarbons, do not dissolve in water. The non-polar molecules can easily separate because their intermolecular forces are weak. But the strong hydrogen bonding between water molecules acts as a barrier that keeps out molecules that cannot themselves form hydrogen bonds.

Organic molecules that can form hydrogen bonds, such as alcohols, do dissolve and mix with water. Ethanol molecules, for example, can break into the hydrogen-bonded structure of water by forming new hydrogen bonds between ethanol and water molecules. Alcohols with longer hydrocarbon chains do not mix and dissolve in water so easily. The longer the chain, the less soluble the alcohol.

Solutions of ionic salts in water

It is not obvious why the charged ions in a solid like sodium chloride separate and dissolve in water so easily. The explanation for the solubility of ionic salts, such as sodium chloride, in water is that the charged ions are strongly attracted to polar water molecules. The water molecules cluster around the ions and bind to them. This binding of water molecules to ions is called hydration. In the case of sodium chloride, the energy released as water molecules bind to the ions is enough to compensate for the energy needed to overcome the ionic bonding between the ions.

Other salts are insoluble in water because the energy that would be released when the ions are hydrated is not large enough to balance the energy needed to separate the ions.

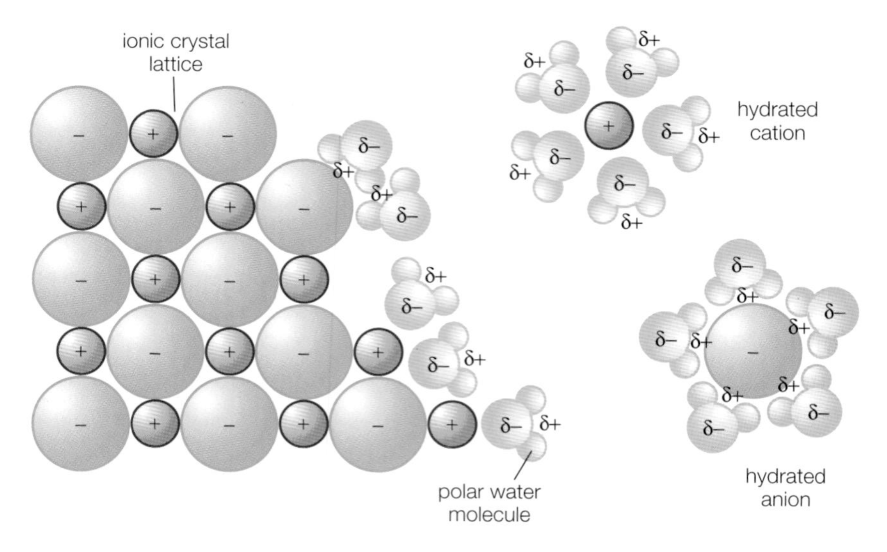

Figure 6.50 ▶ Sodium and chloride ions leaving a crystal lattice and becoming hydrated as they dissolve in water. The bond between the ions and the polar water molecules is an electrostatic attraction.

REVIEW QUESTIONS

1 This question is about calcium and calcium oxide.

a) i) Describe the bonding in calcium. (3)

ii) Explain why calcium is a good conductor of electricity. (2)

b) Draw '*dot-and-cross*' diagrams for the ions in calcium oxide showing **all** the electrons and the ionic charges. (4)

c) Under what conditions does calcium oxide conduct electricity? Explain your answer. (4)

2 The following table shows the melting points of the elements in Period 3 of the periodic table.

Element	Na	Mg	Al	Si	P	S	Cl	Ar
Melting point/K	371	922	933	1683	317	386	172	84

a) Explain why the melting point of sodium is much lower than that of magnesium. (3)

b) Phosphorus and sulfur exist as molecules of P_4 and S_8 respectively. Explain their difference in melting points. (2)

c) State the type of structure and the nature of the bonding in each of the following elements:

i) aluminium

ii) silicon

iii) chlorine. (6)

3 a) Use sodium chloride, hydrogen chloride and copper to explain what is meant by covalent, ionic and metallic bonding. (9)

b) Compare and explain the conduction of electricity by sodium chloride and copper in terms of structure and bonding. (6)

4 a) Phosphorus forms the chloride PCl_3. Draw a '*dot-and-cross*' diagram for PCl_3. (2)

b) Draw and name the shape of the PCl_3 molecule. (2)

c) Explain why PCl_3 has this shape. (2)

d) Why does PCl_3 form a stable compound with BCl_3? (3)

5 When space travel was being pioneered, one of the first rocket fuels was hydrazine, H_2NNH_2.

a) Draw a '*dot-and-cross*' diagram to show the electron structure of a hydrazine molecule. (2)

b) Predict the value of the H–N–H bond angle in a hydrazine molecule and explain your reasoning. (3)

c) Suggest a possible equation for the reaction that occurs when hydrazine vapour burns in oxygen. (2)

d) When 1.00 g of hydrazine burns in excess oxygen, 18.3 kJ of thermal energy is released. Calculate the energy released when 1 mole of hydrazine burns in oxygen. (2)

6 Three of the hydrides of Group 6 elements are H_2O, H_2S and H_2Se.

a) Explain why all three hydrides have polar molecules. (2)

b) State and explain the trend in electronegativity from O to Se down Group 6. (3)

c) In which of the hydrides of Group 6 are the bonds most polar? (1)

7 Thin streams of some liquids are attracted towards a charged rod, but there is no such effect with other liquids.

a) Explain why some liquids are attracted while others are not. (2)

b) i) Which of the liquids, water, hexane, bromoethane and tetrachloromethane, are attracted towards a charged rod? (2)

ii) Explain your predictions. (2)

c) Why are the affected liquids always attracted towards the charged rod and never repelled by it? (2)

8 Explain the following statements about oil refining in terms of intermolecular forces and covalent bonds.

a) It is possible to separate the hydrocarbons in crude oil into fractions by distillation. (3)

b) Cracking turns a mixture of liquids separated from crude oil into a mixture of gases. (2)

c) Isomerisation turns straight chain alkanes into alkanes with lower boiling points. (2)

9 a) Name the strongest of the intermolecular forces between water molecules and describe the bonding with the help of a diagram. (3)

b) Explain why:

i) the boiling point of water is higher than the boiling points of the other hydrides of Group 6 elements

ii) ice is less dense than water at 0 °C

iii) water and pentane do not mix. (6)

7 Periodicity

Inorganic chemistry is the study of the hundred or so chemical elements and their compounds. The amount of information can be bewildering, hence the importance of the periodic table. This table helps to reveal the many similarities and trends in the properties and reactions of chemicals.

7.1 Discovering periodic patterns

The periodic table was a triumph for nineteenth-century chemistry at a time when scientists were discovering many new elements. In the first 20 years of the century, scientists found new elements at an average rate of one a year. They noticed that some of the elements had similar properties. They knew about Dalton's work (Section 1.3) on atomic theory and began to wonder if there might be a connection between the chemical properties of elements and their relative masses.

One of the first attempts to investigate the problem was made by the German chemist Johan Döbereiner (1780–1849). He studied groups of three elements that were chemically similar, such as calcium, strontium and barium. He noted that the properties of strontium seemed to be just halfway between those of the two other metals. He also noticed that the atomic mass of strontium was midway between the values for magnesium and barium. He found other examples and suggested that there are groups of three elements in which a middle element has properties that are an average of the properties of the other two. He called these groups 'triads'.

Figure 7.1 ▶ Humphry Davy (1778–1829) used electrolysis to discover six elements in two years. This cartoon by James Gillray shows him pumping the bellows during a demonstration of laughing gas at the Royal Institution.

Test yourself

1 Use modern data to test whether these groups of similar elements fit with Döbereiner's 'law of triads'.
 a) Lithium, sodium and potassium
 b) Chlorine, bromine and iodine
 c) Carbon, silicon and germanium

Döbereiner published his idea in 1829. It was taken up by other scientists but progress was slow because accurate values of atomic masses were not available for most elements. It was not until 1864 that the British chemist John Newlands (1838–1898) made another attempt to find a pattern among the elements. In 1860, a major conference of chemists at Karlsruhe in Germany had established a much more accurate list of relative atomic masses than had ever been available before. Newlands used this data to arrange all of the known elements in order of atomic mass. He wrote: 'The 8th element starting from a given one, is a kind of repetition of the first, like the eighth note in an octave of music.' He called his rule the 'law of octaves'. Even so, Newlands could not persuade others to accept his ideas, and the Chemical Society in London refused to publish his paper.

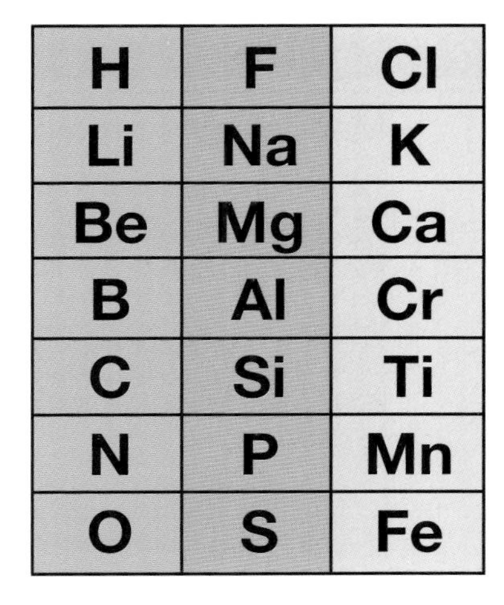

H	F	Cl
Li	Na	K
Be	Mg	Ca
B	Al	Cr
C	Si	Ti
N	P	Mn
O	S	Fe

Figure 7.2 ▲
Version of part of Newlands' table of elements using modern chemical symbols.

Test yourself

2 Refer to Figure 7.2 and show that Newland's law of octaves worked for the three alkali metals shown.
3 One of the reasons that Newlands' theory was rejected was that his table had no spaces for undiscovered elements. Was this a justifiable criticism? Give your reasons.
4 How does the conference in Karlsruhe illustrate the importance of scientists publishing and checking results?

In 1869, a Russian chemist, Dmitri Mendeleev, published the table on which all later versions of the periodic table have been based. He succeeded where others failed because he realised that Newlands had made two mistakes. One mistake was not to leave space for elements that had yet to be discovered. The other was to think that the 'law of octaves' would work throughout the table with seven elements in every row. Mendeleev recognised that later rows in the table had to be longer than those at the beginning. He stated his periodic law as follows: 'When the elements are arranged in order of atomic mass, similar properties recur at intervals.'

Mendeleev made the bold move of predicting the properties of the elements he expected to fill the gaps he had left in his table. For example, there was a gap in his table between silicon and tin in Group 4 of the table. Mendeleev called the missing element eka-silicon and successfully predicted its properties. He was equally successful in predicting properties of other missing elements, including gallium (discovered in 1875) and scandium (discovered in 1879). Within a few years, it was impossible to doubt the usefulness of Mendeleev's table.

Mendeleev organised his table so that elements with similar properties appeared in the same group. To do this, he sometimes found that he had to break the rule of putting the elements in order of atomic mass. Despite the values of their atomic masses, he put iodine in Group 7 and tellurium in Group 6. At the time he could not explain why there was this break in the pattern. It was not until much more was known about atomic structure that it was possible to understand why some elements seemed to break the rules.

Figure 7.3 ▲
Two elements that did not seem to fit Mendeleev's periodic law:
a) tellurium and b) iodine.

Test yourself

5 a) Use data to show that tellurium and iodine break Mendeleev's periodic law.
b) Identify another pair of 'law breakers' in the modern periodic table.
6 a) Which whole group of elements had not been discovered when Mendeleev drew up his periodic table?
b) Why had the elements not been discovered by chemists at the time?
7 Identify two properties of iodine that mean that Mendeleev placed iodine in Group 7 with bromine and chlorine and not in Group 6.

Activity

Mendeleev's predictions

Table 7.1 compares Mendeleev's predictions of the properties of eka-silicon with the properties of germanium, which was found to fill the gap in 1886.

Property	Mendeleev's predictions about eka-silicon	Observed properties of germanium
Relative atomic mass	72	72.6
Density/g cm^{-3}	5.5	5.35
Formula and melting point of oxide	EsO_2 with a high melting point and density 4.7 g cm^{-3}	GeO_2, melting at 1116 °C and density 4.7 g cm^{-3}
Properties of chloride	Chloride boiling below 100 °C and with a density of 1.9 g cm^{-3}	$GeCl_4$ boiling at 86.5 °C and with a density of 1.89 g cm^{-3}

Table 7.1 ▲
Comparison of Mendeleev's predictions about eka-silicon with the properties of germanium.

When the German chemist Clemens Winkler discovered germanium in argyrodite, he had to separate it from arsenic and antimony minerals. He first thought he had discovered eka-antimony. He later realised that the element had to be eka-silicon when he found that it formed two sulfides: a red sulfide, GeS, and a white sulfide, GeS_2.

1. What are the patterns that make it possible to predict the formulae of the oxide and chloride of an element from the element's position in the periodic table?
2. Explain why the oxide of germanium has a high melting point while the chloride boils below 100 °C.
3. Winkler first thought that the element's 'place in the periodic system should be between antimony and bismuth'. Which element could have been what he thought was eka-antimony?
4. Why could Winkler be sure that his element was eka-silicon once he had identified the two sulfides?
5. Why was Mendeleev more successful than others in persuading scientists of the truth of his periodic law?

7.2 Explaining periodic patterns

The connection between the periodic table and the electronic structure of atoms is described in Topic 5. The next stage of the story began in the twentieth century. In 1913, the young English physicist Henry Moseley was studying the wavelength of X-rays given off by elements bombarded by electrons. He worked out from his results that the atomic nucleus of each element must have a characteristic positive charge. He called this its atomic number. He found that when he arranged the elements in order of their atomic numbers, the sequence was almost identical to the sequence when they were arranged in order of their relative atomic masses. The elements that did not fit the pattern were the misfits in Mendeleev's periodic table.

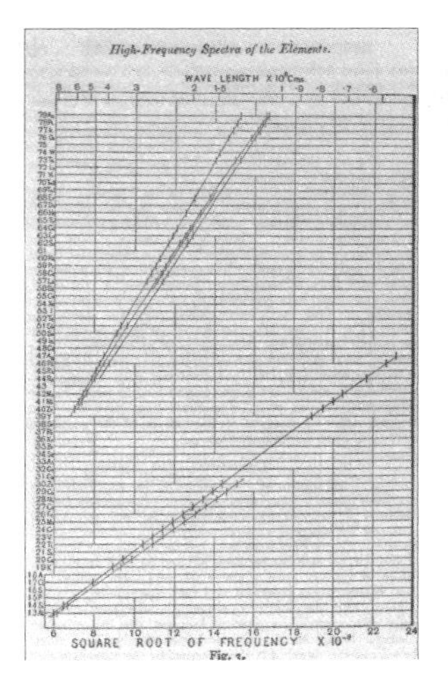

Figure 7.4 ◀
Henry Moseley showed that data from the X-ray spectra of elements produced a periodic pattern when plotted against atomic number.

The discovery of isotopes, at about the same time, helped to confirm that the atomic number is the fundamental property of atoms and not the relative atomic mass. Why this is so could be explained when James Chadwick discovered neutrons in 1932 (Section 1.4).

Understanding of the connection between atomic number and the properties of elements was complete once chemists appreciated that the chemical properties of all elements are determined by the number and arrangement of electrons in atoms. The patterns of electron structures are closely related to the positions of elements in the periodic table (Section 5.5).

Test yourself

8 The name 'isotope' comes from Greek words meaning 'same place'. How did the discovery of neutrons help to explain why the isotopes of an element all belong in the same place in the periodic table?

9 Outline why the chemical properties of an element are determined by its atomic number and not by its relative atomic mass.

7.3 Modern periodic table

The American scientist Glenn Seaborg (1912–1999) was responsible for the last big change to the periodic table. He introduced the actinide series of elements. Most of these elements do not exist naturally but have been created by bombarding atoms with fast neutrons. They include the first 11 elements beyond uranium – the so-called transuranium elements.

The first transuranium element that Seaborg and his team identified was plutonium (atomic number 94). They produced atoms of the element by bombarding uranium with neutrons. Their main difficulty was to prove that they had created new atoms. The challenge was to extract the tiny amount of the artificial element from the mass of uranium.

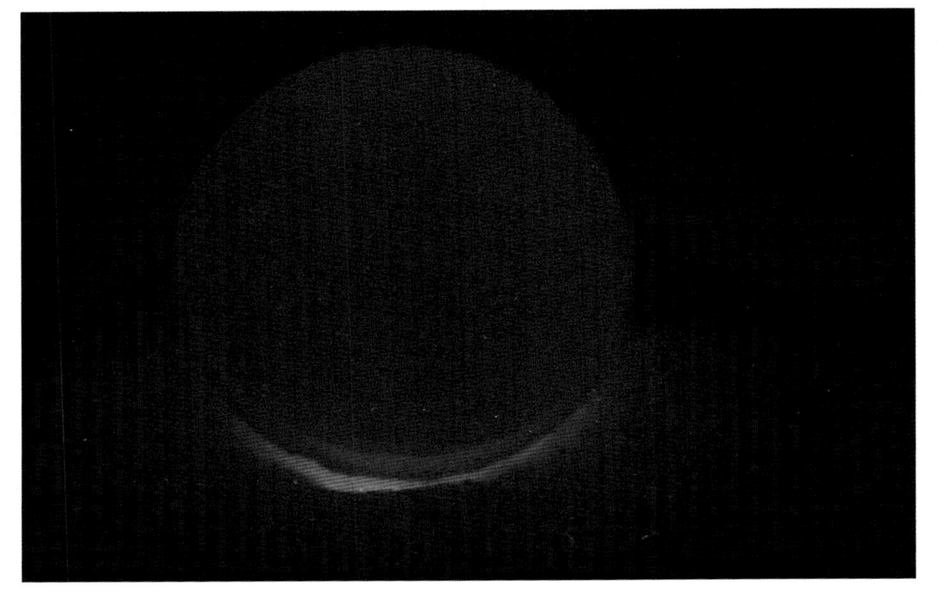

Figure 7.5 ◀
Pellet of plutonium lit up by the glow of its own radioactivity.

Test yourself

10 How did the periodic table help Seaborg to isolate new elements?

11 Give examples of discoveries related to atomic structure or the periodic table that have:
- a) been predicted in advance
- b) been the result of a planned investigation
- c) come about by chance.

12 Suggest two reasons why scientists can be reluctant to accept new ideas when they conflict with established ways of thinking.

Definitions

Physical properties are the properties that describe how a material behaves when it is chemically unchanged. Examples are colour, strength, electrical conductivity, melting point and boiling point.

In the periodic table, **periodicity** describes any pattern in the physical or chemical properties of elements that repeats itself from one period to the next.

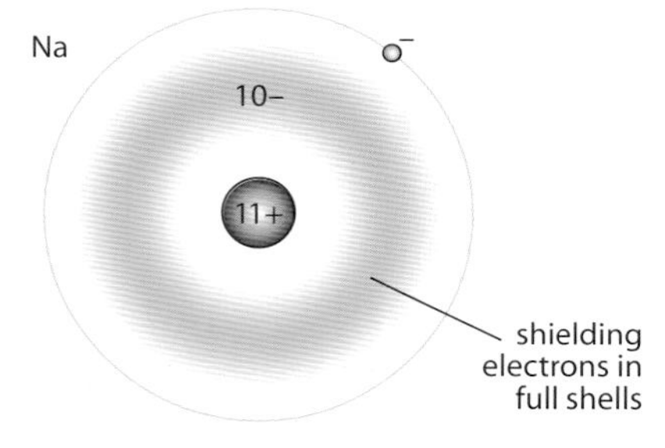

Figure 7.6 ▲
Shielding effect of electrons in inner shells reduces the pull of the nucleus on the electrons in the outer shell.

Note

The International Union of Pure and Applied Chemistry (IUPAC) now recommends that the groups in the periodic table should be numbered from 1 to 18. Groups 1 and 2 are the same as before. Groups 3–12 are the vertical families of d-block elements, and the groups traditionally numbered 3–7 and 0 become Groups 13–18.

Seaborg and his co-workers proved the existence of plutonium in 1940. Making and identifying elements 95 and 96 turned out to be much more difficult. For two years they worked hard and thought deeply about the problem but without success. Then Seaborg had a brainwave. It occurred to him that these heavy elements, starting with actinium (element 89), might form a separate series on the periodic table. If so, their chemistry might be different. This would mean that he needed to use different separation techniques to find them. His team quickly tried new methods and they had soon identified curium (element 96) and americium (element 95).

Despite this success, many scientists doubted that these elements really formed a separate series. It was some years before the idea was generally accepted. As Seaborg said, 'Most ideas go through two stages: at first people think that you're crazy; then, a long time afterwards, they think you were slow not to have thought of it before.'

7.4 Periodic properties

When Mendeleev arranged the elements in order of atomic mass, he saw a repeating pattern in their properties. A repeating pattern is a periodic pattern, hence the terms 'periodic properties' and 'periodicity'.

Perhaps the most obvious repeating pattern in the periodic table is from metals on the left through elements with intermediate properties (called metalloids) to non-metals on the right. Plots of the physical properties of the elements, such as melting points, electrical conductivities and first ionisation energies, also show repeating patterns. The models of bonding between atoms and molecules can be used to explain the properties of elements and the repeating patterns in the periodic table.

First ionisation energies of the elements

Figure 7.7 shows the clear periodic trend in the first ionisation energies of the elements. The general trend is that first ionisation energies increase from left to right across a period.

The ionisation energy of an atom is determined by three atomic properties.

- **Distance of the outermost electron from the nucleus** – As the distance increases, the attraction of the positive nucleus for the negative electron decreases, and this tends to reduce the ionisation energy.
- **Size of the positive nuclear charge** – As the positive nuclear charge increases, its attraction for outermost electrons increases, and this tends to increase the ionisation energy.
- **Shielding effect of electrons** – Electrons in inner shells exert a repelling effect on electrons in the outer shell of an atom, which reduces the pull of the nucleus on the electrons in the outer shell. Shielding means that the 'effective nuclear charge' that attracts electrons in the outer shell is much less than the full positive charge of the nucleus. As expected, the shielding effect increases as the number of inner shells increases (Figure 7.6).

Moving from left to right across any period, the nuclear charge increases as electrons are added to the same outer shell. The increasing nuclear charge tends to pull the outer electrons closer to the nucleus, while the shielding effect of full inner shells is constant. The increased nuclear charge and the reduced distance between the outer electrons and the nucleus makes the outer electrons more difficult to remove, so the first ionisation energy increases.

Notice that the rising trend in ionisation energies across a period in Figure 7.7 is not smooth. There is a 2-3-3 pattern, which reflects the way in which electrons feed into s and p orbitals (Section 5.4).

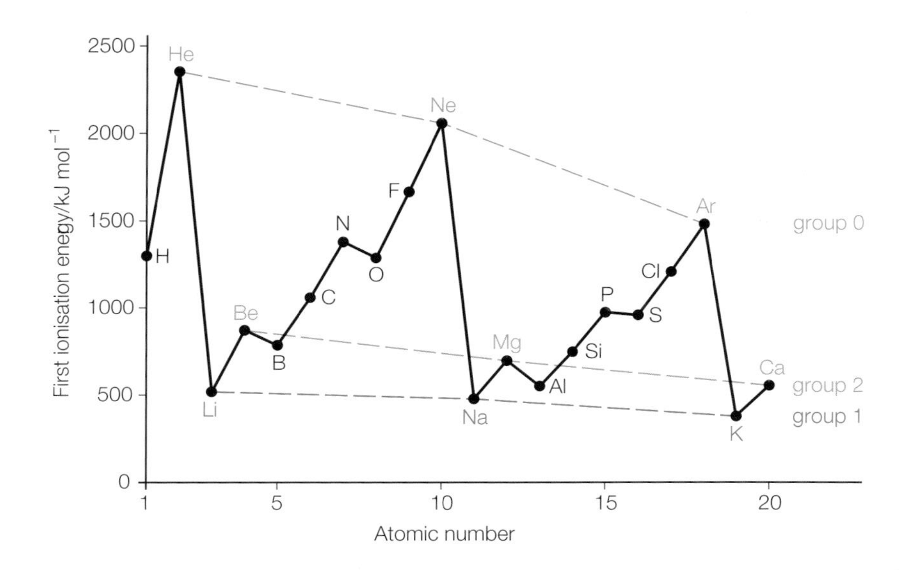

Figure 7.7 ◀
Periodicity in the first ionisation energies of the elements.

13 Use data on the CD-ROM to explore whether there is any pattern in the boiling points of the chlorides of elements in Periods 2 and 3.

14 Why do you think the first ionisation energies of elements decrease with atomic number in every group of the periodic table?

Data

Atomic radii

Atomic radii measure the size of atoms in crystals and molecules. Chemists use X-ray diffraction and other techniques to measure the distance between the nuclei of atoms. The atomic radius of an atom cannot be defined precisely because it depends on the type of bonding and on the number of bonds.

Atomic radii for metals are calculated from the distances between atoms in metal crystals (metallic radii). The atomic radii for non-metals are calculated from the lengths of covalent bonds in crystals or molecules (covalent radii).

Atomic radii decrease from left to right across a period. Across the period Na to Ar, atomic radii fall from 0.191 nm for sodium to 0.099 nm for chlorine. From one element to the next across a period the charge on the nucleus increases by one as the number of electrons in the same outer shell increases by one. Shielding by electrons in the same shell is limited, so the 'effective nuclear charge' increases and the electrons are drawn more tightly to the nucleus.

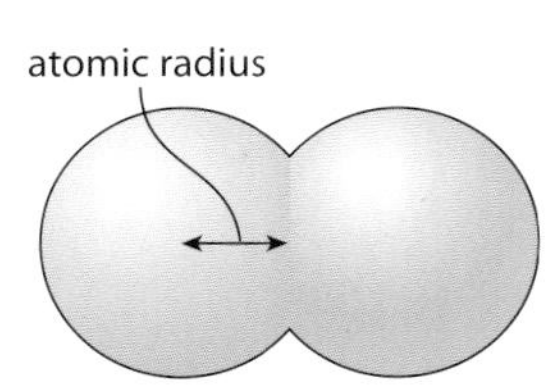

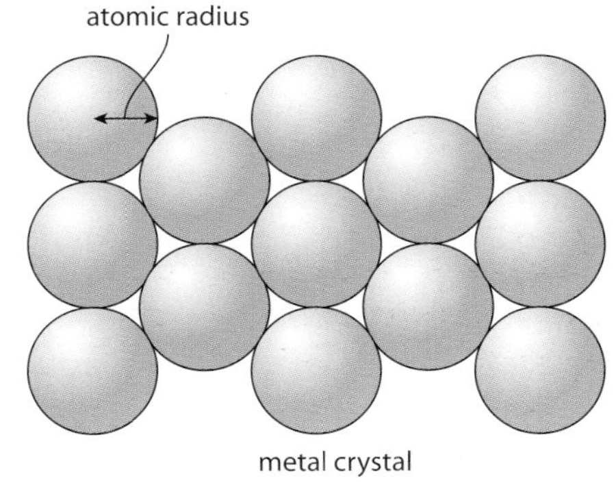

Figure 7.8 ▲
Atomic radii and the internuclear distance in a molecule and a crystal.

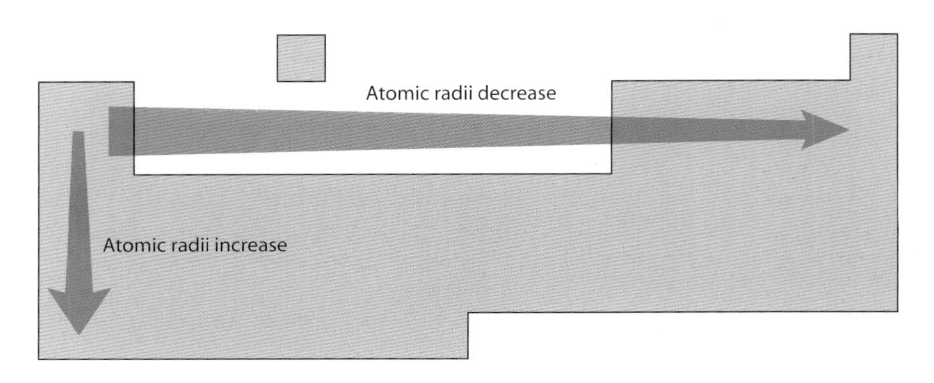

Figure 7.9 ▲
Periodicity of atomic radii in the periodic table.

Test yourself

15 Arrange these elements in order of atomic radius: Al, B, C, K and Na.

16 Which atom or ion in each of these pairs has the larger radius?
a) Cl or Cl^-
b) Al or N

Melting and boiling points of the elements

Figure 7.10 shows the periodic pattern revealed by plotting the melting points of elements against atomic number. The melting point of an element depends on both its structure and the type of bonding between its atoms. In metals, the bonding between atoms is strong (Section 6.3), so their melting points are usually high. The more electrons an atom contributes from its outermost shell to the shared delocalised electrons, the stronger the bonding and the higher the melting point.

Melting points therefore rise from Group 1 to Group 2 to Group 3. In Group 4, the elements carbon and silicon have giant covalent structures. The bonds in these structures are highly directional, so most of the bonds must break before the solid melts. This means that the melting points of the elements in Group 4 are very high and at the peak of the graph.

The non-metal elements in Groups 5, 6 and 7 form simple molecules. Elements in Group 0 are mono-atomic. The intermolecular forces between single molecules, and between uncombined atoms, are weak, so these elements have low melting points.

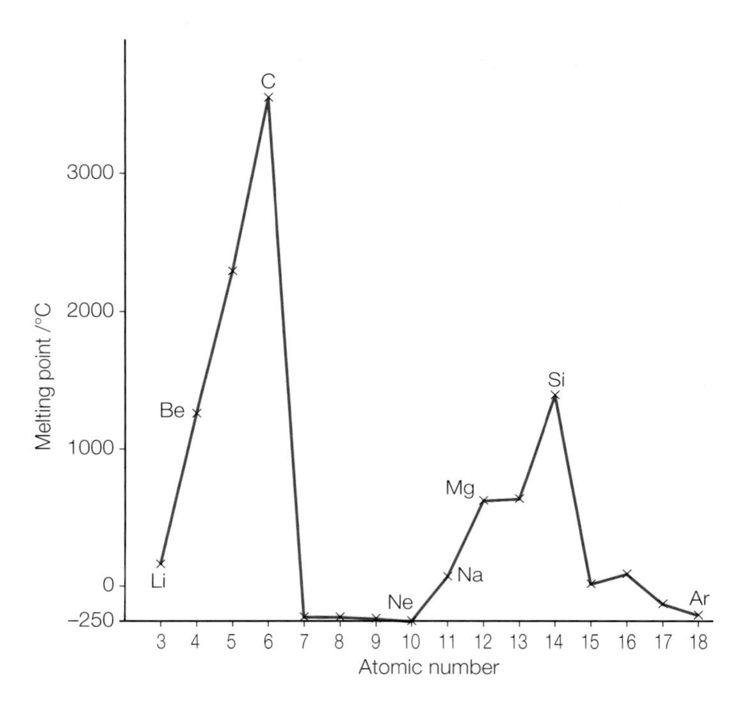

Figure 7.10 ▶ Periodicity in the melting points of the elements.

Test yourself

17 a) Use data on the CD-ROM to explore whether there is a periodic pattern in the boiling points of the elements in Periods 2 and 3.

b) Explain the data on the boiling points of the elements in terms of structure and bonding.

REVIEW QUESTIONS

Extension questions

1 The table shows the formulae of chlorides for the elements in periods 2 and 3.

Groups 1	2	3	4	5	6	7
LiCl	$BeCl_2$	BCl_3	CCl_4	NCl_3	OCl_2	FCl
NaCl	$MgCl_2$	$AlCl_3$	$SiCl_4$	PCl_3	SCl_2	ClCl

a) Why are there no entries for Group 0 in the table? (1)

b) What periodic pattern is shown by the formulae of these chlorides? (2)

c) i) Draw up a similar table to show the formulae of the oxides of the elements in Periods 2 and 3.

ii) Is there a periodic pattern in the formulae of these oxides? (3)

d) i) Draw up a similar table to show the charges on simple ions formed by elements in Periods 2 and 3.

ii) Is there a periodic pattern in the charges on simple ions? (3)

Data

2 The table below shows the melting points of the elements in Period 3 of the periodic table.

Element	Na	Mg	Al	Si	P	S	Cl	Ar
Melting point/°C	98	649	660	1410	44	119	−101	−189

The trend in the melting temperatures across Period 3 and other periods is described as a 'periodic property'.

a) What is the general pattern in melting temperatures across periods in the periodic table? (2)

b) How is this general trend related to the different types of elements? (1)

c) What do you understand by the term 'periodic property'? (2)

d) State two other physical properties that can be described as periodic in relation to the periodic table. (2)

8 Group 2 – the alkaline earth metals

Group 2 elements belong to the family of alkaline earth metals. Many of the compounds of these elements occur as minerals in rocks – hence the name 'earth metals'. Chalk, marble and limestone, for example, are forms of calcium carbonate. Dolomite consists of a mixture of calcium and magnesium carbonates. Fluorspar is a form of calcium fluoride that is mined as the ornamental mineral Blue John in Derbyshire. These Group 2 compounds are insoluble, unlike the equivalent Group 1 compounds, so they do not dissolve in water.

Figure 8.1 ▶
Dolomites in north Italy. Rocks made of calcium and magnesium carbonates.

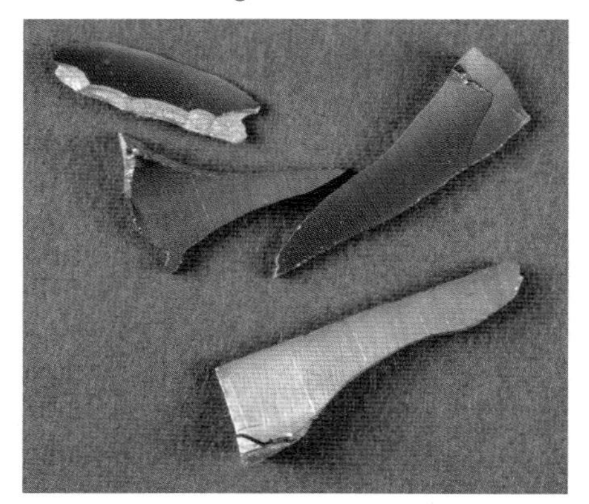

Figure 8.2 ▲
Sample of beryllium metal. Beryllium is light and strong and has a high melting point, so it is used for the construction of high-speed aircraft, missiles and space rockets. Thin sheets of the metal are transparent to X-rays and neutrons.

Figure 8.3 ▲
Samples of the silvery metal calcium usually look grey because they are covered with a layer of calcium oxide.

Data

8.1 Elements

The Group 2 metals are harder and denser than Group 1 metals such as sodium or potassium. In air, the surface of the metals is covered with a layer of oxide.

The first member of the group, beryllium, Be, is a strong metal with a high melting point, but its density is much less than that of transition metals such as iron. The element makes useful alloys with other metals.

Magnesium is manufactured by the electrolysis of molten magnesium chloride obtained from sea water or salt deposits. The low density of the metal helps to make light alloys, especially with aluminium. These alloys, which are strong for their weight, are especially valuable for cars and aircraft.

Barium is a soft, silvery-white metal. It is so reactive with air and moisture that it is generally stored under oil, like the alkali metals.

Atoms and ions

The Group 2 elements have similar chemical properties because their atoms have similar electron structures with two electrons in the outer s orbital (Section 5.5). Even so, there are trends in properties down the group from beryllium to barium.

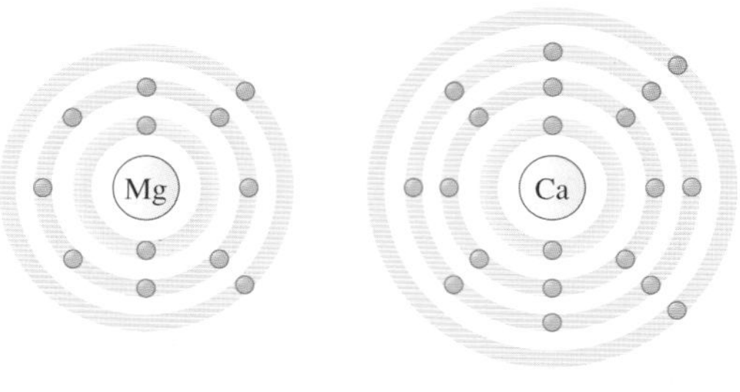

Figure 8.4 ▲
Diagrams to represent the electron configurations of magnesium and calcium atoms.

Metal	Electron configuration
Beryllium, Be	$[He]2s^2$
Magnesium, Mg	$[Ne]3s^2$
Calcium, Ca	$[Ar]4s^2$
Strontium, Sr	$[Kr]5s^2$
Barium, Ba	$[Xe]6s^2$

Table 8.1 ▲
Shortened forms of the electron configurations of Group 2 metals (Section 5.4).

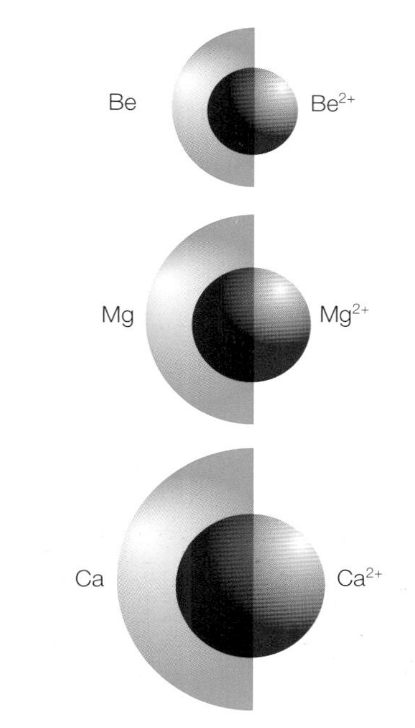

Figure 8.5 ▲
Relative sizes of the atoms and ions of Group 2 elements.

The atoms change in two ways down the group: the charge on the nucleus increases and the number of filled inner shells also increases as a result. The increasing number of filled inner shells means that atomic and ionic radii increase down the group. For each element, the 2+ ion is smaller than the atom because of the loss of the outer shell of electrons.

The first and second ionisation energies decrease down the group. The shielding effect of the inner electrons means that the effective nuclear charge that attracts the outer electron is 2+. Down the group, the outer s electrons get further and further away from the same effective nuclear charge and are held less strongly, so the ionisation energies decrease. This trend helps to account for the increasing reactivity of the elements down the group.

The removal of a third electron to form a 3+ ion takes much more energy, because that third electron has to be removed against the attraction of a much larger effective nuclear charge. This means that it is never energetically favourable for the metals to form M^{3+} ions.

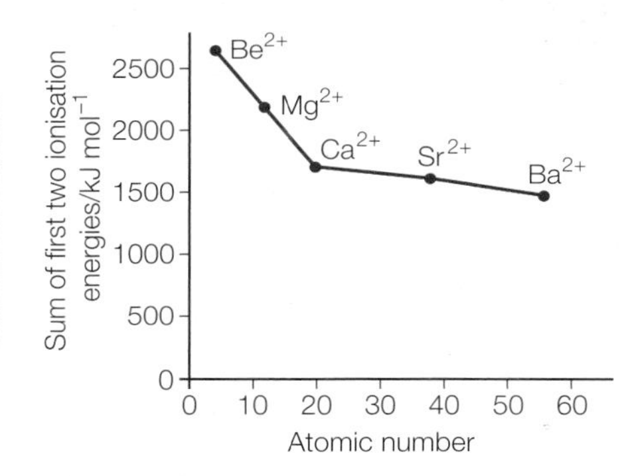

Figure 8.6 ▲
Graph to show the trend in the sum of the first two ionisation energies of Group 2 metals: $M(g) \rightarrow M^{2+}(g) + 2e^-$.

Test yourself

1 Write the full electron configurations of the atoms and ions of Mg, Ca and Sr showing the numbers of s, p and d electrons (Section 5.4).
2 Explain why Group 2 ions in any period are smaller than the Group 1 ion in that period.

Oxidation states

All the Group 2 metals have similar chemical properties because they have similar electron structures with two electrons in an outer s orbital. When the metal atoms react to form ions, they lose the two outer electrons to give ions with a 2+ charge: Mg^{2+}, Ca^{2+}, Sr^{2+} and Ba^{2+}. These elements therefore exist in the +2 oxidation state in all their compounds.

8.2 Reactions of the Group 2 elements

Group 2 metals are reducing agents. They readily give up their two s electrons to form M^{2+} ions (where M represents Mg, Ca, Sr or Ba).

$$M \rightarrow M^{2+} + 2e^-$$

Reactions with oxygen

Apart from beryllium, the Group 2 metals burn brightly on heating in oxygen to form white ionic oxides, $M^{2+}O^{2-}$.

Magnesium burns very brightly in air, with an intense white flame, to form the white solid magnesium oxide, MgO. For this reason, magnesium powder is an ingredient of fireworks and flares.

Calcium also burns brightly in air, but with a red flame, to form the white solid calcium oxide, CaO. Strontium reacts in a similar way.

Barium burns in excess air or oxygen, with a green flame, to form a peroxide, BaO_2, that contains the peroxide ion, O_2^{2-}.

Figure 8.7 ▲
Calcium burning in air.

Reactions with water

The metals Mg to Ba in Group 2 react with water. The reactions are not as vigorous as the reactions of the Group 1 metals, but, as in Group 1, the rate of reaction increases down the group.

Magnesium reacts very slowly with cold water but much more rapidly on heating in steam.

$$Mg(s) + H_2O(g) \rightarrow MgO(s) + H_2(g)$$

Calcium reacts with cold water to produce hydrogen and a white precipitate of calcium hydroxide.

$$Ca(s) + 2H_2O(l) \rightarrow Ca(OH)_2(s) + H_2(g)$$

Barium reacts even faster with cold water, but its hydroxide is more soluble.

Reactions with chlorine

All the metals react with chlorine on heating to form white chlorides, MCl_2.

$$Mg(s) + Cl_2(g) \rightarrow MgCl_2(s)$$

Test yourself

3 Write balanced equations for the reactions of:
   a) calcium with oxygen
   b) barium with oxygen
   c) strontium with chlorine
   d) barium with water.

4 Show that magnesium acts as a reducing agent when it reacts with:
   a) oxygen
   b) water.

8.3 Properties of the compounds of Group 2 metals

Oxides

Apart from beryllium oxide, the oxides are basic and react with acids to form salts.

$$CaO(s) + 2HNO_3(aq) \rightarrow Ca(NO_3)_2(aq) + H_2O(l)$$

Definition

A **basic oxide** is a metal oxide that reacts with acids to form salts and water. It is the oxide ion that acts as a base by taking a hydrogen ion from the acid. Basic oxides that dissolve in water are alkalis.

Magnesium oxide is a white solid made by heating magnesium carbonate. In water, it turns to magnesium hydroxide, which is slightly soluble. Magnesium oxide has a high melting point and is used as a refractory ceramic to line furnaces.

Calcium oxide is a white solid made by heating calcium carbonate. Calcium oxide reacts very vigorously with cold water – hence its traditional name 'quicklime'. The product is calcium hydroxide.

Hydroxides

The hydroxides of the elements Mg to Ba are:

- similar in that they all have the formula $M(OH)_2$ and are to some degree soluble in water-forming alkaline solutions
- different in that their solubility increases down the group.

Magnesium hydroxide is only slightly soluble in water and gives a solution with a pH of about 10. It is used as an antacid and laxative. As an antacid, it can be taken in tablets or as a suspension in water called 'Milk of Magnesia', and it neutralises some of the hydrochloric acid in the stomach.

Calcium hydroxide, $Ca(OH)_2$, is only sparingly soluble in water and forms an alkaline solution usually called lime water, which has a pH of about 12.

The lime water test for carbon dioxide works because a solution of calcium hydroxide reacts with the gas to form a white, insoluble precipitate of calcium carbonate.

Barium hydroxide is the most soluble of the hydroxides. It is sometimes used as an alkali in chemical analysis. It has an advantage over sodium and potassium hydroxides in that it cannot be contaminated by its carbonate because barium carbonate is insoluble in water.

Test yourself

5 Write balanced equations for the reactions of:
 a) magnesium oxide with dilute hydrochloric acid
 b) calcium oxide with water
 c) lime water with carbon dioxide.

Carbonates

The carbonates of Group 2 metals (Mg to Ba) are:

- similar in that they all have the formula MCO_3, are insoluble in water, react with dilute acids and decompose on heating to give the oxide and carbon dioxide

 $CaCO_3(s) \rightarrow CaO(s) + CO_2(g)$

- different in that they become more difficult to decompose down the group – in other words, they become more thermally stable.

Table 8.2 shows the temperatures at which the carbonates of Group 2 metals begin to decompose. The values indicate that magnesium carbonate is the least stable – it decomposes easily to the oxide and carbon dioxide when heated with a Bunsen flame. Barium carbonate is the most stable.

Carbonate	Decomposition point
$MgCO_3$	540 °C
$CaCO_3$	900 °C
$SrCO_3$	1280 °C
$BaCO_3$	1360 °C

Table 8.2 ▲
Temperature at which Group 2 carbonates begin to decompose.

Sulfates

The sulfates are:

- similar in that they are all colourless solids with the formula MSO_4
- different in that they become less soluble down the group.

Epsom salts consist of hydrated magnesium sulfate, $MgSO_4.7H_2O$, which is a laxative.

Plaster of Paris is the main ingredient of building plasters and much is used to make plasterboard. The white powder is made by heating the mineral gypsum in kilns to remove most of the water of crystallisation. Stirring plaster of Paris with water produces a paste that soon sets as it turns back into interlocking grains of gypsum. Plaster makes good moulds because it expands slightly as it sets so that it fills every crevice.

Figure 8.8 ▲
Crystals of gypsum – a hydrated form of calcium sulfate.

Figure 8.9 ▲
Barium occurs naturally as barites, $BaSO_4$, and also as witherite, $BaCO_3$.

Activity

Limestone and its uses

Figure 8.10 ▲
Limestone cliff in Yorkshire. Limestone gives rise to attractive scenery.

Calcium carbonate occurs naturally as limestone. Limestone is an important mineral. Some of the rock is quarried for road building and construction.

Pure limestone is used in the chemical industry. Heating limestone in a furnace at 1200 K converts it to calcium oxide (quicklime). The reaction of quicklime with water produces calcium hydroxide (slaked lime).

1 Name two other minerals that consist largely of calcium carbonate.

2 Suggest a reason for grinding up limestone lumps before heating them with sodium carbonate and sand to make glass.

3 Write an equation for the decomposition of calcium carbonate on heating in a furnace.

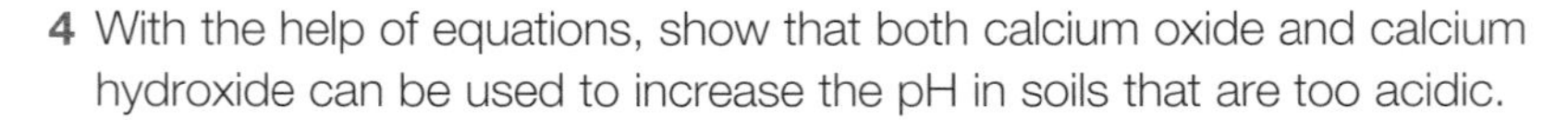

4 With the help of equations, show that both calcium oxide and calcium hydroxide can be used to increase the pH in soils that are too acidic.

5 Lime mortar was used in older buildings. It is a mixture of slaked lime, sand and water. It sets slowly by reacting with carbon dioxide in the air. Identify the main product of the reaction of slaked lime with carbon dioxide.

6 A suspension of calcium hydroxide in water is used as an industrial alkali. Suggest why a suspension and not just a solution of the hydroxide is used?

7 What is the laboratory use of a solution of calcium hydroxide?

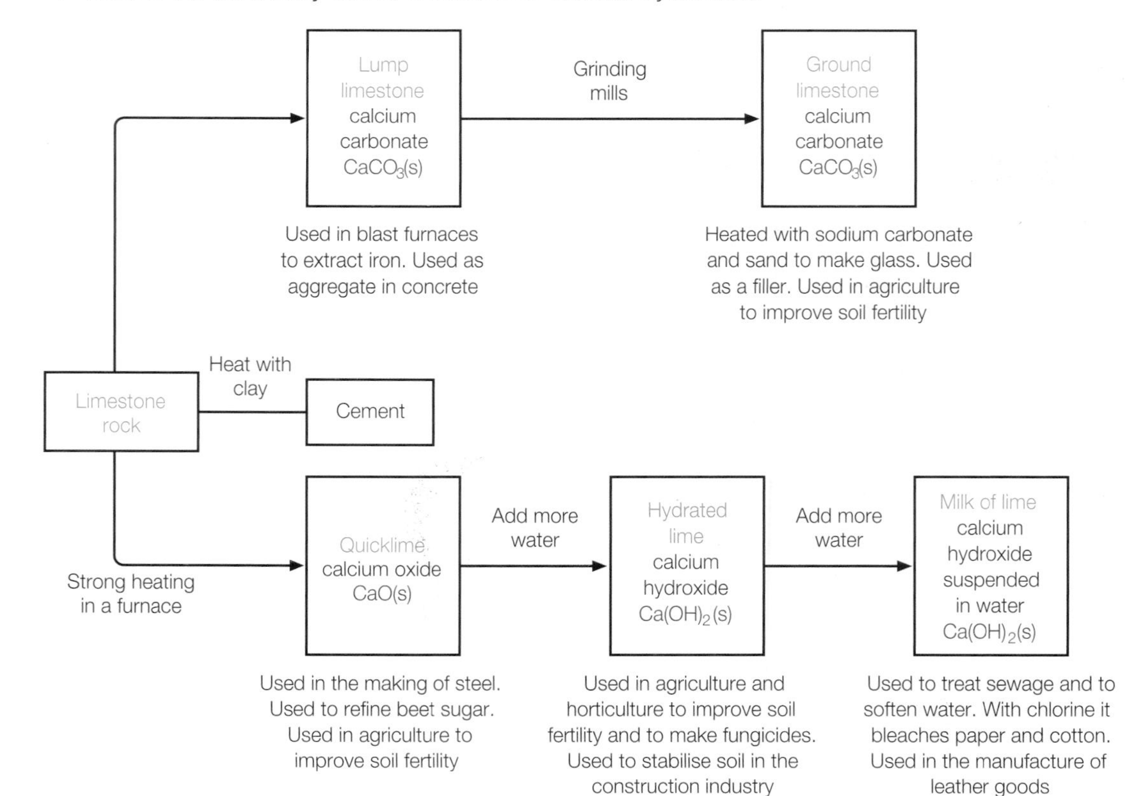

Figure 8.11 ▲
Products from limestone and their uses.

Barium sulfate absorbs X-rays strongly, so it is the main ingredient of 'barium meals' used to diagnose disorders of the stomach or intestines. Soluble barium compounds are toxic, but barium sulfate is very insoluble, so it is not absorbed into the bloodstream from the gut. X-rays cannot pass through the 'barium meal', which therefore creates a shadow on the X-ray film.

A soluble barium salt can be used to test for sulfate ions, because barium sulfate is insoluble even when the solution is acidic. Adding a solution of barium nitrate or barium chloride to an acidified solution produces a white precipitate only if sulfate ions are present.

$$Ba^{2+}(aq) + SO_4^{2-}(aq) \rightarrow \underset{\text{white precipitate}}{BaSO_4(s)}$$

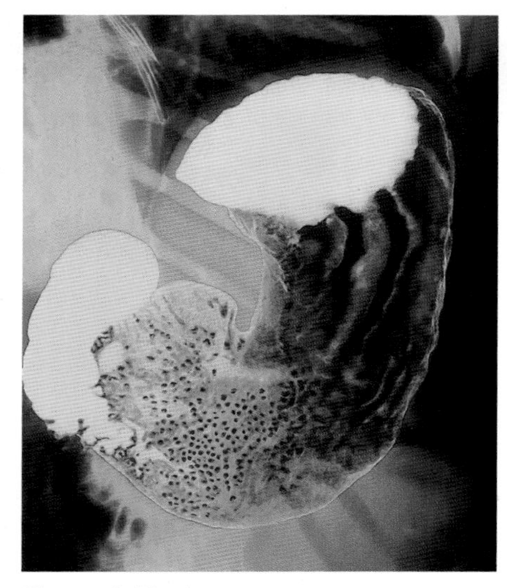

Figure 8.12 ▲
X-ray photograph of the digestive system of a patient who has taken a barium meal.

Nitrates

The nitrates of Group 2 metals (Mg to Ba) are similar in that they all have the formula $M(NO_3)_2$, are colourless crystalline solids, are very soluble in water and decompose to the oxide on heating:

$$2Mg(NO_3)_2 \rightarrow 2MgO(s) + 4NO_2(g) + O_2(g)$$

Test yourself

6 Draw and label a diagram of a simple apparatus to show that magnesium carbonate decomposes on heating.

7 Write equations for:
   a) the thermal decomposition of magnesium carbonate
   b) the reaction of magnesium carbonate with hydrochloric acid
   c) the thermal decomposition of calcium nitrate
   d) the reaction of barium nitrate solution with zinc sulfate solution.

8 With the help of oxidation numbers, identify the elements that are oxidised and reduced during the thermal decomposition of magnesium nitrate.

Flame colours

Flame tests help to identify compounds of calcium, strontium and barium.

Metal ion	Flame
Beryllium	No colour
Magnesium	No colour
Calcium	Brick red
Strontium	Bright red
Barium	Pale green

Table 8.3 ▲
Flame colours of Group 2 metal compounds.

REVIEW QUESTIONS

1 The elements Mg to Ba in Group 2, and their compounds, can be used to show the trends in properties down a group of the periodic table. State and explain the trend down the group in:

a) atomic radius (3)

b) first ionisation energy (3)

c) reactivity of the metals. (3)

2 Radium is a highly radioactive element. Use your knowledge of the chemistry of the elements Mg to Ba in Group 2 to predict properties of radium and its compounds in terms of the following. Include descriptions of any chemicals changes, equations and the appearance of the products in your predictions.

a) Reaction of radium with oxygen (3)

b) Reaction of radium with water (3)

c) Reaction of radium oxide with water (3)

d) Reaction of radium hydroxide with dilute hydrochloric acid (3)

e) Solubility of radium sulfate in water (2)

f) Effect of heating radium carbonate. (3)

3 a) Explain the use of calcium hydroxide in agriculture to neutralise soils. (3)

b) Explain the use of magnesium hydroxide in some indigestion tablets as an antacid. (3)

9 Group 7 – the halogens

Fluorine, chlorine, bromine and iodine belong to the family of halogens. All four are reactive non-metals – fluorine and chlorine extremely so. The elements are hazardous because they are so reactive. For the same reason, they are never found free in nature. However, they do occur as compounds with metals. Many of the compounds of Group 7 elements are salts – hence the name 'halogen' meaning 'salt-former'. The halogens are important economically as the ingredients of plastics, pharmaceuticals, anaesthetics, dyestuffs and chemicals for water treatment.

9.1 Elements

Under laboratory conditions, chlorine is a yellow-green gas, bromine is a dark red liquid and iodine is a dark grey solid.

Fluorine is much too dangerous to be used in normal laboratories. Astatine, the final member of the group, is the rarest naturally occurring element. It is highly radioactive. Its most stable isotope has a half-life of just over eight hours.

The halogens have similar chemical properties because they all have seven electrons in the outer shell – one less than the next noble gas in Group 0.

Halogen	Electron configuration
Fluorine, F	$[He]2s^2 2p^5$
Chlorine, Cl	$[Ne]3s^2 3p^5$
Bromine, Br	$[Ar]3d^{10} 4s^2 4p^5$
Iodine, I	$[Kr]4d^{10} 5s^2 5p^5$

Table 9.1 ▲
Shortened form of the electron configurations of the halogens.

Fluorine is the most electronegative of all elements (Section 6.11). Its oxidation state is −1 in all its compounds. Uses of fluorine include the manufacture of a wide range of compounds consisting of only carbon and fluorine (fluorocarbons). The most familiar of these is the very slippery, non-stick polymer poly(tetrafluorethene).

Chlorine reacts directly with most elements. In its compounds, chlorine is usually present in the −1 oxidation state, but it can be oxidised to positive oxidation states by oxygen and fluorine. Most chlorine is used in the production of polymers such as polyvinylchloride (PVC). Water companies use chlorine to kill bacteria in drinking water, while another important use of this element is to bleach paper and textiles.

Bromine, like the other halogens, is a reactive element, but it is a less powerful oxidising agent than chlorine. Bromine is used to make a range of products, including flame retardants, medicines and dyes.

Figure 9.1 ▲
Chlorine gas.

Figure 9.2 ▲
Bromine is a dark red liquid at room temperature – it is very volatile and gives off a choking, orange vapour.

Iodine is also an oxidising agent but less powerful than bromine. Iodine and its compounds are used to make many products, including medicines, dyes and catalysts. Iodine is needed in our diet so that the thyroid gland can make the hormone thyroxine, which regulates growth and metabolism. In many regions,

Figure 9.3 ▲
Iodine is a lustrous grey-black solid at room temperature, which sublimes when gently warmed to give a purple vapour.

sodium iodide is added to table salt to supplement iodine in the diet and drinking water to prevent goitre – a swelling of the thyroid gland in the neck.

Test yourself

1 Predict the state of the following at room temperature and pressure and give your reasons:
  a) fluorine
  b) astatine.
2 Write down the full electron configurations of:
  a) a chlorine atom
  b) a chloride ion
  c) a bromine atom
  d) a bromide ion.
3 Explain why:
  a) the atomic radii of halogen atoms increase down the group
  b) the ionic radii of halides are larger than their corresponding atomic radii.
4 Draw '*dot-and-cross*' diagrams to show the bonding in:
  a) a fluorine molecule
  b) a molecule of hydrogen bromide
  c) a molecule of iodine monochloride, ICl.

9.2 Halogen molecules

Boiling points

All the halogens consist of diatomic molecules, X_2, linked by a single covalent bond. They are all volatile. The van der Waals' forces between the molecules arise from the attractions between temporary dipoles and the fleeting dipoles they induce in neighbouring molecules (Section 6.13). The greater the number of electrons in a molecule, the more polarisable it is and the greater the possibility for temporary induced dipoles. This explains why the boiling temperatures increase down the group of halogens.

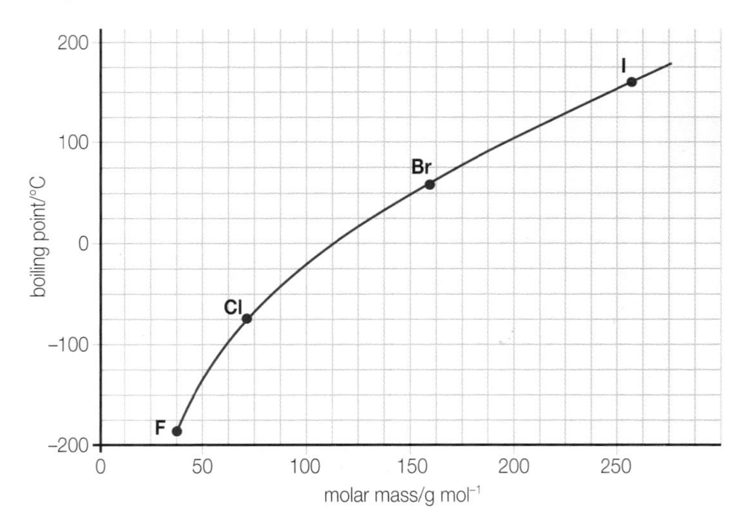

Figure 9.4 ▶
Boiling temperatures of the halogens plotted against molar mass.

Solutions of the halogens

The halogens dissolve freely in hydrocarbon solvents such as hexane. When dissolved in hexane, the solutions have a very similar colour to the free halogen vapours. Iodine in hexane, for example, has an attractive violet colour.

The halogens are less soluble in water. Aqueous chlorine and bromine are useful reagents. Their colours are similar to the colours of their vapours. These elements also react with water (Section 9.5).

Iodine does not dissolve in water, but it does dissolve in aqueous potassium iodide. Iodine dissolves in this way because iodine molecules react with iodide ions to form tri-iodide ions, I_3^-. The solution is yellow when very dilute but dark yellow-brown when more concentrated.

9.3 Reactions of the Group 7 elements

The halogens are powerful oxidising agents. Apart from fluorine, chlorine is the strongest oxidising agent among these elements.

The tendency to form negative ions decreases down the group. The shielding effect of the inner electrons means that the effective nuclear charge that attracts the outer electron is 7+ for all of the atoms. Down the group the outer electrons get further and further away from the same effective nuclear charge and so they are held less strongly and the tendency to gain an extra electron decreases.

With other non-metals, the halogens form molecules with covalent bonds. Halogen atoms are all highly electronegative (Section 6.11), but electronegativity decreases down the group. As a result, the covalent bonds formed by iodine are less polar than the bonds formed by chlorine.

Reactions with metals

Chlorine and bromine react with s-block metals to form ionic halides in which the halogen atoms gain one electron to fill the outermost p sub-shell. The halogens also react with most metals in the d-block. Hot iron, for example, burns brightly in chlorine to form iron(III) chloride. The reaction with bromine is similar but much less exothermic. Iodine(III) iodide does not exist, because iodide ions reduce iron(III) ions to iron(II) ions. Heating iron with iodine vapour produces iron(II) iodide.

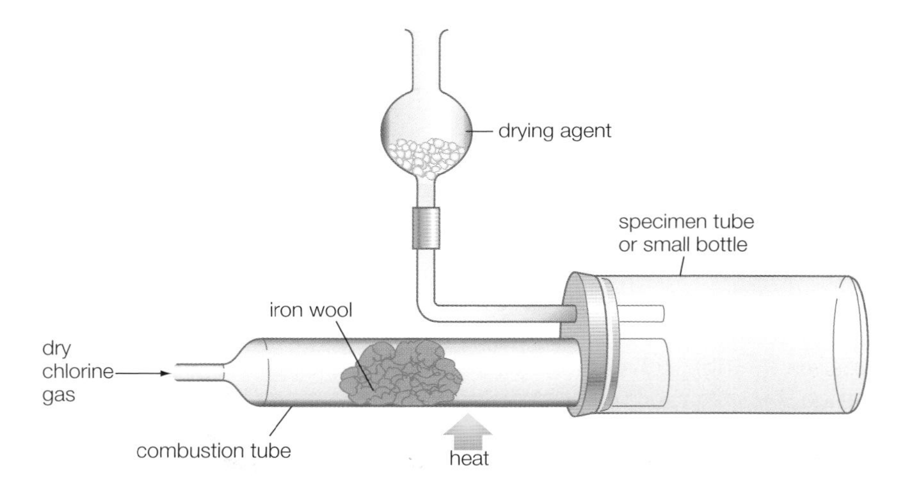

Figure 9.5 Laboratory apparatus for making anhydrous iron(III) chloride.

Test yourself

5 Write balanced equations for the following reactions and show the changes in oxidation states:
- a) bromine with magnesium
- b) chlorine with iron
- c) iodine with iron.

6 In Figure 9.5, explain the reason for:
- a) drying the chlorine gas
- b) using iron wool instead of small lumps of iron
- c) collecting the product in a specimen tube
- d) allowing excess gas to escape through a tube with a drying agent.

Test yourself

7 a) Show that the reactions of halogens with hydrogen illustrate a trend in reactivity down Group 7.
   b) Predict the formula of the product and the vigour of the reaction when fluorine reacts with hydrogen.

8 Write an equation for the reaction between phosphorus and chlorine and use oxidation numbers to show that phosphorus is oxidised in this reaction.

Reactions with non-metals

Chlorine reacts with most non-metals to form molecular chlorides. Hot silicon, for example, reacts to form silicon tetrachloride, $SiCl_4(l)$, and phosphorus reacts to produce phosphorus trichloride, $PCl_3(l)$. Chlorine, however, does not react directly with carbon, oxygen or nitrogen.

Hydrogen burns in chlorine to produce the colourless, acidic gas hydrogen chloride, HCl. Igniting a mixture of chlorine and hydrogen gases leads to a violent explosion.

Bromine also oxidises non-metals such as sulfur and hydrogen on heating to form molecular bromides. A mixture of bromine vapour and hydrogen gas reacts smoothly with a pale bluish flame.

$$H_2(g) + Br_2(g) \rightarrow 2HBr(g)$$

Iodine oxidises hydrogen on heating to form hydrogen iodide. Unlike the reactions of chlorine and bromine, this is a reversible reaction.

$$H_2(g) + I_2(g) \rightleftharpoons 2HI(g)$$

9.4 Halogens in oxidation state –1

Halide ions

Halide ions are the ions of the halogen elements in oxidation state –1. They include the fluoride, F^-, chloride, Cl^-, bromide, Br^-, and iodide, I^-, ions.

In Group 7, a more reactive halogen oxidises the ions of a less reactive halogen. For example, bromine reacts with a solution of an iodide to produce iodine and a bromide. This is because bromine has a stronger tendency to gain electrons and turn into ions than iodine.

$$Br_2(aq) + 2I^-(aq) \rightarrow 2Br^-(aq) + I_2(s)$$

Silver nitrate solution can be used to distinguish between halides. Silver halides are insoluble. Adding silver nitrate to a solution of one of these halide ions produces a precipitate.

$$Ag^+(aq) + Cl^-(aq) \rightarrow AgCl(s)$$

Silver chloride is white and quickly turns purple-grey in sunlight. This distinguishes it from silver bromide, which is a creamy colour, and silver iodide, which is a brighter yellow.

The colour changes are not very distinct, but a further test with ammonia helps to distinguish the precipitates. Silver chloride easily dissolves in dilute ammonia solution. Silver bromide dissolves in concentrated ammonia solution, but silver iodide does not dissolve in ammonia solution at all.

Figure 9.6 ▲
Adding a solution of chlorine in water to aqueous potassium iodide. Adding a hydrocarbon solvent to the mixture and shaking produces a violet colour in the organic solvent, which shows that iodine has been formed.

Test yourself

9 Describe the colour changes on adding:
   a) a solution of chlorine in water to aqueous sodium bromide
   b) a solution of bromine in water to aqueous potassium iodide.

10 Write ionic equations for the reactions of aqueous chlorine with:
   a) bromide ions
   b) iodide ions.

11 Put the chloride, bromide, and iodide ions in order of their strength as reducing agents, with the strongest reducing agent first. Explain your answer.

12 Write ionic equations for the reactions of silver nitrate solution with:
   a) potassium iodide solution
   b) sodium bromide solution.

Practical guidance

Data

Hydrogen halides

The hydrogen halides are compounds of hydrogen with the halogens. They are all colourless molecular compounds with the formula HX, where X stands for Cl, Br or I. The bonds between hydrogen and the halogens are polar.

Hydrogen chloride, hydrogen bromide and hydrogen iodide are similar in that they are:

- colourless gases at room temperature, which fume in moist air
- very soluble in water to form acidic solutions (hydrochloric, hydrobromic and hydriodic acids) that ionise completely in water
- strong acids, which means that they ionise completely in water to form hydrogen ions and halide ions.

Mixing any of the hydrogen halides with ammonia produces a white smoke of an ammonium salt. Ammonia molecules turn into ammonium ions, NH_4^+, in this reaction.

$$NH_3(g) + HBr(g) \rightarrow NH_4Br(s)$$

This reaction can be used as a test to detect hydrogen halides.

Test yourself

13 Explain why the compound of hydrogen and fluorine is a liquid at room temperature on a cool day when the other hydrogen halides are gases.

14 a) Show that the reaction of ammonia with hydrogen chloride gas involves proton transfer.

b) Explain why the product of the reaction is a solid.

9.5 Halogens in oxidation states +1 and +5

Chlorine oxoanions form when chlorine reacts with water and alkalis.

When chlorine dissolves in water, it reacts reversibly to form a mixture of weak chloric(I) acid and strong hydrochloric acid.

$$H_2O(l) + Cl_2(g) \rightarrow HClO(aq) + HCl(aq)$$

This is an example of a reaction in which the same element both increases and decreases its oxidation number. In other words, some of the element is oxidised while the rest of the element is reduced. Reactions of this type are called disproportionation reactions.

Bromine reacts in a similar way but to a much lesser extent. Iodine is almost insoluble in water and hardly reacts at all.

When chlorine dissolves in sodium (or potassium) hydroxide solution at room temperature, it produces chlorate(I) and chloride ions.

$$Cl_2(aq) + 2OH^-(aq) \rightarrow ClO^-(aq) + Cl^-(aq) + H_2O(l)$$

(arrow from Cl_2 to ClO^- labelled +1; arrow from Cl_2 to Cl^- labelled −1)

The active ingredient in household bleach is sodium chlorate(I), which is made by dissolving chlorine in sodium hydroxide solution. On heating, the chlorate(I) ions disproportionate to chlorate(V) and chloride ions:

$$3ClO^-(aq) \rightarrow ClO_3^-(aq) + 2Cl^-(aq)$$

+1	+5	−1	oxidation states

Bromine and iodine react in a similar way to chlorine with alkalis. The BrO^- and IO^- ions are less stable, so they disproportionate at a lower temperature. A hot solution of iodine in potassium hydroxide produces a solution that contains potassium iodate(V) and potassium iodide.

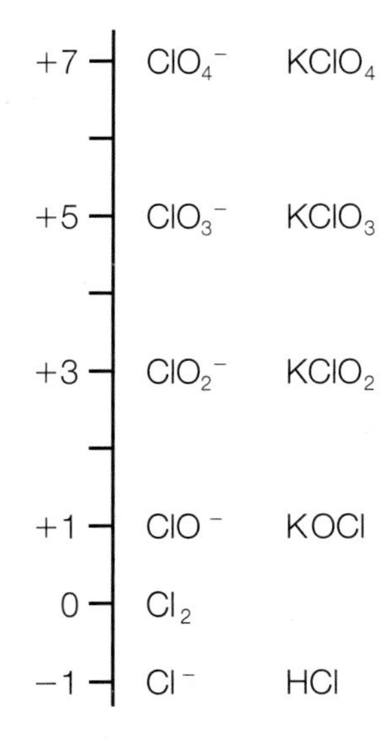

Figure 9.7 ▲ Oxidation states of chlorine.

Test yourself

15 a) Use oxidation numbers to write a balanced equation for the reaction of iodine with hot aqueous hydroxide ions to form IO_3^- and I^- ions.

b) Show that this is a disproportionation reaction.

Definition

A **disproportionation reaction** is a reaction in which the same element both increases and decreases its oxidation number, so the element is both oxidised and reduced.

9.6 Benefits and risks of water treatment

Chlorine is added to drinking water to kill micro-organisms such as bacteria and viruses. This greatly reduces the risk of waterborne diseases such as cholera and typhoid fever.

Chlorine disinfects tap water by forming chloric(I) acid, HClO, when it reacts with water. Chloric(I) acid is a powerful oxidising agent and a weak acid. It is an effective disinfectant because the molecule can pass through the cell walls of bacteria, unlike ClO^- ions. Once inside the bacterial cells, the HClO molecules break them open and kill the organism by oxidising and chlorinating molecules that make up the structure of the cells.

Disinfection with chlorine can produce by-products that may be harmful. Chlorine can also react with organic matter in the water. This produces traces of chlorinated hydrocarbons (Section 11.5). The main products are the trihalomethanes, which include trichloromethane, $CHCl_3$.

Tests with laboratory animals suggest that exposure to high levels of trihalomethanes can increase the risk of cancer. At low levels, the risks are very small, but they have to be investigated because millions of people drink treated water for many years. One study suggested an increased risk of bladder cancer and possibly colon cancer in people who drank chlorinated water for 35 years or more. A study in California found that pregnant women who drank large amounts of water with relatively high levels of trihalomethanes had an increased risk of miscarriage. The interpretation of the evidence remains controversial, and the links between trichloromethanes and cancer or miscarriage are not firmly established. Whatever these risks, the benefits of chlorination to prevent waterborne disease are widely accepted to be much greater than the small health risks from trihalomethanes.

Test yourself

16 Suggest why levels of chlorinated hydrocarbons in drinking water are generally lower:

- **a)** in winter than in summer
- **b)** when deep wells or large lakes rather than rivers are the source of drinking water.

17 Why is it easier to show the benefits of chlorinating drinking water than to assess the risks from the by-products of chlorination?

Activity

Water treatment in swimming pools

Swimming pools can be sterilised with chlorine compounds that produce chloric(I) acid when they dissolve in water. Chloric(I) acid is a weak acid so it does not easily ionise. The concentration of unionised HClO in a solution depends on the pH as shown in the graph in Figure 9.8.

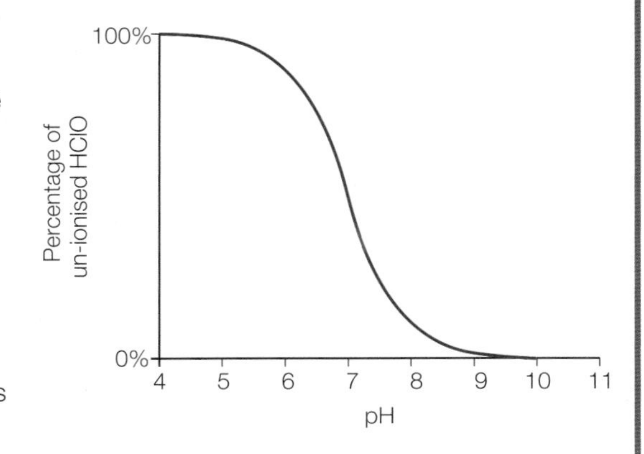

Figure 9.8 ▶
Concentration of chloric(I) acid, HClO, varies over a range of pH values at 20 °C.

Swimming pool managers have to check the pH of the water carefully. They aim to keep the pH in the range 7.2–7.8.

1 Write an equation to show the formation of chloric(I) acid when chlorine reacts with water.

2 Swimming pools were once treated with chlorine gas from cylinders that contained the liquefied gas under pressure. Nowadays they are usually treated with chemicals that produce chloric(I) acid when they are added to water. Suggest reasons for this change.

3 Explain why sodium chlorate(I) produces chloric(I) acid when it is added to water at pH 7–8.

4 Explain why the pH of water in swimming pools must not be allowed to increase above 7.8.

5 Suggest why pool water must not become acidic, even though this would increase the concentration of unionised HClO.

Figure 9.9 ▲
Chlorine compounds treat the water in swimming pools.

6 Nitrogen compounds, including ammonia, urea and proteins, react with HClO to form chloramines, which are irritating to skin and eyes. Chloramines formed from ammonia can react with themselves to form nitrogen and hydrogen chloride, which gets rid of the problem. However, if there is excess HClO, another reaction produces nitrogen trichloride, which is responsible for the so-called 'swimming pool smell' and is very irritating to the eyes. Write equations to show:

a) the formation of the chloramine, NH_2Cl, from ammonia and chloric(I) acid

b) the removal of chloramines by the reaction of NH_2Cl with $NHCl_2$

c) the formation of nitrogen trichloride from chloramine, NH_2Cl.

7 Explain why swimming pools do not smell of chlorine if properly maintained.

REVIEW QUESTIONS

Extension questions

1 Astatine, At, is the element below iodine in Group 7. Predict, giving your reasons:

a) the physical state of astatine at room temperature **(1)**

b) the effect of bubbling chorine through an aqueous solution of sodium astatide **(2)**

c) the colour of silver astatide and its solubility in concentrated ammonia solution. **(2)**

2 a) Write an equation for the reaction of chlorine with aqueous sodium hydroxide and use this example to explain what is meant by disproportionation. **(2)**

b) On heating, chlorate(I) ions in solution disproportionate to chlorate(V) ions and chloride ions. Write an ionic equation for this reaction. **(2)**

c) On heating to just above its melting temperature, $KClO_3$ reacts to form $KClO_4$ and KCl. Write a balanced equation for the reaction and show that it is a disproportionation reaction. **(3)**

3 Identify the element that disproportionates in each of the following reactions by giving the oxidation states of the element before and after the reaction.

a) $2H_2O_2(aq) \rightarrow 2H_2O(l) + O_2(g)$ **(1)**

b) $Cl_2(aq) + 2NaOH(aq) \rightarrow NaCl(aq) + NaClO(aq) + H_2O(l)$ **(1)**

c) $Cu_2O(s) + H_2SO_4(aq) \rightarrow CuSO_4(aq) + Cu(s) + H_2O(l)$ **(1)**

d) $3MnO_4^{2-}(aq) + 4H^+(aq) \rightarrow 2MnO_4^-(aq) + MnO_2(s) + 2H_2O(l)$ **(1)**

4 State the trend in the power of chloride, bromide and iodide ions as reducing agents. Describe how you could demonstrate the trend by carrying out experiments in the laboratory. Include in your description the main observations that illustrate the differences between the ions and equations for the reactions. **(10)**

Unit 2

Chains, energy and resources

10 Introduction to organic chemistry
11 Alkanes
12 Alkenes
13 Alcohols
14 Halogenoalkanes
15 Instrumental analysis
16 Enthalpy changes and energetics
17 Rates and equilibria
18 Chemistry of the air
19 Green chemistry

10 Introduction to organic chemistry

Carbon is an amazing element. The number of compounds containing carbon and hydrogen is well over ten million. This is far more than the number of compounds of all the other elements put together.

Most compounds containing carbon also contain hydrogen. The main sources of these compounds are organic living or once-living materials in animals and plants. Because of this, the term 'organic chemistry' is used to describe the branch of chemistry concerned with the study of compounds containing C–H bonds. This covers most of the compounds of carbon. Simple carbon compounds that do not contain C–H bonds, such as carbon dioxide and carbonates, are usually included in the study of inorganic chemistry.

Figure 10.1 ▶
Children playing with their toys. From the cells in their bodies to the fibres in their clothes and the plastics in their toys, almost everything in this photograph consists of organic chemicals – molecules containing carbon–hydrogen bonds.

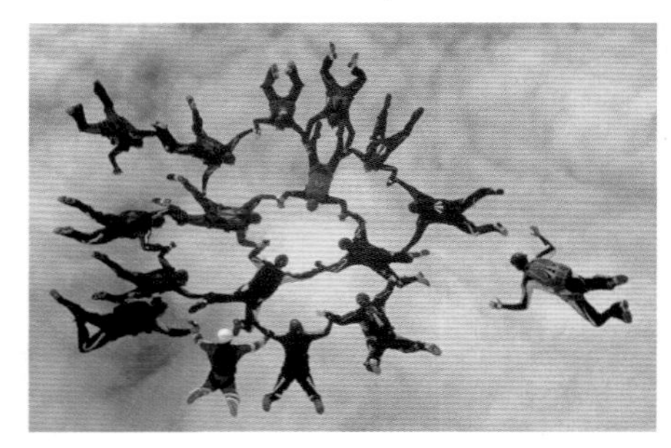

Figure 10.2 ▲
Sky divers can use their arms and legs to form four links to one another. Like carbon atoms, they can form chains and rings.

10.1 Carbon – a special element

There are two main reasons why carbon can form so many compounds.

The first reason is that carbon atoms have an exceptional ability to form chains, branched chains and rings of varying sizes. No other element can form long chains of its atoms in the same way as carbon.

The second reason why carbon can form so many compounds is the relative inertness and unreactive nature of the C–C and C–H bonds (Section 11.5).

When carbon atoms form a chain or ring linked by single covalent bonds, no more than two of the bonds on each atom are used. This means that at least two other bonds on each carbon atom can bond with other atoms. In fact, carbon often forms bonds with hydrogen, oxygen, nitrogen and halogen atoms (Figure 10.3).

A knowledge of organic chemicals enables chemists to extract, synthesise and manufacture a wide range of important products including fuels, plastics, medicines, anaesthetics and antibiotics.

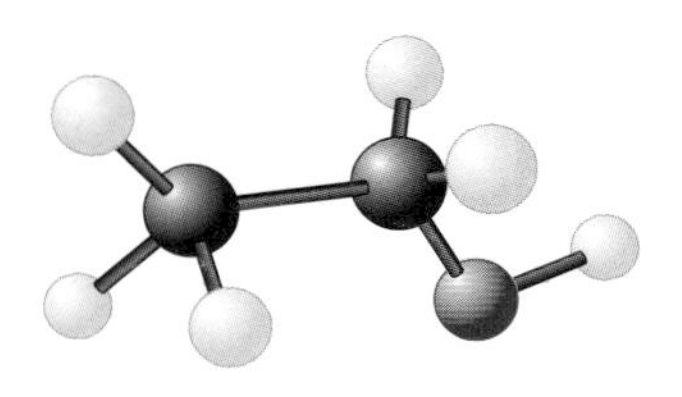

Figure 10.3 ▲
The structure of ethanol, CH_3CH_2OH. Ethanol is commonly called alcohol. It has two carbon atoms linked to each other and to hydrogen and oxygen atoms by covalent bonds.

Big molecules

Most molecules in living things are carbon compounds, so biochemistry and molecular biology are important applications of organic chemistry. The compounds found in living cells include carbohydrates, fats, proteins and nucleic acids. The molecules of these compounds are large. Some of them are very large. The carbohydrate cellulose in cotton, for example, is a natural polymer made of very long chains of glucose units linked together. Its relative molecular mass is about one million.

Organic chemists can synthesise other long-chain molecules by linking together thousands of small molecules to make polymers. These synthetic polymers include polythene, pvc, polystyrene and nylon (Figure 10.4).

With so many organic compounds to study and understand, chemists need something to simplify and organise their knowledge. They have found a method of classifying organic compounds into families or series, each of which has a distinctive group of atoms called a functional group (Section 10.2). Examples of these families include alkanes (Topic 11), alkenes (Topic 12), alcohols (Topic 13) and halogenoalkanes (Topic 14). These families of similar compounds with the same functional group are called homologous series.

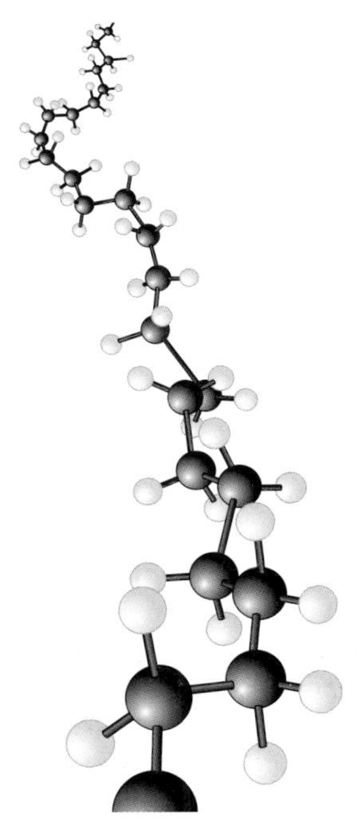

Figure 10.4 ▲
A short section of a polythene molecule.

Test yourself

1. What is organic chemistry?
2. State two reasons why carbon can form so many compounds.
3. Use the data section on your CD-ROM to look up the average bond enthalpies (bond energies) of various single bonds between similar elements.
   a) How does the strength of the single C–C bond compare with other single bonds between two atoms of the same non-metal?
   b) How do you think the relative strength of the C–C bond will affect the number of carbon compounds?

10.2 Functional groups

Ethane, CH_3CH_3, and ethanol, CH_3CH_2OH, have very different properties despite their similar structures (Figure 10.5). Ethane is a gas and ethanol is a liquid. Ethane does not react with sodium, but ethanol reacts vigorously to form hydrogen. Clearly, the –OH group in ethanol has a big effect on its properties.

The –OH group in ethanol is an example of a functional group. The functional group in a molecule is responsible for most of its reactions, while the hydrocarbon chain that makes up the rest of the organic compound is relatively unreactive (Figure 10.6).

Definitions

A functional group is the group of atoms that gives an organic compound its characteristic reactions.

A homologous series is a series of organic compounds with the same functional group in which the formula of each successive member differs by CH_2.

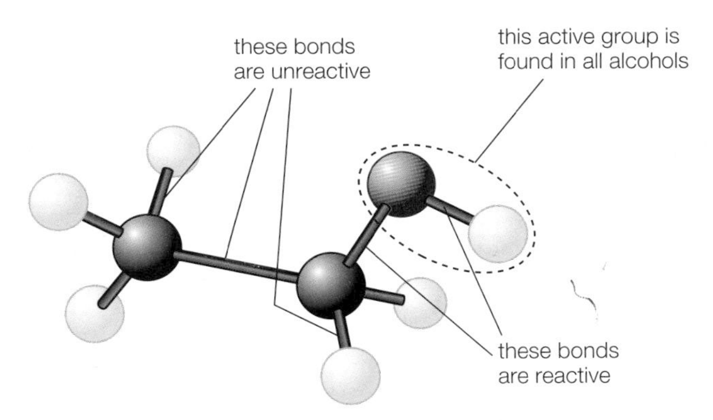

Figure 10.6 ▲
The structure of ethanol labelled to show the reactive functional group and the unreactive hydrocarbon skeleton.

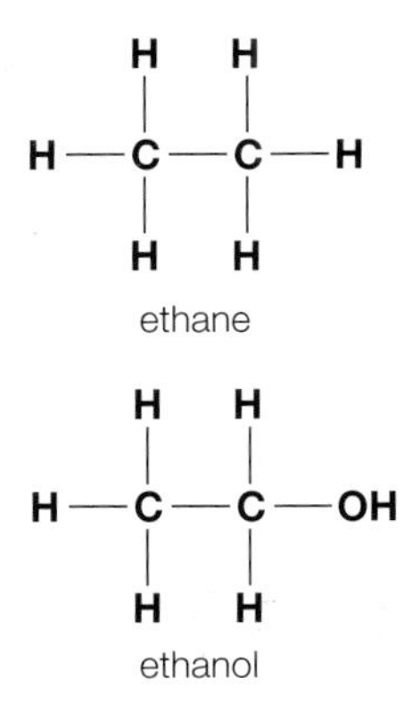

Figure 10.5 ▲
The structures of ethane and ethanol.

Functional groups, such as –OH, have more or less the same effect whatever the size of the hydrocarbon skeleton to which they are attached. This makes the study of organic compounds much simpler because all molecules containing the same functional group will have similar properties. In this respect, molecules with the same functional group can be regarded as a chemical family and compared to a group of elements in the periodic table.

Ethanol is a member of the homologous series of alcohols, all of which contain the –OH functional group. Ethene, $CH_2{=}CH_2$, is a member of the homologous series of alkenes that contain the >C=C< functional group. The common functional groups and homologous series that we will study in this AS course are shown in Table 10.1.

Table 10.1 ▶
Some common functional groups and homologous series.

Functional group	Name of the homologous series	Example
–OH	Alcohols	CH_3CH_2OH Ethanol
>C=C<	Alkenes	$CH_2{=}CH_2$ Ethene
–Hal	Halogenoalkanes	CH_3CH_2Cl Chloroethane
C–O–C	Ethers	CH_3OCH_3 Methoxymethane
–C(=O)H	Aldehydes	CH_3CHO Ethanal
>C=O	Ketones	CH_3COCH_3 Propanone
–C(=O)OH	Carboxylic acids	CH_3COOH Ethanoic acid
–O–C(=O)–	Esters	CH_3OCOCH_3 Methylethanoate
$-NH_2$	Primary amines	$CH_3CH_2NH_2$ Aminoethane (Ethylamine)

Some organic molecules have two or more functional groups. Lactic acid in sour milk, for example, has both an –OH group and a –COOH group (Figure 10.7). In its reactions, lactic acid sometimes acts like an alcohol, sometimes acts like an acid and sometimes shows the properties of both types of compound.

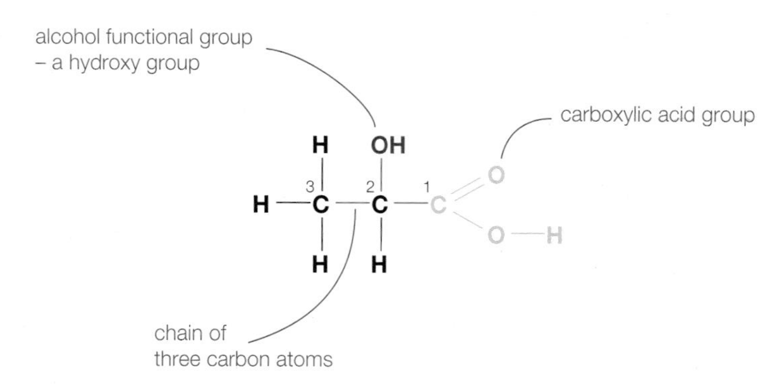

Figure 10.7 ▲
The structure of lactic acid, 2-hydroxypropanoic acid.

Test yourself

4 a) Why do all alcohols have similar properties?
   b) Why is there a gradation of the same property in alcohols from ethanol, C_2H_5OH, to hexanol, $C_6H_{13}OH$?

5 Identify the functional groups in the following compounds and the homologous series to which they belong.
   a) CH_3CH_2CHO
   b) CH_3CH_2I
   c) $CH_2{=}CHCH_2Cl$

10.3 Empirical, molecular and structural formulae

Empirical formulae

The empirical formula of a compound is the formula found by experiment. In Section 2.2, we used the masses of elements combined in a compound to calculate its empirical formula. The formulae we obtain by this method show the simplest whole number ratio of the atoms of different elements in a compound. The following example will help you to recall our calculations.

Worked example

0.15 g of a liquid was analysed and found to contain 0.06 g of carbon, 0.01 g of hydrogen and 0.08 g of oxygen. What is the empirical formula of the liquid?

	C	H	O
Masses of elements combined	0.06 g	0.01 g	0.08 g
Molar masses	$12\,g\,mol^{-1}$	$1\,g\,mol^{-1}$	$16\,g\,mol^{-1}$
Amounts of elements combined	$\frac{0.06\,g}{12\,g\,mol^{-1}}$ = 0.005 mol	$\frac{0.01\,g}{1\,g\,mol^{-1}}$ = 0.01 mol	$\frac{0.08\,g}{16\,g\,mol^{-1}}$ = 0.005 mol
Ratio of moles of elements	1	2	1
∴ Ratio of atoms of elements	1	2	1

So, the empirical formula of the compound is CH_2O.

Figure 10.8 ◀ Chemists involved in medical research studying the computer model of a protein. The purple ribbon shows the general structure of the molecule. The green pattern shows the atoms at the active site of the molecule which helps cells respond to hormones.

Molecular formulae

The molecular formula of a compound shows the actual number of atoms of each element in one molecule. The molecular formula of ammonia is NH_3 and that of ethanol is C_2H_6O. The term 'molecular formula' applies only to substances that consist of molecules.

For molecular compounds, the relative molecular mass shows whether or not the molecular formula is the same as the empirical formula. A molecular formula is always a simple multiple of the empirical formula.

For example, analysis shows that the empirical formula of octane in petrol is C_4H_9, but the mass spectrum of octane shows that its relative molecular mass is 114.

The relative mass of the empirical formula,

$$M_r(C_4H_9) = (4 \times 12) + (9 \times 1) = 57$$

The relative molecular mass is twice this value, so the molecular formula is twice the empirical formula,

$\therefore$ the molecular formula of octane = C_8H_{18}

Although the empirical and molecular formulae of organic compounds can be determined by analysis along the lines just described, the modern way to find molecular formulae is by mass spectrometry (Section 15.2).

Test yourself

6 A sample of a hydrocarbon was burned completely in oxygen. All of the carbon in the sample was converted to 1.69 g of carbon dioxide and all of the hydrogen was converted to 0.346 g of water.
   a) What is the percentage of carbon in carbon dioxide?
   b) What is the mass of carbon in 1.69 g of carbon dioxide?
   c) What is the percentage of hydrogen in water?
   d) What is the mass of hydrogen in 0.346 g of water?
   e) Use the masses of carbon and hydrogen you calculated in parts **b)** and **d)** to calculate the empirical formula of the hydrocarbon.

7 A compound that contains only carbon, hydrogen and oxygen was analysed. It consisted of 38.7% carbon and 9.68% hydrogen.
   a) What percentage of oxygen did it contain?
   b) What is its empirical formula?
   c) The relative molecular mass of the compound is 62. What is its molecular formula?

8 A hydrocarbon that consists of 82.8% carbon has an approximate relative molecular mass of 55.
   a) What is its empirical formula?
   b) What is its molecular formula?

Structural formulae

The molecular formula of a compound shows the number of atoms of the different elements in one molecule, but it does not show how the atoms are arranged. To understand the properties of a compound, we need to know its structural formula. This shows which atoms, or groups of atoms, are attached to each other. For example, the molecular formula of ethanol is C_2H_6O, but this does not show how the two carbon atoms, six hydrogen atoms and one oxygen atom are arranged. Its structural formula, however, is written as CH_3CH_2OH. This shows that ethanol has a CH_3 group attached to a CH_2 group which, in turn, is attached to an OH group.

It is often clearer to draw a more detailed structure showing all the atoms and all the bonds between them in one molecule of a compound (Figure 10.9). This type of formula is called a displayed formula.

Skeletal formulae are sometimes used – these show only the carbon–carbon bonds and functional groups in a compound (Figure 10.10). Skeletal formulae are outline formulae only. They provide a useful shorthand for large and complex molecules. Skeletal formulae need careful study, however, because they show the hydrocarbon part of a molecule with nothing more than lines for the bonds between carbon atoms and for the bonds from carbon atoms to functional groups. The symbols of carbon and hydrogen atoms in the carbon skeleton are omitted. In contrast, functional groups are shown in full.

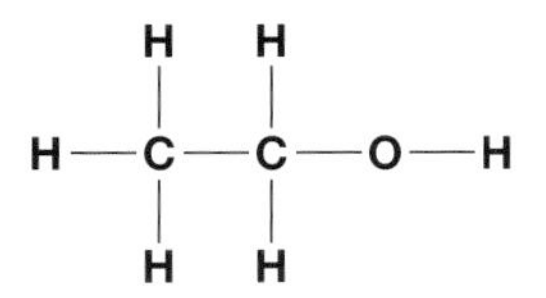

Figure 10.9 ▲
The displayed formula of ethanol.

Molecular formula	Structural formula	Displayed formula	Skeletal formula
C_3H_8	$CH_3CH_2CH_3$	H H H H—C—C—C—H H H H	∧
C_2H_6O	CH_3CH_2OH	H H H—C—C—O—H H H	∧OH

Figure 10.10 ▲
Alternative formulae for propane and ethanol.

Definitions

A **structural formula** provides the minimal detail to show the arrangement of atoms in one molecule of a compound.

A **displayed formula** shows all the atoms and all the bonds between them in one molecule of a compound.

A **skeletal formula** shows the hydrocarbon part of a molecule simply as lines between carbon atoms, omitting the symbols for carbon and hydrogen atoms, but showing the functional groups in full.

Activity

The alkanes – an important series of organic compounds

Methane, CH_4, ethane, C_2H_6, propane, C_3H_8 and butane, C_4H_{10}, are the first four members of the homologous series of alkanes. Most of their empirical, molecular, structural and displayed formulae are shown in Table 10.2.

Name	Methane	Ethane	Propane	Butane
Empirical formula	CH_4		C_3H_8	C_2H_5
Molecular formula	CH_4	C_2H_6		C_4H_{10}
Structural formula		CH_3CH_3	$CH_3CH_2CH_3$	
Displayed formula	H H—C—H H		H H H H—C—C—C—H H H H	H H H H H—C—C—C—C—H H H H H

Table 10.2 ▲
Empirical, molecular, structural and displayed formulae of the simplest alkanes.

1 Copy and complete the table by adding the missing formulae.

2 The names of all alkanes end in -ane. The names of the first four alkanes in Table 10.2 do not follow a logical system. All other straight-chain alkanes are named using a Greek numerical prefix for the number of carbon atoms in one molecule with the ending -ane. So, C_5H_{12} is pentane, C_7H_{16} is heptane and C_9H_{20} is nonane. The prefixes are the same as those used for geometrical figures (pentagon, heptagon, etc.).

What is the name for:

a) $CH_3CH_2CH_2CH_2CH_2CH_3$
b) $CH_3CH_2CH_2CH_2CH_2CH_2CH_2CH_3$
c) $CH_3CH_2CH_2CH_2CH_2CH_2CH_2CH_2CH_2CH_3$?

3 Which of the following molecular formulae are alkanes?

a) C_2H_2
b) C_3H_8
c) C_4H_8
d) C_8H_{18}
e) $C_{10}H_{20}$

4 Look at the formulae of methane, ethane, propane and butane in Table 10.2. What is the difference in carbon and hydrogen atoms between:

a) one molecule of methane and one molecule of ethane
b) one molecule of ethane and one molecule of propane
c) one molecule of propane and one molecule of butane?

5 It is possible to write a general formula for alkanes in the form of C_xH_y.

a) Suppose x equals n. If an alkane has n carbon atoms, how many hydrogen atoms will it have? (Hint: In a long-chain alkane, every carbon atom has two hydrogen atoms except the end carbon atoms, each of which has one extra hydrogen atom.)
b) What is the value of y in terms of n?
c) What is the general formula for alkanes in terms of C, H and n?
d) Predict the number of hydrogen atoms in one molecule and the molecular formula of the alkane with 12 carbon atoms in one molecule.

6 a) Draw the skeletal formula of butane.
b) Why is it not possible to draw a skeletal formula of methane?

Definition

The **general formula** of a homologous series is the simplest algebraic formula for its members.

10.4 Naming simple organic compounds

The International Union of Pure and Applied Chemistry (IUPAC) is the recognised authority for naming chemical compounds. IUPAC has developed systematic names based on a set of rules. These IUPAC rules make it possible to work out the structure of a compound from its name and to work out its name from its structure.

Figure 10.11 ▶
Four ways of representing the structure of the alkane with an unbranched chain of six carbon atoms. This is hexane.

CH_3 – CH_2 – CH_2 – CH_2 – CH_2 – CH_3

$CH_3CH_2CH_2CH_2CH_2CH_3$

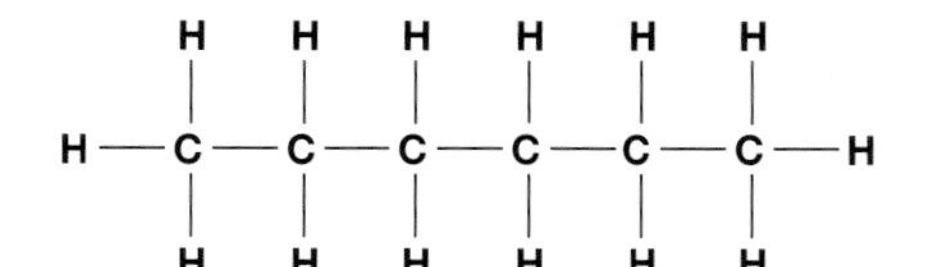

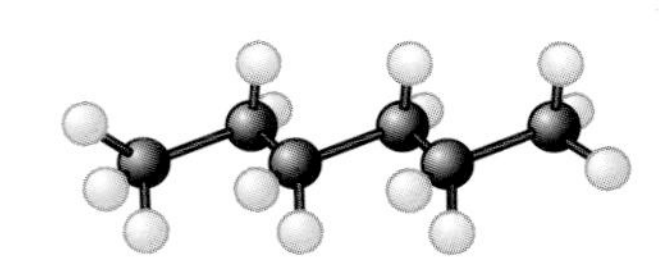

Naming alkanes

In naming an alkane, it is important to follow the IUPAC rules.

1 Look for the longest unbranched chain of carbon atoms in the carbon skeleton of the molecule and name that part of the compound.

So, $CH_3-CH_2-CH_2-CH_2-CH_3$ is pentane,

$CH_3-CH_2-CH_2-CH(CH_3)-CH_3$ is pentane with a CH_3 group attached and

$CH_3-CH_2-CH_2-CH(CH_2CH_3)-CH(CH_3)-CH_2-CH_3$ is heptane with one CH_3 and one CH_3CH_2 group attached.

2 Identify the alkyl groups that are attached to the longest unbranched chain. The simplest alkyl group is the methyl group, CH_3^-, which is methane with one hydrogen atom removed. Alkyl groups are alkane molecules minus one hydrogen atom (Table 10.3). So,

$CH_3-CH_2-CH_2-CH(CH_3)-CH_3$ has a methyl side group and

$CH_3-CH_2-CH_2-CH(CH_2CH_3)-CH(CH_3)-CH_2-CH_3$ has an ethyl and a methyl side group.

3 Number the carbon atoms in the main chain to identify which carbon atoms the side groups are attached to.

4 Name the compound using the name of the longest unbranched chain, prefixed by the names of the side groups and the numbers of the carbon atoms to which they are attached. The numbering of the carbon atoms can be from either the left or the right to give the name with the lowest numbers. So,

$CH_3-CH_2-CH_2-CH(CH_3)-CH_3$ is 2-methylpentane not 4-methylpentane and

$CH_3-CH_2-CH_2-CH(CH_2CH_3)-CH(CH_3)-CH_2-CH_3$ is 4-ethyl-3-methylheptane not 4-ethyl-5-methylheptane.

5 When there is more than one type of side group, these should be arranged alphabetically. So,

$CH_3-CH_2-CH_2-CH(CH_2CH_3)-CH(CH_3)-CH_2-CH_3$ is 4-ethyl-3-methylheptane not 3-methyl-4-ethylheptane.

6 When there are two or more of the same side group, add the prefix 'di', 'tri', 'tetra' and so on. So,

$CH_3-CH_2-CH(CH_3)-CH(CH_3)-CH_3$ is 2,3-dimethylpentane not 2-methyl-3-methylpentane and not 3,4-dimethylpentane.

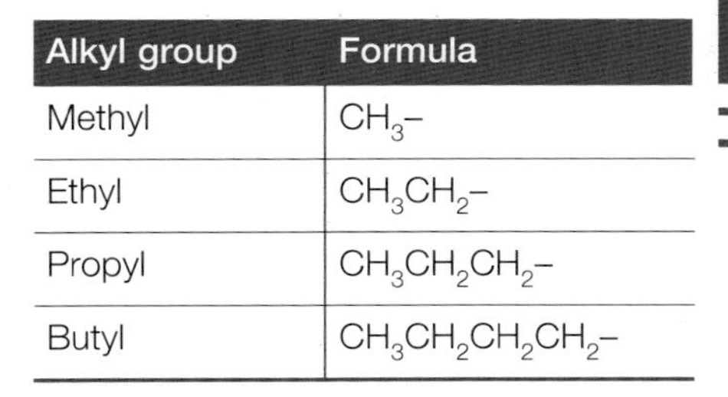

Alkyl group	Formula
Methyl	CH_3-
Ethyl	CH_3CH_2-
Propyl	$CH_3CH_2CH_2-$
Butyl	$CH_3CH_2CH_2CH_2-$

Table 10.3 ▲
The structures of alkyl groups.

Tutorial

Test yourself

9 Name the following alkanes.

a) $CH_3(CH_2)_6CH_3$

b)
```
CH3 - CH - CH - CH - CH2 - CH3
      |    |    |
      CH3  CH3  CH3
```

c)
```
CH3CH2CH2CHCH3
          |
          CH2
          |
          CH3
```

d)
```
        CH3
        |
CH3CH2CCH3
        |
        CH3
```

10 Draw the displayed formulae of the following alkanes:

a) 3,3,4-trimethylheptane

b) 2-methylbutane

c) 3-ethyl-2-methyl-5,5-dipropyldecane.

Test yourself

11 Name the following alkenes.

a)
```
CH3 — C = CH2
      |
      CH3
```

b) $CH_3CH_2CH_2CH{=}CHCH_3$

c)
```
CH3 - CH2 - C = C - CH3
            |   |
            CH3 CH3
```

12 Draw the displayed formulae of the following alkenes.

a) 2-methylbut-2-ene

b) 3,4-dimethylpent-1-ene

Naming alkenes

Ethene, $CH_2{=}CH_2$, and propene, $CH_3CH{=}CH_2$, are the first two members of the homologous series of alkenes with the functional group $>C{=}C<$.

Alkenes are named using the same general rules as alkanes, but with the suffix -ene instead of -ane. This suffix is sometimes prefixed by a number to indicate the position of the double bond in the chain.

With ethene and propene, there is no need to number the carbon atoms because the double bond must be between carbon atoms 1 and 2. With a chain of four or more carbon atoms, however, the double bond may be in more than one position. The molecule $CH_2{=}CHCH_2CH_3$ is therefore named but-1-ene. Although the double bond links carbon atoms 1 and 2, the number 1 is used because it is the lower. Using the same rule, $CH_3CH{=}CHCH_3$ is named but-2-ene.

The methods of naming organic compounds with other functional groups will be explained as they arise during this AS course.

10.5 Isomerism

Another reason why carbon forms so many compounds is because it is sometimes possible to join the same atoms together in different ways. Consider, for example, the molecular formula C_4H_{10}. You probably already realise that this could be butane, but another compound, 2-methylpropane, also has the molecular formula C_4H_{10} (Figure 10.12). Compounds like butane and 2-methylpropane, which have the same molecular formula but different structural formulae, are called structural isomers.

Definition

Structural isomers are compounds with the same molecular formula but different structural formulae.

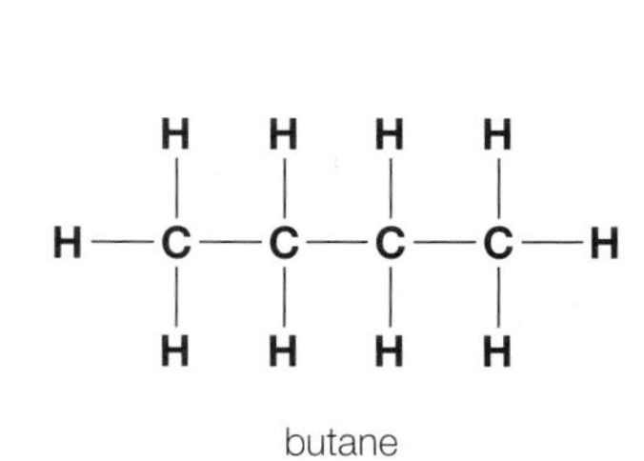

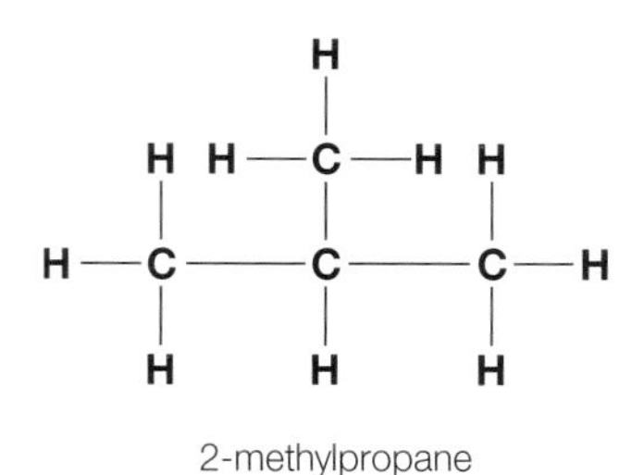

Figure 10.12 ▶ Isomers of C_4H_{10}.

It is useful to divide structural isomers into three different types: chain isomers, position isomers and functional group isomers.

- **Chain isomers** have different chains of carbon atoms (Figure 10.12).
- **Position isomers** have different positions of the same functional group (Figure 10.13).

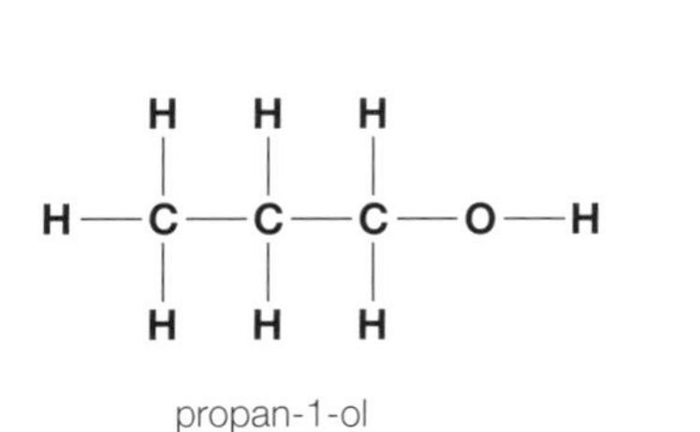

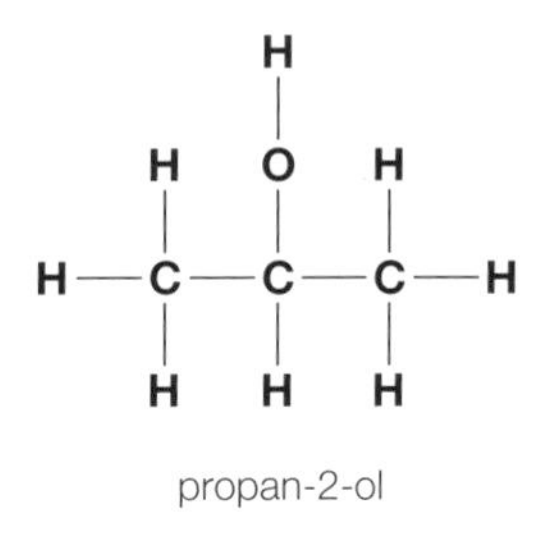

Figure 10.13 ◀
Propan-1-ol and propan-2-ol are both alcohols like ethanol, CH_3CH_2OH. All alcohols contain the –OH group. In propan-1-ol and propan-2-ol, the –OH groups are in different positions on the carbon chain.

- Functional group isomers have different functional groups (Figure 10.14).

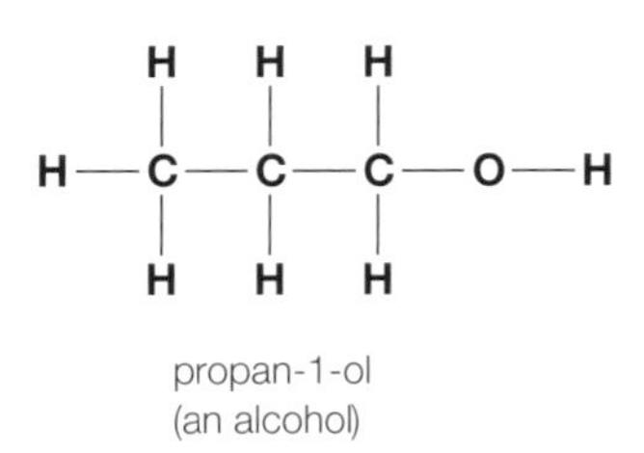

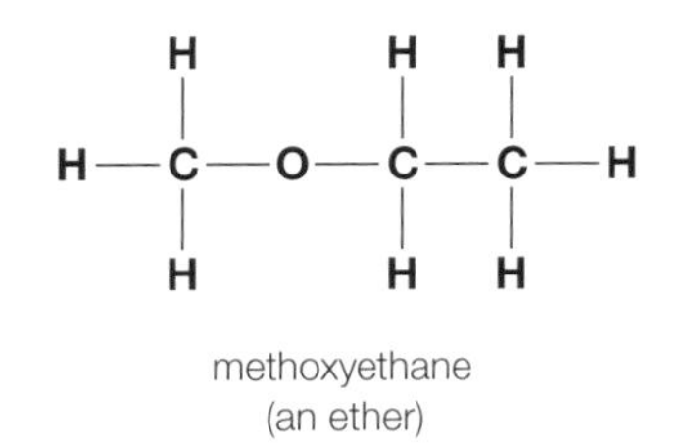

Figure 10.14 ◀
Propan-1-ol is an alcohol with the –OH functional group. Methoxyethane is an ether with the C–O–C functional group. Both of these compounds have the same molecular formula, C_3H_8O.

Notice that the word describing the type of structural isomers (that is, chain, position and functional group) tells you how they differ from each other.

Figure 10.15 ▲
Most vehicles rely on petrol (gasoline) for fuel. Petrol is a mixture of more than 100 different alkanes with between 5 and 10 carbon atoms. There are three isomers of C_5H_{12}, five of C_6H_{14}, 18 of C_8H_{18} and 75 of $C_{10}H_{22}$.

Test yourself

13 a) Draw the carbon chains of the three chain isomers with the molecular formula C_5H_{12}. (Hint: One obvious chain is C–C–C–C–C.)

b) Use a molecular model kit to construct models of the three isomers and look at the rotatable models of their structures on the CD-ROM.

c) Draw displayed formulae of the three isomers and name them.

10.6 Introduction to the mechanisms of organic reactions

The mechanism of a reaction shows, step by step, the bonds that break and the new bonds that form as reactants turn into products. Chemists have used great ingenuity in working out mechanisms. They began with simple reactions but are now using their knowledge to explain what happens during industrial processes and in living cells, in which enzymes control biochemical processes.

During chemical reactions, bonds can break in two ways: homolytically or heterolytically. These two ways are often described as homolytic fission and heterolytic fission.

Note

The prefix 'homo' means 'the same' or 'similar'. Chemical terms that include this prefix include homolytic fission, homogeneous catalyst and homologous series.

The prefix 'hetero' means 'different'. Chemical terms that include this prefix include heterolytic fission and heterogeneous catalyst.

Homolytic fission

A covalent bond consists of a shared pair of electrons. If the bond breaks homolytically, each atom keeps one electron.

When chlorine is exposed to ultraviolet radiation, homolytic fission occurs. This produces free chlorine atoms with unpaired electrons.

$$Cl : Cl \rightarrow Cl\bullet + Cl\bullet$$

In equations like this, a single dot is used to represent the unpaired electron. Homolytic fission also results in groups of atoms with unpaired electrons, like $CH_3\bullet$ and $CH_3CH_2\bullet$. These atoms or groups of atoms with unpaired electrons are called free radicals or simply radicals. They exist for a short time during a reaction but quickly react to form new products. So, radicals are usually very short-lived. They are intermediates that form during a reaction but then disappear as the reaction is completed.

Radicals often occur in reactions that take place in the gas phase or in non-polar solvents. Ultraviolet light can speed up radical reactions.

Examples of radical processes include the thermal cracking of hydrocarbons (Section 11.3), the burning of petrol and other alkanes (Section 11.3) and the substitution reactions of alkanes with halogens (Section 11.5). Radical reactions are important high in the atmosphere where gases are exposed to intense ultraviolet radiation from the Sun. The reactions that form and destroy the ozone layer involve radicals (Section 18.6).

Figure 10.16 ▶ Radicals are produced from oxygen and hydrocarbons when petrol burns and explodes.

Heterolytic fission

When a covalent bond breaks heterolytically, one atom takes both of the electrons from the bond. This results in the formation of an anion and a cation.

$$H_3C \overset{\bullet}{\underset{\times}{\ }} Br \longrightarrow H_3C^+ + \overset{\bullet}{\underset{\times}{\ }}Br^-$$

$$H_3C{-}Br \longrightarrow H_3C^+ + Br^-$$

Figure 10.17 ▶ Heterolytic fission. Note the use of a curly arrow to show what happens to the electrons as the bond breaks. The covalent bond breaks and the atoms separate, with one atom taking both electrons in the shared pair.

Heterolytic fission produces ionic intermediates such as CH_3^+ and Br^- in reactions like that shown in Figure 10.17. This type of bond breaking is favoured when reactions take place in polar solvents such as water. Often the bond that breaks is already polar (Section 6.10), with a δ+ end and a δ– end like the C–Br bond in Figure 10.18. Notice that the curly arrows in Figures 10.17 and 10.18 start from either a bond, a pair of electrons or a negative charge and that the head of the arrow points to where the bond, the electrons or the negative charge will be after the change.

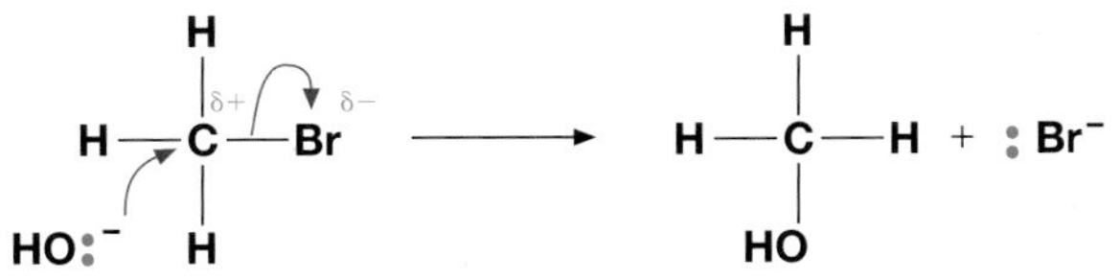

Figure 10.18 ▲
An ionic reagent attacking a polar bond leading to heterolytic bond breaking.

Some reagents that start reactions seek out the δ+ end of polar bonds. These are called nucleophiles. This is the way in which OH^- acts in Figure 10.18. Other reagents seek out the δ– end of polar bonds or the electron dense regions in molecules. These are called electrophiles.

Nucleophiles

Nucleophiles are molecules or ions with a lone pair of electrons that can form a new covalent bond. They are 'electron-pair donors'. Nucleophiles are reagents that attack molecules in which there is a partial positive charge, δ+, so they seek out positive charges – they are 'nucleus loving'. The substitution reactions of halogenoalkanes involve nucleophiles (Section 14.4).

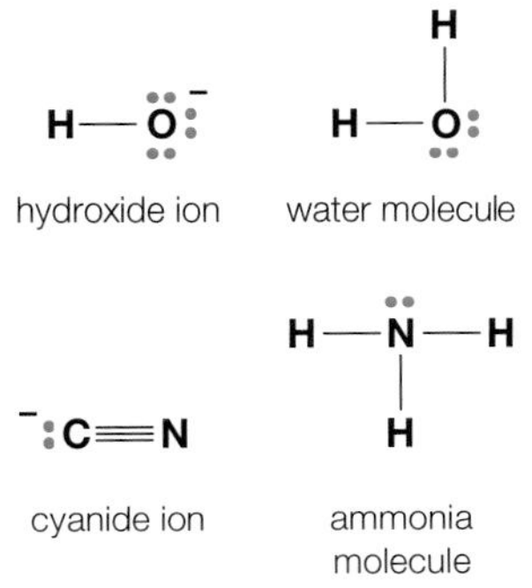

hydroxide ion
water molecule
cyanide ion
ammonia molecule

Figure 10.19 ▲
Examples of nucleophiles.

Electrophiles

Electrophiles are reactive ions and molecules that attack parts of molecules that are rich in electrons – negative ions or negative centres in molecules. They are 'electron-loving' reagents. Electrophiles form a new bond by accepting a pair of electrons from the molecule attacked during a reaction.

An example of an electrophile is the H atom at the δ+ end of the H–Br bond in hydrogen bromide. Electrophiles play an important part in the addition reactions of alkenes (Section 12.4).

Definitions

Radicals are reactive single atoms or groups of atoms with an unpaired electron.

Nucleophiles are electron-pair donors – anions or molecules with a lone pair of electrons that attack positive ions or positive centres in molecules.

Electrophiles are electron-pair acceptors – cations and molecules that attack negative ions or negative centres in molecules.

Test yourself

14 Write equations showing how:
- a) a bromine molecule breaks homolytically
- b) a hydrogen bromide molecule breaks heterolytically
- c) a C–H bond in a methane molecule breaks homolytically.

15 In each of the following examples, decide whether the reagent attacking the carbon compound is a radical, a nucleophile or an electrophile.
- a) $CH_3CH_2I + H_2O \rightarrow CH_3CH_2OH + HI$
- b) $CH_2{=}CH_2 + HBr \rightarrow CH_3CH_2Br$
- c) $Cl\bullet + CH_4 \rightarrow CH_3Cl + H\bullet$
- d) $CH_3CH_2Br + CN^- \rightarrow CH_3CH_2CN + Br^-$

10.7 Theoretical and percentage yields

If a chemical reaction is totally efficient, all the starting reactant is converted to the required product. This gives a 100% yield. Few reactions are completely efficient, however, and many reactions give low yields, especially organic reactions. There are various reasons why yields are low.

Definitions

The **actual yield** is the mass of product obtained.

The **theoretical yield** is the mass of product obtained if the reaction goes according to the equation.

The **percentage yield** = $\frac{\text{actual yield}}{\text{theoretical yield}} \times 100\%$

A **limiting reactant** is a substance which is present in an amount that limits the theoretical yield. Often some reactants are added in excess to ensure that the most valuable reactant is converted to as much product as possible.

- The reactants may not be totally pure.
- Some of the product may be lost during transfer of the chemicals from one container to another when the product is separated and purified.
- There may be side reactions in which the reactants form different products.
- Some of the reactants may not react because the reaction is so slow (Section 17.1) or because it comes to equilibrium (Section 17.7).

Worked example

Ethene, C_2H_4, can be obtained by dehydration of ethanol, C_2H_6O. In an experiment, 0.50 g of ethene was obtained from 5.00 g of ethanol.

What is the theoretical yield and percentage yield of ethene?

The equation for the dehydration of ethanol is:

$$C_2H_6O(l) \rightarrow C_2H_4(g) + H_2O(l)$$
$$1 \text{ mol of } C_2H_6O \rightarrow 1 \text{ mol of } C_2H_4$$
$$\therefore (2 \times 12) + (6 \times 1) + 16 \text{ g } C_2H_6O \rightarrow (2 \times 12) + (4 \times 1) \text{ g } C_2H_4$$
$$46 \text{ g } C_2H_6O \rightarrow 28 \text{ g } C_2H_4$$
$$\therefore 1 \text{ g } C_2H_6O \rightarrow \frac{28}{46} \text{ g } C_2H_4$$
$$\text{So } 5 \text{ g } C_2H_6O \rightarrow \frac{28}{46} \times 5 \text{ g } C_2H_4 = 3.04 \text{ g } C_2H_4$$

$\therefore$ Theoretical yield of ethene = 3.04 g

The actual yield of ethene = 0.50 g

$\therefore$ Percentage yield $= \frac{0.50}{3.04} \times 100\% = 16\%$

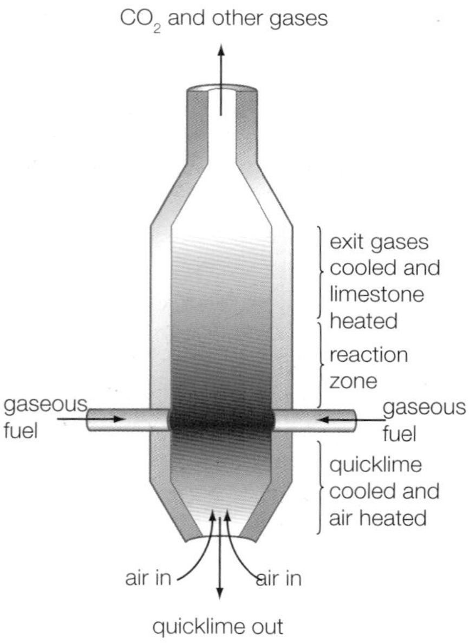

Figure 10.20 ▲ A gas-fuelled lime kiln.

Test yourself

16 A modern gas-fuelled lime kiln produces 500 kg of calcium oxide, CaO (quicklime) from 1000 kg of crushed calcium carbonate, $CaCO_3$ (limestone).

a) Write an equation for the decomposition of calcium carbonate to form calcium oxide.

b) What is the theoretical yield of calcium oxide from 1000 kg of calcium carbonate? (Ca = 40, C = 12, O = 16)

c) What is the percentage yield of calcium oxide in the gas-fuelled kiln?

17 When ethene, C_2H_4, reacts with hydrogen, it undergoes an addition reaction to give a very high percentage yield of ethane, C_2H_6.

a) Write an equation for the reaction.

b) Suggest a reason why the percentage yield is very high.

c) What is the theoretical yield of ethane from 140 kg of ethene?

18 500 kg of calcium oxide (pure quicklime) was reacted with water to produce calcium hydroxide, $Ca(OH)_2$ (slaked lime). 620 kg of calcium hydroxide was produced. Calculate the theoretical and percentage yields.

Atom economy

The yields in most laboratory and industrial processes focus on the desired product. However, many atoms in the reactants do not end up in the desired product, and this leads to a huge waste of material. For example, when glucose, $C_6H_{12}O_6$, is fermented to produce ethanol, C_2H_5OH, part of the glucose is lost as carbon dioxide. This waste in many reactions has led scientists and industrialists to use the term atom economy when calculating the overall efficiency of a chemical process. The atom economy of a reaction is the mass of the atoms in the desired product expressed as a percentage of the mass of the atoms in all the products shown in the equation.

Definition

Atom economy = $\frac{\text{mass of atoms in desired product}}{\text{mass of atoms in all reactants}} \times 100$

Worked example

The equation for the fermentation of glucose, $C_6H_{12}O_6$, to produce ethanol is:

$$C_6H_{12}O_6(aq) \rightarrow 2C_2H_5OH(aq) + 2CO_2(g)$$

What is the atom economy of this process?

$$\begin{aligned}\text{Molecular mass of desired product} &= 2 \times M(C_2H_5OH)\\ &= 2 \times 46\\ &= 2\,\text{mol} \times 46\,\text{g mol}^{-1}\\ &= 92\,\text{g}\end{aligned}$$

$$\begin{aligned}\text{Sum of molecular masses of all products} &= 2\,\text{mol} \times 46\,\text{g mol}^{-1} + 2\,\text{mol} \times 44\,\text{g mol}^{-1}\\ &= 180\,\text{g}\end{aligned}$$

$$\rightarrow \text{atom economy} = \frac{92}{180} \times 100\% = 51.1\%$$

Almost half the mass of the products is 'wasted' in this process as carbon dioxide is lost to the atmosphere. If we are to use raw materials as efficiently as possible, chemists must look for high atom economies as well as high percentage yields, especially in industrial processes. In particular, chemists must look towards addition reactions for the commercial production of bulk materials, as these have an atom economy of 100%.

Until the 1970s, ethanoic acid, CH_3COOH, was manufactured by the reaction of hydrocarbons from crude oil with oxygen. The process operated at a temperature of 200 °C and a pressure of 45 atmospheres. The atom economy of the process was only 35%. Since 1970, most ethanoic acid has been produced by reacting methanol, CH_3OH, with carbon monoxide using a catalyst of rhodium or iridium metal and iodide ions:

$$CH_3OH(l) + CO(g) \rightarrow CH_3COOH(l)$$

This new process has an atom economy of 100%. There is no waste and much less energy is needed to separate and purify the ethanoic acid. The catalysed reaction is very fast, and the yield of ethanoic acid is more than 95%.

At one time, industrial chemists looked for processes with a high percentage yield. However, some processes with a high percentage yield have a low atom economy. This is particularly true of processes involving substitution reactions, such as the manufacture of titanium by heating titanium(IV) chloride with magnesium:

$$TiCl_4(g) + 2Mg(l) \rightarrow Ti(s) + 2MgCl_2(s)$$

In this process, almost 80% of the reactants are wasted, as magnesium and chlorine atoms are lost as magnesium chloride. In the long term, processes as wasteful as this are just not sustainable.

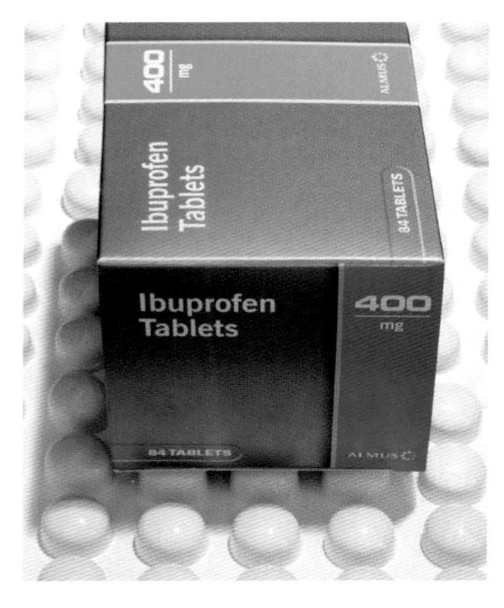

Figure 10.21 ▲
The production of Ibuprofen is an excellent example of atom economy. Ibuprofen is an important medicine that reduces swelling and pain in the joints. In the 1960s, Boots made Ibuprofen in five steps with an atom economy of only about 40%. This left 60% of the mass of products as waste material. When the patent expired, another company developed a new process that involved just two steps with an atom economy of 100%. The benefits and advantages of this new process which leaves no waste material are very significant.

Test yourself

19 Calculate the atom economy for:
  a) the manufacture of ethanol, C_2H_5OH, by hydration of ethene, C_2H_4:

  $$C_2H_4(g) + H_2O(g) \rightarrow C_2H_5OH(g)$$

  b) the manufacture of tin, Sn, from tinstone, tin(IV) oxide, SnO_2:

  $$SnO_2(s) + 2C(s) \rightarrow Sn(s) + 2CO(g)$$

20 Why are addition reactions preferable to substitution reactions in industrial processes?

Figure 10.22 ◄
Titanium is an exceptionally strong metal, yet its density is relatively low. This has led to the use of titanium alloys in aircraft construction. The fuselage of this stealth fighter plane is made from an alloy with a very high percentage of titanium.

REVIEW QUESTIONS

Extension questions

1 a) Carbon is able to form a vast number of chemical compounds. Suggest two reasons for this. (2)

b) Petrol is a mixture of hydrocarbons containing between six and ten carbon atoms. Some of these hydrocarbons are structural isomers.

i) Explain the term structural isomers. (2)

ii) Some of the hydrocarbons in petrol are alkanes. Write the molecular formula of an alkane that could be present in petrol. (1)

c) Petrol also contains cycloalkanes. Draw the structure of cyclohexane, and write the general formula of cycloalkanes. (2)

2 a) The compound X below, which contains two functional groups, can be extracted from oil of violets.

CH_3CH_2 and H on one C; CH_2OH and H on the other C; C=C

X

Figure 10.23 ▲

State the empirical and molecular formula of X and draw its skeletal formula.

Explain the term 'functional group', and name the functional groups present in X. (7)

b) X reacts with hydrogen and a nickel catalyst in the gas phase to produce compound Y, with the formula $CH_3(CH_2)_3CH_2OH$.

i) What is the systematic name of Y? (1)

ii) Write an equation, including state symbols, for the reaction of X with hydrogen to form Y. (2)

iii) Y has two structural isomers that are also position isomers. Name these two position isomers of Y. (2)

iv) Y also has structural isomers with a different functional group. Write the structural formula of one of these isomers. (1)

3 a) Crude oil is a mixture of many hydrocarbons. Fractional distillation can be used to separate it into fractions which can be refined to produce hydrocarbons such as dodecane.

i) What is meant by the term 'hydrocarbon'? (1)

ii) One molecule of dodecane contains 12 carbon atoms. What is the molecular formula of dodecane? (1)

iii) What is the empirical formula of dodecane? (1)

b) Decane, $C_{10}H_{22}$, is a straight-chain alkane. It reacts with chlorine in a radical reaction to form the compound $C_{10}H_{21}Cl$.

i) Explain the term 'radical'. (2)

ii) Write an equation for the formation of chlorine radicals from Cl_2. (1)

iii) What type of bond fission is involved in the formation of chlorine radicals? (1)

iv) How many different structural isomers can be produced when decane reacts with chlorine to form $C_{10}H_{21}Cl$? (1)

v) Draw and name the structural formula of one of these structural isomers. (2)

4 Cyclohexene, C_6H_{10}, is manufactured by dehydration of cyclohexanol, $C_6H_{11}OH$, in the gas phase.

a) Write an equation with state symbols for this process. (2)

b) Draw the structural formula of cyclohexene. (1)

c) Calculate the theoretical yield of cyclohexene from 10 kg of cyclohexanol. (3)

d) The actual yield of cyclohexene from 10 kg of cyclohexanol was 7.1 kg. What was the percentage yield of cyclohexene? (2)

e) What is the atom economy for the manufacture of cyclohexene by dehydration of cyclohexanol? (2)

11 Alkanes

Alkanes are important because they make up most of crude oil and natural gas, which are the source of most fuels and the main raw materials for the chemical industry. Alkanes are hydrocarbons – compounds containing only hydrogen and carbon. The carbon atoms in alkane molecules may be joined together in straight chains, branched chains or rings, but all of the bonds are single bonds.

11.1 Types of hydrocarbons

There are three important types of hydrocarbons.

- Aliphatic hydrocarbons have no rings of carbon atoms. Their chains of carbon atoms may be branched or unbranched. Alkanes and alkenes (Topic 12) are examples of aliphatic compounds.
- Alicyclic hydrocarbons have rings of carbon atoms, such as cycloalkanes and cycloalkenes (Figure 11.1). Alicyclic hydrocarbons are usually just called cyclic hydrocarbons.
- Arenes are ring compounds such as benzene, methylbenzene and naphthalene in which there are delocalised electrons (Figure 11.2). Arenes are sometimes called 'aromatic hydrocarbons' because of their smell (aroma).

In addition to this classification of hydrocarbons as aliphatics, alicyclics and arenes, two other terms are often used to describe hydrocarbons: saturated and unsaturated.

In saturated compounds, all the bonds between the atoms in their molecules are single bonds. Alkanes and cycloalkanes are examples of saturated compounds.

The term 'saturated' is also used to describe compounds with saturated hydrocarbon chains, such as saturated fats and fatty acids in food.

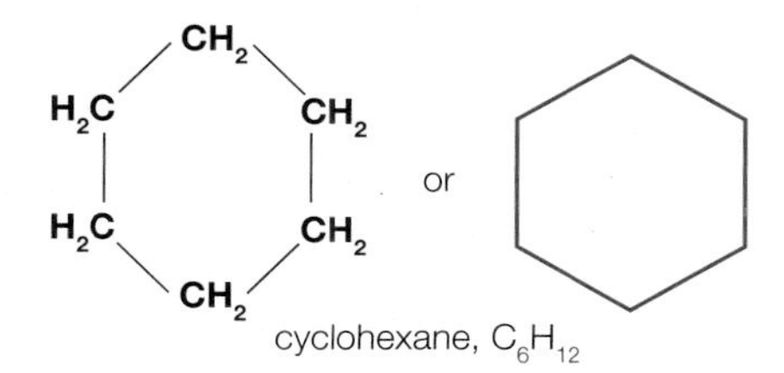

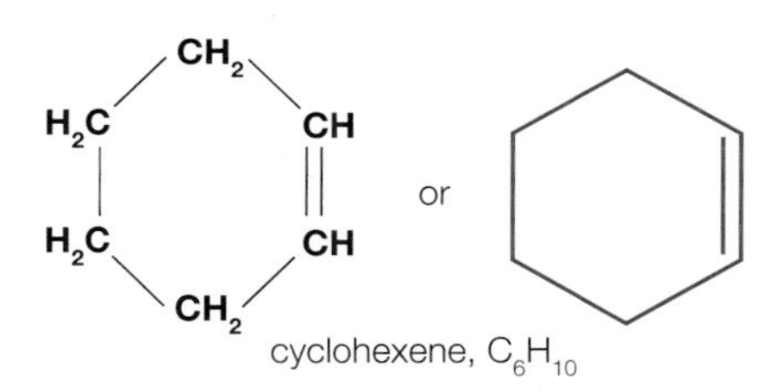

Figure 11.1 ▲
Structures of the cyclic hydrocarbons cyclohexane and cyclohexene.

Figure 11.3 ▲
Saturated fats in foods such as fish and chips, beef burgers and doughnuts contain alkyl groups with long chains of carbon atoms. These substances, if taken in excess, lead to high levels of cholesterol in the blood, which causes furring and blocking of the arteries (arteriosclerosis).

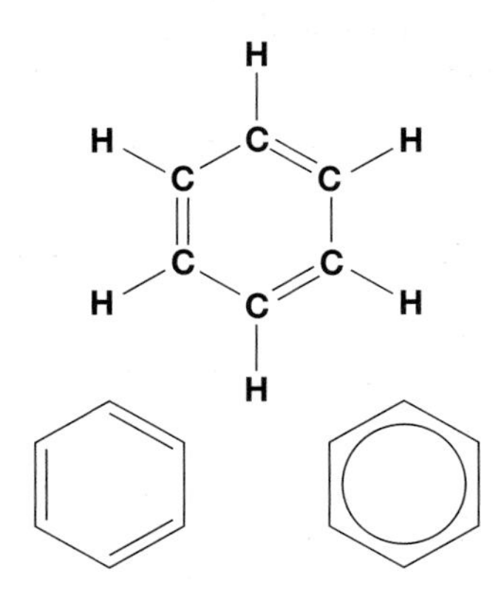

Figure 11.2 ▲
Representations of the structure of benzene. At one time, chemists thought that the ring structure in benzene had three double and three single bonds. However, X-ray studies showed that all six bonds in the ring are identical and that each carbon atom contributes one electron to a cloud of delocalised electrons. This has led to the third structure with a ring inside a hexagon. The strength and length of carbon–carbon bonds in arenes are between those of single and double bonds.

Unsaturated compounds have one or more double or triple bonds between the atoms in their molecules. The term 'unsaturated' is often applied to alkenes

Definition

Hydrocarbons are compounds containing only hydrogen and carbon.

Aliphatic hydrocarbons are straight-chain and branched hydrocarbons with no rings of carbon atoms.

Alicyclic hydrocarbons are hydrocarbons with a ring or rings of carbon atoms.

Arenes are hydrocarbons with a ring or rings of carbon atoms in which there are delocalised electrons.

such as ethene and cyclohexene which readily undergo addition reactions. The term is also used to describe foods containing unsaturated fats which have carbon–carbon double bonds (C=C).

11.2 Alkanes

Alkanes are the hydrocarbons that make up most of crude oil and natural gas. Alkanes form a homologous series of organic compounds with the general formula C_nH_{2n+2} (Section 10.3).

In alkane molecules, every carbon atom forms four single covalent bonds by using the four electrons in its outer shell. The electrons in these four bonds repel each other to tetrahedral positions around each carbon atom (Section 6.8). The tetrahedral shape around each carbon atom is shown neatly in the models of propane in Figure 11.4.

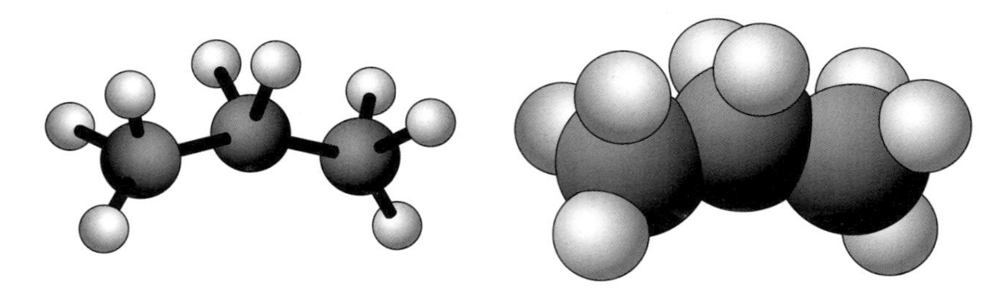

Figure 11.4 ▶
Models of the structure of propane showing the tetrahedral shape around each carbon atom in alkanes.

Test yourself

1 Classify the following compounds as aliphatic, alicyclic, arene, saturated and unsaturated.
  a) hexane
  b) propene
  c) cyclobutene
  d) HN=NH
2 a) Write the molecular formulae and names for the first four members of the alkanes.
  b) What is the difference in molecular formula from one alkane to the next?
3 Why is the general formula of alkanes C_nH_{2n+2}?

Physical properties

The molecules in alkanes are only held together by weak intermolecular forces (Section 6.13). As the molecules get longer and larger, the intermolecular forces increase, so the melting and boiling points rise as the number of carbon atoms per molecule increases.

At room temperature and pressure, alkanes in the range C_1 to C_4 are gases, those in the range C_5 to C_{17} are liquids and those from C_{18} upwards are solids.

As the electronegativities of carbon and hydrogen are so similar, alkane molecules are non-polar. This means that they do not mix with or dissolve in polar solvents like water.

11.3 Fuels from crude oil

Crude oil (petroleum) is arguably the most important naturally occurring raw material. It provides a vast source of hydrocarbons, which can be separated into fractions with different boiling points and used as fuels or feedstock for the petrochemical industry.

Crude oil is a complex mixture of hydrocarbons, most of which are alkanes. In its raw form, crude oil has no uses. The challenge for refineries is to produce the various products in the proportions required by industrial and domestic users.

Activity

Intermolecular forces and the properties of alkanes

In this activity you will be looking for patterns in the boiling points of alkanes. You will need to refer to the data table for alkanes on your CD-ROM and explain the patterns you find in terms of intermolecular forces.

Boiling points of the straight-chain alkanes

Plot the boiling points of straight-chain alkanes against their number of carbon atoms for the range C_1 to C_{10}.

1 What is the approximate increase in boiling point:
- **a)** for each $-CH_2-$ added to the chain for the range C_6 to C_{10}?
- **b)** from pentane to hexane?
- **c)** from nonane to decane?

2 Estimate the boiling point of pentadecane, $C_{15}H_{32}$.

3 What type of intermolecular forces act between alkane molecules?

4 Explain the trend in boiling points in your graph.

Boiling points of the branched alkanes

Add the boiling points of the 2-methylalkanes from 2-methylpropane to 2-methylheptane to your graph.

5 What is the effect of chain branching on the boiling points of alkanes?

6 Explain the effect you found in question 5.

7 Write down the names and boiling points of the three isomers with the molecular formula C_5H_{12} and explain the trend in values.

Generally, crude oil contains too much of the high boiling fractions with larger molecules and not enough of the low boiling fractions with smaller molecules. These smaller molecules are needed for fuels such as petrol. In order to satisfy the demand for very different products, crude oil undergoes fractional distillation, cracking and reforming.

Fractional distillation

Fractional distillation is the first stage in refining crude oil. The continuous process operates on a large scale and separates crude oil into different fractions. This produces fuels and lubricants, as well as feedstocks for the petrochemical industry.

A furnace heats the crude oil to about 400 °C, which then flows into a fractionating tower containing about 40 horizontal 'trays' pierced with small holes.

The column is hotter at the bottom and cooler at the top. Rising vapour condenses when it reaches the tray containing liquid at a temperature just below the vapour's boiling point. The condensing vapour releases energy, which heats the liquid on the tray. This evaporates the more volatile compounds in the mixture.

With a series of trays, the outcome is that hydrocarbons with small molecules and low boiling points rise to the top of the column, while larger molecules stay at the bottom. Fractions are drawn off from the column at various levels.

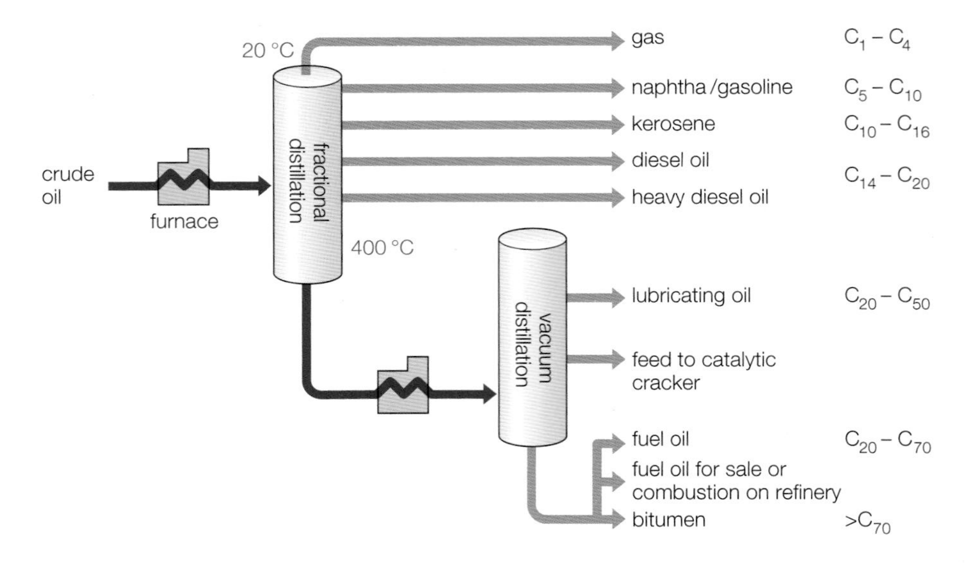

Figure 11.5 ▶
Fractional distillation of crude oil.

Some components of crude oil have boiling points too high for them to vaporise at the furnace temperature and atmospheric pressure. Lowering the pressure in a separate vacuum distillation column reduces the boiling points of these hydrocarbons and makes it possible to separate them.

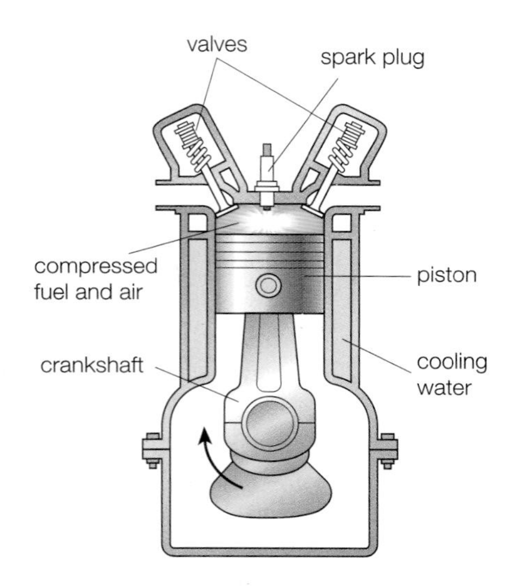

Figure 11.6 ▲
The working parts of a cylinder of an internal combustion engine that runs on petrol. Sparks from the plugs cause the compressed fuel and air to ignite. This produces more gas molecules, increasing the pressure and forcing the piston down. The product gases are then allowed to escape and, as the pressure falls, the piston rises ready for the next ignition.

Fuel fractions

Petrol is a blend of hydrocarbons based on the gasoline fraction (hydrocarbons with 5–10 carbon atoms), whereas jet fuel is produced from the kerosene or paraffin fraction (hydrocarbons with 10–16 carbon atoms). Fuel for diesel engines is made from diesel oil (hydrocarbons with 14–20 carbon atoms).

All these fuels must be refined to remove sulfur compounds that would cause air pollution when they burn. In addition, petrol must be blended carefully if modern engines are to start reliably and run smoothly. The proportion of volatile hydrocarbons added to petrol is higher in winter to help cold starts and lower in summer to prevent vapour forming too readily.

For smooth running, petrol must burn smoothly in the engines of vehicles and not in fits and starts. To promote smoother, more efficient combustion, companies produce more volatile 'higher-octane' fuel by increasing the proportions of branched alkanes and cyclic hydrocarbons.

The two main processes for doing this are:

- cracking, which makes smaller molecules and converts straight-chain hydrocarbons into branched and cyclic hydrocarbons
- reforming, which turns straight-chain cyclic alkanes into arenes such as benzene and methylbenzene.

Catalytic cracking

Catalytic cracking converts heavier fractions from the fractional distillation, such as diesel oil and fuel oil, into more useful hydrocarbon fuels by breaking up larger molecules into smaller ones. Cracking converts longer chain alkanes with 12 or more carbon atoms into smaller, more useful molecules in a mixture of branched alkanes, cycloalkanes, alkenes and branched alkenes.

The catalyst is synthetic sodium aluminium silicate, which belongs to a class of compounds called zeolites. Cracking takes place on the surface of the catalyst at about 500 °C. Synthetic zeolites make excellent catalysts because they can be designed with active sites that favour the shape and size of those molecules that react to give the desired product.

Catalytic cracking is a continuous process. The finely powdered catalyst gradually becomes coated with carbon, so it circulates through a regenerator in which the carbon burns away in a stream of air.

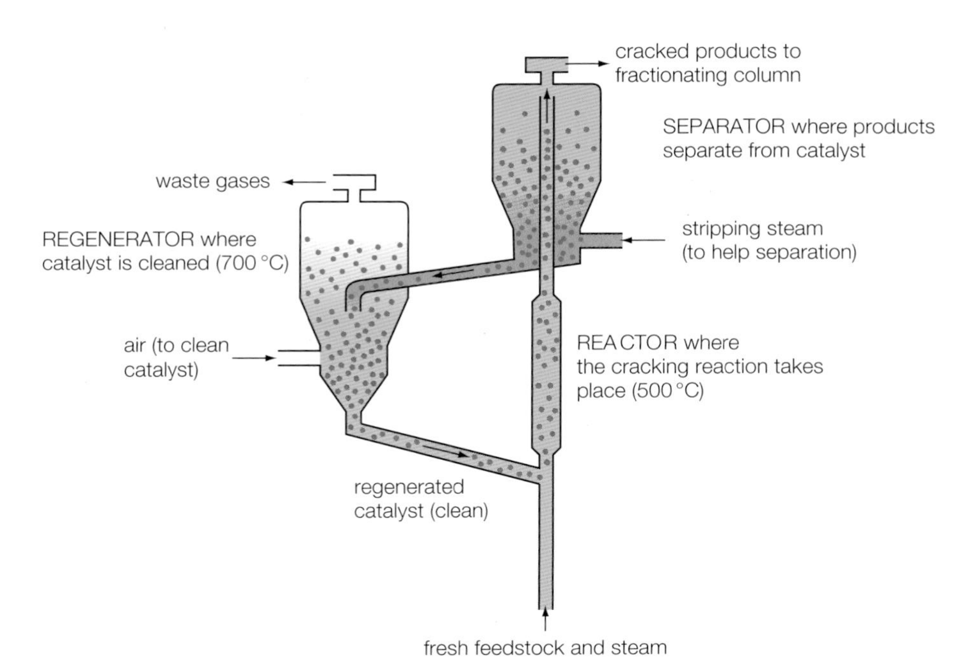

Figure 11.7 ◀ Catalytic cracking. Fresh feedstock and catalyst powder flow to the vertical reactor where cracking takes place. The cracked products pass to a fractionating column while the catalyst flows to the regenerator.

Reforming

Reforming converts straight-chain and cyclic alkanes to arenes (aromatic hydrocarbons) such as benzene and methylbenzene (Section 11.1). Hydrogen is a valuable by-product of the process, which can be used in processes elsewhere at the refinery.

The catalyst for this process is often one or more of the precious metals, such as platinum and rhodium, supported on an inert material such as aluminium oxide. The process operates at about 500 °C (Figure 11.8).

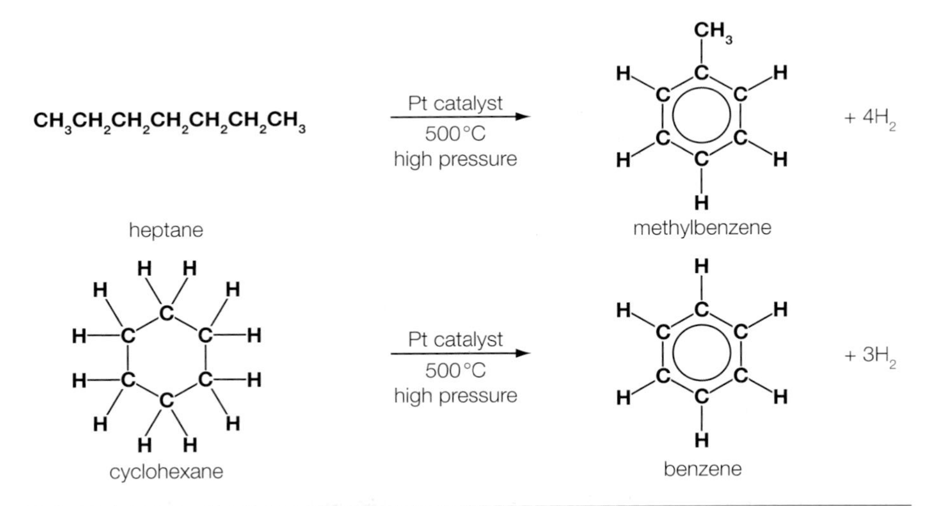

Figure 11.8 ◀ Examples of reforming.

Test yourself

4 a) Why is crude oil so important?
   b) Why should we try to conserve our reserves of crude oil?
5 What methods are used to increase the octane number of petrol?
6 a) Why is cracking important?
   b) What conditions are used for cracking?
7 Write structural formulae to show how:
   a) catalytic cracking converts hexane to butane and ethene
   b) catalytic cracking converts decane, $C_{10}H_{22}$, to 2,3-dimethylpentane and propene
   c) reforming converts hexane to benzene.

11.4 Alternative fuels

Crude oil is not the only source of organic raw materials and fuels. Industries are increasingly turning to plants as the source of chemicals and fuels because of the impact on the environment from our use of fossil fuels.

The concentrations of greenhouse gases, particularly carbon dioxide, in the air are rising. This is primarily due to human activity, especially the burning of fossil fuels in power stations, aviation, industries, our homes and our vehicles. The carbon dioxide produced is enhancing the greenhouse effect, and there is mounting evidence that this is responsible for global warming and climate change (Section 18.2).

Concern about the enhanced greenhouse effect is encouraging people and governments to reduce their CO_2 emissions and seek a more **sustainable lifestyle**. This involves planning to live within the means of the environment so that the Earth's natural resources are not destroyed but remain available for future generations (Section 19.2).

Any attempt to reduce CO_2 emissions means seeking alternatives to fossil fuels, such as biofuels like bioethanol and biodiesel.

Figure 11.9 ▶ Farmers throughout the world are turning to crops such as rapeseed, which can be used to produce biodiesel and reduce our reliance on fossil fuels.

> **Definition**
>
> A **sustainable lifestyle** involves living within the means of the environment so that the Earth's natural resources are available for future generations.

Crops take in carbon dioxide from the air as they photosynthesise and make sugars and vegetable oils that can be used to produce biofuels. When biofuels burn, the carbon dioxide taken up during photosynthesis is returned to the air. This analysis suggests that the use of biofuels should have no overall effect on the level of carbon dioxide in the atmosphere. Because of this, biofuels are sometimes described as carbon neutral (Section 19.6). This, however, ignores the carbon dioxide released from fossil fuels during the mechanical planting, harvesting and processing of the crop and the manufacture of fertilisers applied to the crop during the growing season.

The principal biofuels now being used are bioethanol and biodiesel. Bioethanol is manufactured by fermenting carbohydrates such as starch and sugar in crops like sugar cane. Fermentation converts starch to glucose and then glucose to ethanol and carbon dioxide. The reaction is catalysed by an enzyme in yeast:

$$\underset{\text{glucose}}{C_6H_{12}O_6(aq)} \rightarrow \underset{\text{carbon dioxide}}{2CO_2(g)} + \underset{\text{ethanol}}{2C_2H_5OH(aq)}$$

Biodiesel is produced by extracting and processing the oils from crops such as rapeseed.

The increasing desire for renewable fuels, such as ethanol and biodiesel in 'rich countries' has clear benefits. However, if poorer countries are tempted to grow 'crops for fuel' in place of 'crops for food', the situation may lead to serious problems of food supply and malnutrition.

Figure 11.10 ◄
Brazil is the leading country to use bioethanol. This photo shows a petrol station in Sao Paulo with separate pumps for ethanol (alcool), petrol (gasolina) and an ethanol/petrol mixture.

Test yourself

8 Why are governments around the world becoming increasingly concerned about our use of fossil fuels?
9 Suggest three ways in which our use of fossil fuels might be reduced.
10 Biofuels are sometimes described as 'carbon neutral'. Why is this?
11 Explain what is meant by the term 'sustainable development' using examples from the manufacture of fuels.

11.5 Chemical reactions of the alkanes

Practical guidance

The bond enthalpies of C–C and C–H bonds are relatively high, so the bonds in alkanes are difficult to break. In addition, these bonds in alkanes are non-polar (Section 6.10). This means that alkanes are very unreactive with ionic reagents in water, such as acids, alkalis, oxidising agents and reducing agents. There are, however, three important reactions of alkanes involving homolytic fission and radicals (Section 10.6). These three important reactions are burning (combustion), halogenation and cracking (Section 11.3).

Burning (combustion)

In a plentiful supply of air or oxygen, alkanes are completely oxidised to carbon dioxide and water, and the reaction is highly exothermic. When one mole (58 g) of butane (camping gas) burns completely in oxygen, it releases 2876 kJ of energy (heat):

$$\underset{\substack{\text{butane}\\\text{in GAZ}}}{C_4H_{10}(g)} + 6\tfrac{1}{2}O_2(g) \rightarrow 4CO_2(g) + 5H_2O(g) \qquad \Delta H = -2876\,\text{kJ mol}^{-1}$$

Figure 11.11 ▲
Red Calor gas cylinders contain propane for use as a fuel.

The exothermic nature of the combustion of alkanes has led to their widespread use as fuels in industry, homes and vehicles.

The combustion of alkanes involves a radical mechanism which occurs rapidly in the gas phase. This means that liquid and solid alkanes must vaporise before they burn and it explains why less volatile alkanes burn less easily.

The burning of alkanes is immensely important in any advanced technological society. It is used to generate energy of one kind or another in power stations, furnaces, domestic heaters, cookers, candles and vehicles. Unfortunately, this mass burning of alkanes is now accepted to be the major cause of global warming and the increased greenhouse effect (Section 18.3).

In a limited supply of air or oxygen, alkanes burn incompletely forming highly toxic carbon monoxide, as well as carbon dioxide and water. If the supply of air or oxygen is even more restricted, the major products will be soot (carbon) and water.

$$C_4H_{10}(g) + 2\tfrac{1}{2}O_2(g) \rightarrow 4C(s) + 5H_2O(g)$$

$$C_4H_{10}(g) + 4\tfrac{1}{2}O_2(g) \rightarrow 4CO(g) + 5H_2O(g)$$

Figure 11.12 ▲
Serious faults with the engine of this lorry means that it is emitting dangerous and illegal levels of carbon (soot) and toxic carbon monoxide from its exhaust system.

Carbon monoxide is colourless, it has no smell and it is very poisonous. It reacts irreversibly with haemoglobin in the blood to form carboxyhaemoglobin, and this reduces the capacity of the blood to carry oxygen. Because of this, it is dangerous to burn hydrocarbon fuels in a poor supply of air or in faulty gas appliances.

Reactions with chlorine and bromine (halogenation)

Alkanes react with chlorine and bromine on exposure to ultraviolet light. During the reactions, hydrogen atoms in the alkane molecules are replaced (substituted) by halogen atoms. These are described as substitution reactions.

Any or all of the hydrogen atoms in an alkane may be replaced and the reaction can continue until all of the hydrogen atoms have been substituted by halogen atoms. Consequently, the product is a mixture of similar compounds that are difficult to separate and purify. Because of this, the reactions of alkanes with halogens are limited as methods of synthesising halogenoalkanes, even though these products can be used in various ways. Alternative ways of synthesising different halogenoalkanes and the uses of halogenoalkanes are discussed in Topic 14.

Definition

A **substitution reaction** is one in which an atom or group of atoms is replaced (substituted) by another atom or group of atoms.

In strong sunlight, methane and chlorine react explosively. The first products are chloromethane, CH_3Cl, and hydrogen chloride (Figure 11.13).

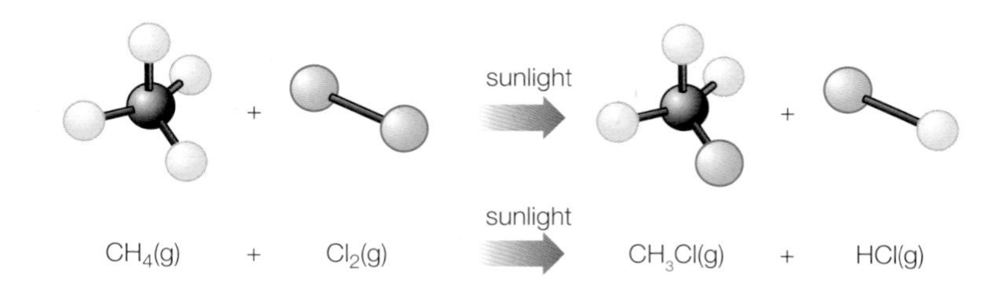

Figure 11.13 ▶
The equation and models representing the initial substitution reaction of methane with chlorine.

The reaction involves breaking certain bonds for which energy must be supplied, and then making new bonds, when energy is released.

The reaction between methane and chlorine will not occur in the dark because the molecules do not have enough energy for bonds to break when they collide. The energy provided by photons in ultraviolet (UV) light is sufficient to cause homolytic fission of chlorine molecules. This produces free chlorine atoms (highly reactive radicals) to start the reaction, and this step in the reaction mechanism is described as initiation.

$$Cl_2 \xrightarrow{\text{UV light}} Cl\bullet + Cl\bullet$$

In the next stage, chlorine atoms with an unpaired electron first remove hydrogen atoms from methane molecules to form hydrogen chloride and a new methyl radical, $CH_3\bullet$:

$$Cl\bullet + CH_4 \rightarrow HCl + CH_3\bullet$$

The $CH_3\bullet$ radical then reacts with more chlorine to form chloromethane, CH_3Cl, and generate another chlorine radical:

$$CH_3\bullet + Cl_2 \rightarrow CH_3Cl + Cl\bullet$$

This new $Cl\bullet$ radical can react with another CH_4 molecule and the last two reactions can be repeated again and again until either all the Cl_2 or all the CH_4 is used up. These two repeated reactions create a chain reaction and are described as propagation stages.

Propagation ends when two free radicals combine. This is the termination stage of the reaction which is very exothermic because covalent bonds form.

$$Cl\bullet + Cl\bullet \rightarrow Cl_2$$
$$CH_3\bullet + Cl\bullet \rightarrow CH_3Cl$$
$$CH_3\bullet + CH_3\bullet \rightarrow CH_3CH_3$$

The three stages in the radical substitution of methane with chlorine (initiation, propagation and termination) are summarised in Figure 11.14.

Stage 1	Initiation	$Cl : Cl$ $\xrightarrow{\text{light}}$	$Cl\bullet + Cl\bullet$
Stage 2	Propagation	$Cl\bullet + CH_4 \rightarrow$	$HCl + CH_3\bullet$
		$CH_3\bullet + Cl_2 \rightarrow$	$CH_3Cl + Cl\bullet$
		chain reaction	
Stage 3	Termination	$Cl\bullet + Cl\bullet \rightarrow$	Cl_2
		$CH_3\bullet + Cl\bullet \rightarrow$	CH_3Cl
		$CH_3\bullet + CH_3\bullet \rightarrow$	CH_3CH_3

Figure 11.14 ◄
Stages in the radical substitution of methane with chlorine.

Definition

Free-radical chain reactions involve three stages:

- **initiation** – the step that produces free radicals
- **propagation** – steps that form products and more free radicals
- **termination** – steps that remove free radicals by turning them into molecules.

Figure 11.15 ▲
The effect of light on a mixture of bromine and hexane after 0, 10 and 16 seconds.

Test yourself

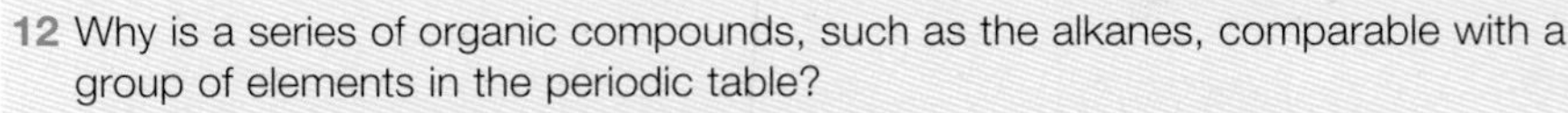

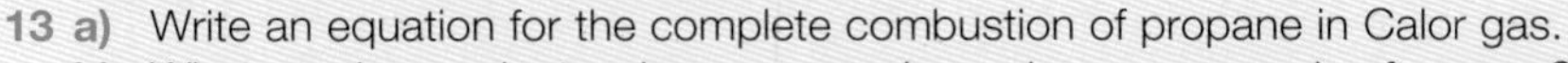

12 Why is a series of organic compounds, such as the alkanes, comparable with a group of elements in the periodic table?

13 a) Write an equation for the complete combustion of propane in Calor gas.
b) What are the products when propane burns in a poor supply of oxygen?

14 Why is it dangerous to allow a car engine to run in a garage with the door closed?

Data

15 a) Use the data section of your CD-ROM to find the boiling points of propane and butane.
b) Why is it wise for campers and caravanners to use Calor gas (propane) rather than GAZ (butane) for cooking and heating during the winter?

16 Write equations for all four possible substitution reactions when chlorine reacts with methane and name the products.

17 a) Explain why a mixture of bromine in hexane remains orange in the dark but fades and becomes colourless in sunlight.
b) Write an equation for the reaction in part a).
c) Why can acid fumes be detected above the solution once the colour has faded?

REVIEW QUESTIONS

Extension questions

1 Crude oil is an important source of materials for the petrochemical industry. Various products are obtained from crude oil by fractional distillation followed by processes involving cracking and reforming.

a) How is the crude oil separated into different fractions by fractional distillation? (6)

b) i) What is meant by 'cracking'? (3)

ii) Dodecane, $C_{12}H_{26}$, can be cracked into ethene and a straight-chain alkane so that the molar ratio of ethene:straight-chain alkane is 2:1.

Write a balanced equation for this reaction and name the straight-chain alkane. (2)

iii) Heat alone could be used to crack alkanes, but oil companies normally use catalysts as well. Suggest two reasons why oil companies use catalysts. (2)

c) Straight-chain alkanes such as heptane can be reformed into cyclic compounds.

Write a balanced equation to show how heptane can be reformed into methylcyclohexane. (2)

2 a) What is an alkane? (2)

b) Why is C_8H_{18} called octane? (4)

c) Write an equation for the complete combustion of octane in petrol. (2)

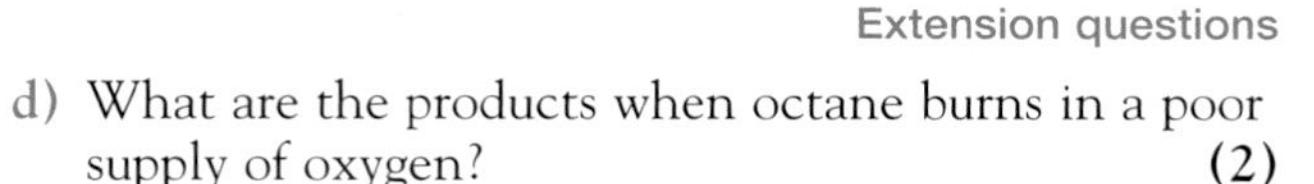

d) What are the products when octane burns in a poor supply of oxygen? (2)

e) Why is it dangerous to burn hydrocarbon fuels in faulty gas appliances? (2)

3 When a few drops of bromine are added to hexane, a deep red solution is formed. No reaction occurs if the mixture is kept in the dark, but in bright sunlight the red colour slowly disappears and a misty gas is produced.

a) What is the misty gas produced in the reaction? (1)

b) Write an equation for the overall reaction of 1 mole of hexane with 1 mole of bromine in bright sunlight. (1)

c) Why is there no reaction in the dark? (2)

d) Why does a reaction occur in sunlight? (3)

e) The reaction involves radicals in a three-stage process.

i) Write an equation for the initiation stage of the reaction. (1)

ii) Show in an equation how hexyl radicals are produced in the propagation stage of the reaction. (2)

12 Alkenes

Alkenes such as ethene and propene are invaluable to chemists because they react with a wide range of chemicals to make useful products. Both ethene and propene are produced during the cracking of heavier fractions obtained from crude oil. They are important starting points for the synthesis of other chemicals because of the reactivity of their double bonds. Alkenes, like alkanes, occur naturally in plant and animal tissues as oils, fats, pigments and perfumes.

12.1 Alkenes

Alkenes form a homologous series in which the functional group is a double bond between two carbon atoms (C=C). Because of this, their molecules have two atoms of hydrogen less than the corresponding alkane, and their general formula is C_nH_{2n}.

Alkenes and cycloalkenes contain at least one C=C bond. This means that they contain fewer hydrogen atoms than they would if all the bonds between carbon atoms were single and each molecule was saturated with hydrogen atoms like alkanes. Because of this, alkenes and cycloalkenes are described as unsaturated.

Definitions

Unsaturated compounds contain one or more double or triple bonds between atoms in their molecules. The term is often applied to alkenes and used to describe unsaturated fats that have double bonds between carbon atoms in their hydrocarbon chains.

Saturated compounds have only single bonds between atoms in their molecules.

ORGANIC SUNFLOWER OIL
Produced under organic standards

NUTRITIONAL INFORMATION:
Typical values per 100ml

Energy:	3382kJ/823kcal
Protein:	Nil
Carbohydrate:	Nil
of which sugars:	Nil
Fat:	91.4g
of which saturates:	13.1g
of which monounsaturates:	66.7g
of which polyunsaturates:	7.5g
Fibre:	Nil
Sodium:	Nil

Napolina is a registered trademark of Napolina Ltd.
Packed in the EU for Napolina Limited,
Royal Liver Building, Pier Head, Liverpool L3 1NX.
www.napolina.com

Figure 12.1 ◀ Olive oil and sunflower oil contain unsaturated fats with double bonds between their carbon atoms like alkenes.

Names and structures

The name of an alkene is based on the name of the corresponding alkane with the ending -ene in place of -ane (Figure 12.2).

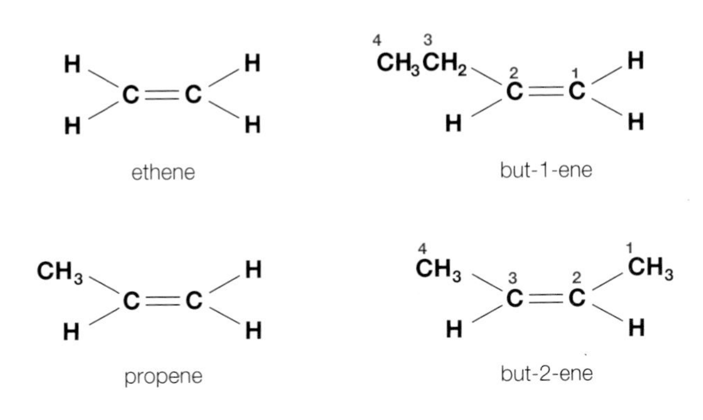

Figure 12.2 ◀ The names and structures of the four simplest alkenes.

Where necessary, a number in the name shows the position of the double bond as in the structural isomers but-1-ene and but-2-ene. Counting starts from the end of the chain that gives the lowest possible number in the name. The number in the name indicates the first of the two atoms connected by the double bond. In but-1-ene, for example, the double bond is between the first and second carbon atoms in the chain.

Each carbon atom attached to the carbon–carbon double bond in an alkene is surrounded by three negative centres – one double bond and two single bonds. The electrons in these bonds repel each other as far as possible. So the atoms attached to these bonds take up trigonal planar positions around each carbon atom of the double bond (Section 6.8).

Physical properties

Data

The boiling points of the alkenes increase like alkanes as the number of carbon atoms in the molecules increase. Ethene, propene and the butenes are gases at room temperature.

Alkenes, like other hydrocarbons, do not mix with or dissolve in water.

The double bond in alkenes

Chemists have extended the theory of atomic orbitals (Section 5.4) to describe the distribution of electrons in molecules. This molecular orbital theory is helpful when discussing the bonding and reactivity of alkenes.

Molecular orbitals result when atomic orbitals overlap and interact forming bonds between atoms. The shapes of molecular orbitals show the regions in space where there is a high probability of finding electrons. A sigma (σ) bond is a single covalent bond formed by a pair of electrons in an orbital in a molecule with the electron density concentrated between the two nuclei. Free rotation is possible around single bonds. Sigma bonds can form through the overlap of two s orbitals, an s orbital and a p orbital, or two p orbitals (Figure 12.3).

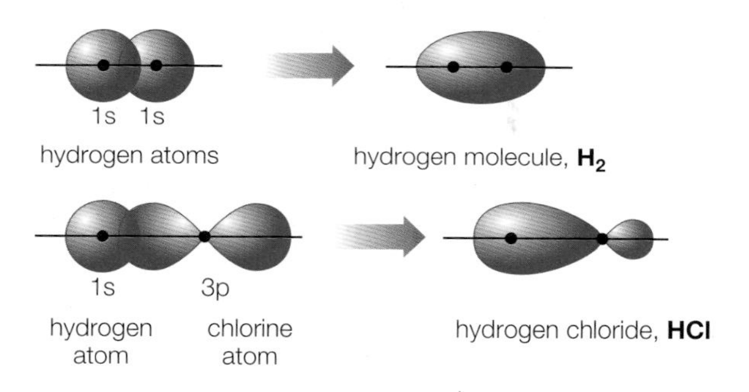

Figure 12.3 ▶ Examples of sigma bonds in molecules.

A pi (π) bond is the type of bond found in molecules with double and triple bonds. The bonding electrons are in a π orbital formed by the *sideways overlap* of two atomic p orbitals. In a π bond, the electron density is concentrated in two regions – one above and one below the plane of the molecule on either side of the line between the nuclei of the two atoms joined by the bond (Figure 12.4).

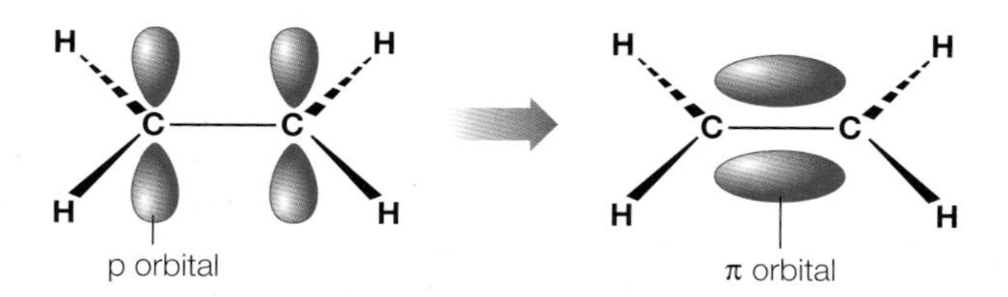

Figure 12.4 ▶ The π bond in ethene.

Test yourself

1 a) Explain why the general formula of straight-chain alkenes is C_nH_{2n}.
b) What is the general formula of cycloalkenes?
2 a) Draw the structure of ethene showing the trigonal planar arrangement around each carbon atom.
b) Why are the bond angles around each carbon in ethene about 120°?
c) Why are the H–C–C bond angles in ethene slightly greater than 120° and the H–C–H bond angles in ethene slightly less than 120°?

12.2 Stereoisomerism in alkenes

X-ray diffraction studies show that the ethene molecule is planar (Figure 12.2) with the three atoms around each carbon atom arranged trigonally at about 120°. In ethane and other alkanes, it is possible to rotate the whole molecule around a single σ bond because this does not affect the overlap of orbitals (Figure 12.5). With a double bond, however, rotation would involve breaking the π bond and the overlap of p orbitals. This requires more energy than the molecules possess at normal temperatures. So, free rotation is not possible around the carbon–carbon double bond in alkenes, and this can give rise to stereoisomers.

These two structures are the same compound because the ends of the molecule can rotate freely around the single bond.

Figure 12.5 ◀ Free rotation can occur around a single σ bond.

Stereoisomers have a different arrangement of their atoms in space. This gives them different shapes and therefore different displayed formulae. Alkenes and other compounds with a C=C bond have stereoisomers if there are two different groups on each carbon atom in the double bond (Figure 12.6).

Definition

Stereoisomers are molecules with the same molecular formula and the same structural formula but with a different arrangement in space.

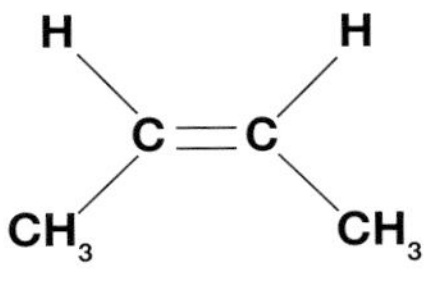

melting point = −139°C
cis-but-2-ene

Different compounds because the double bond stops rotation

melting point = −106°C
trans-but-2-ene

Figure 12.6 ◀ Stereoisomers of but-2-ene are different compounds with different melting points, different boiling points and different densities.

The stereoisomers in certain alkenes, such as but-2-ene, are often labelled *cis* and *trans*, and this type of stereoisomerism is often called *cis–trans* isomerism. In the *cis* isomer, similar groups are on the same side of the double bond (*cis* in Latin means 'on the same side'). In the *trans* isomer, similar groups are on opposite sides of the double bond (*trans* in Latin means 'opposite').

Note

At one time, *cis* and *trans* isomers were called geometric isomers, but this term is no longer recommended.

Test yourself

3 Why is rotation about a carbon–carbon double bond restricted?
4 Which of the following unsaturated compounds have *cis* and *trans* stereoisomers: but-1-ene, 1,1-dichloropropene, pent-2-ene and buta-1,3-diene? (Hint: For *cis–trans* stereoisomers to be possible, there must be two different groups on each carbon atom of the double bond.)
5 a) Draw and name the displayed structures of the three isomers of $C_2H_2Br_2$.
b) What type of structural isomerism do they show?
c) Which two are *cis–trans* stereoisomers?

Activity

Renaming *cis–trans* stereoisomers as E–Z stereoisomers

Cis–trans stereoisomerism arises in compounds containing carbon–carbon double bonds and two different groups on both carbon atoms of the double bond. The different isomers result because of restricted rotation around the double bond.

The existence of a ring structure in a molecule can also restrict rotation and give rise to *cis–trans* stereoisomers. Look at the structure of *trans*-1,2-dichlorocyclobutane in Figure 12.7. Its *cis* isomer is *cis*-1,2-dichlorocyclobutane.

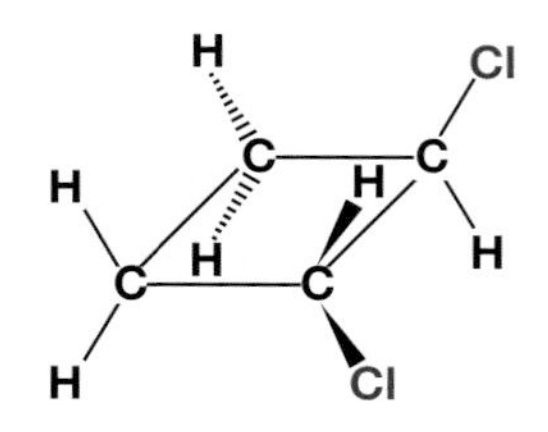

Figure 12.7 ▲ The structure of *trans*-1,2-dichlorocyclobutane.

1 Draw the displayed formula of *cis*-1,2-dichlorocyclobutane.

2 There is another pair of *cis–trans* isomers named dichlorocyclobutane and a separate structural isomer. Name and draw displayed formulae for these three molecules.

Now look at the isomer of 1-bromo-2-chloroprop-1-ene in Figure 12.8.

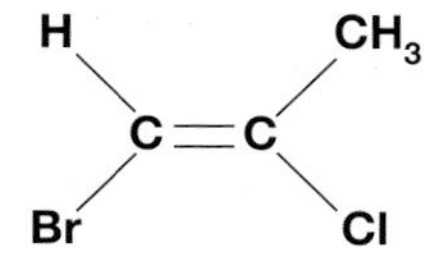

Figure 12.8 ▲ An isomer of 1-bromo-2-chloroprop-1-ene.

3 Draw the other *cis–trans* isomer of 1-bromo-2-chloroprop-1-ene.

But, which one of these isomers is *cis* and which is *trans*? The rule that is normally used to name the isomers as *cis* or *trans* says:

> 'in the *cis* isomer, similar groups are on the same side of the double bond;
>
> in the *trans* isomer, similar groups are on opposite sides of the double bond.'

In 1-bromo-2-chloroprop-1-ene, there are four different groups on the atoms joined by the double bond. So, the normal rule that requires one group to be the same on both carbon atoms cannot be used.

This is where the *cis–trans* naming system breaks down and it becomes necessary to use the E-Z naming system. In fact, the E–Z naming system was developed to enable the systematic naming of complex alkenes in naturally occurring materials, such as the red pigment in tomatoes. One molecule of this pigment has 13 C=C bonds.

The E–Z naming system

- Look at the two atoms bonded to the first carbon atom in the C=C bond. The atom with the highest atomic number takes priority.
- If two atoms with the same atomic number, but in different functional groups, are attached to the first carbon atom, the next bonded atom is taken into account. Thus, CH_3CH_2– has priority over CH_3–.
- This consideration is then repeated with the second carbon atom in the C=C bond.

4 a) What is the order of priority among Br–, C in CH_3–, Cl– and H–?
b) What is the priority among CH_3–, CH_3CH_2–, CH_3O– and $HOCH_2$–?

5 Look again at 1-bromo-2-chloroprop-1-ene in Figure 12.8.

a) Which takes priority between H– and Br– attached to the first carbon atom in the double bond?
b) Which takes priority between Cl– and CH_3– attached to the second carbon atom in the double bond?

- Now, if the two groups of highest priority are on the same side of the double bond, the isomer is designated Z (from the German 'zusammen', which means 'together'). If the two groups of highest priority are on opposite sides of the double bond, the isomer is designated E (from the German 'entgegen', which means 'opposite').

6 Draw the displayed structure of Z-1-bromo-2-chloroprop-1-ene.

7 Use the E-Z system to name the *cis–trans* isomers of:

a) but-2-ene
b) 1,2-dichlorocyclobutane.

Finally, notice from all this that *cis–trans* isomerism is just a special case of E–Z isomerism in which one of the groups attached to each carbon atom of the double bond is the same. In many cases of *cis–trans* isomerism, the same group on each carbon atom is hydrogen – as in but-2-ene.

12.3 Addition reactions of alkenes

The characteristic reactions of alkenes are addition reactions in which small molecules such as H_2, Cl_2, HBr and H_2O add across the double bond to form a single product.

Note

You will not be required to recall and use the four bulleted priority rules to identify which stereoisomer is which in E–Z isomers.

Definition

An **addition reaction** is a reaction in which two molecules add together to form a single product.

Addition of hydrogen

Hydrogen adds to C=C bonds on heating to 150 °C in the presence of a nickel catalyst.

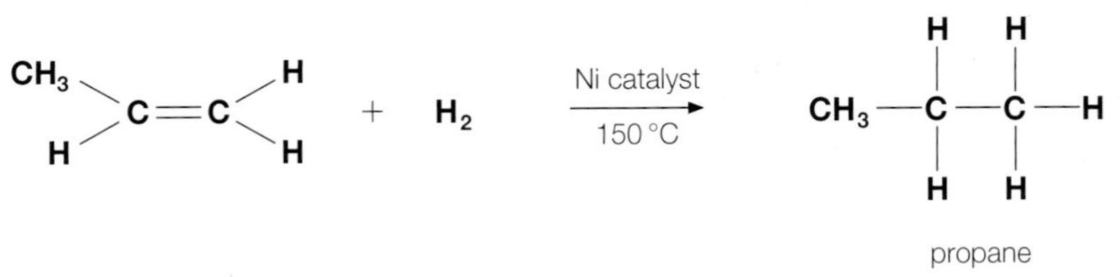

$$CH_3(H)C{=}CH_2 + H_2 \xrightarrow[150\,^\circ C]{\text{Ni catalyst}} CH_3{-}CH_2{-}CH_3$$

propane

Figure 12.9 ▲
The addition of hydrogen to propene forming propane.

This process is known as catalytic hydrogenation. The advantage of using a solid metal catalyst is that it can be held in the reaction vessel as the reactants flow in and the products flow out. In addition, there is no difficulty in separating the products from the catalyst.

At one time, catalytic hydrogenation using hydrogen and a nickel catalyst was important in the manufacture of margarine from unsaturated vegetable oils in palm seeds and sunflower seeds. Vegetable oils are liquids containing carbon–carbon double bonds, nearly all of which are *cis*-double bonds. During hydrogenation, some of these double bonds are converted to carbon–carbon single bonds by the addition of hydrogen. The change in structure turns oily unsaturated liquids into soft saturated fatty solids like margarine.

Note

When describing an organic reaction, always write an equation, name the reagents and products and state the conditions (temperature, pressure and catalysts).

Figure 12.10 ▲
Manufacturers of vegetable fat spreads, such as Bertolli, often claim that they contain virtually no *trans*-fats and that they are rich in unsaturated fat and low in saturated fats compared with butter. These spreads are a healthier option than butter.

Unfortunately, research in the 1960s started to show that saturated fats contribute to heart disease.

Fats that have been partially saturated by hydrogenation often contain *trans*-fats. Evidence found during the 1990s began to suggest that these *trans*-fats could also lead to heart disease. Complete hydrogenation, however, eliminates the double bonds and hence the *trans*-fats. This has led some manufacturers of vegetable fat spreads to use a new production strategy. This involves completely hydrogenating part of the vegetable oil and then blending this with untreated oil to make a spread of the correct texture but with no *trans*-fats.

Addition of halogens

Chlorine and bromine add rapidly to alkenes at room temperature. The products are dichloro- and dibromoalkanes. Fluorine reacts explosively with small alkenes, such as ethene and propene, but the reaction with iodine is slow.

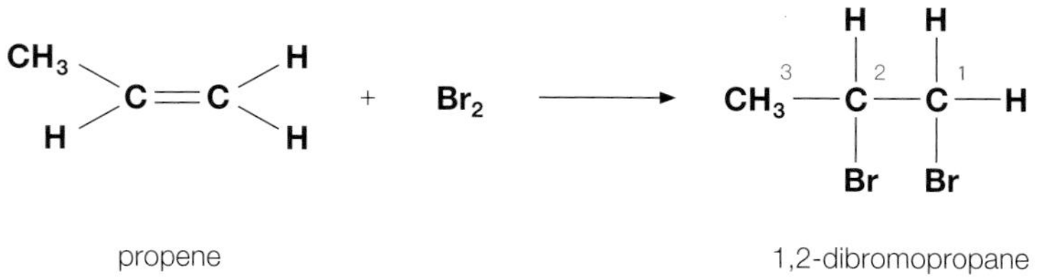

$$CH_3CH{=}CH_2 + Br_2 \longrightarrow CH_3CHBrCH_2Br$$

propene → 1,2-dibromopropane

Figure 12.11 ▲
The addition of bromine to propene forming 1,2-dibromopropane.

Testing for alkenes and the C=C bond

The reaction of bromine water (aqueous bromine, $Br_2(aq)$) with hydrocarbons is a useful test for alkenes and the carbon=carbon double bond.

Unsaturated hydrocarbons with a C=C bond, such as ethene and cyclohexene, quickly decolourise yellow/orange bromine water to produce a colourless mixture.

Saturated hydrocarbons like alkanes do not react with bromine water and the yellow/orange colour of bromine remains.

Addition of hydrogen halides

Hydrogen halides react readily with alkenes at room temperature to form halogenoalkanes. For example, hydrogen bromide reacts with ethene in the gas phase to form bromoethane. The reaction happens at room temperature.

The other hydrogen halides, HCl and HI, react in a similar way.

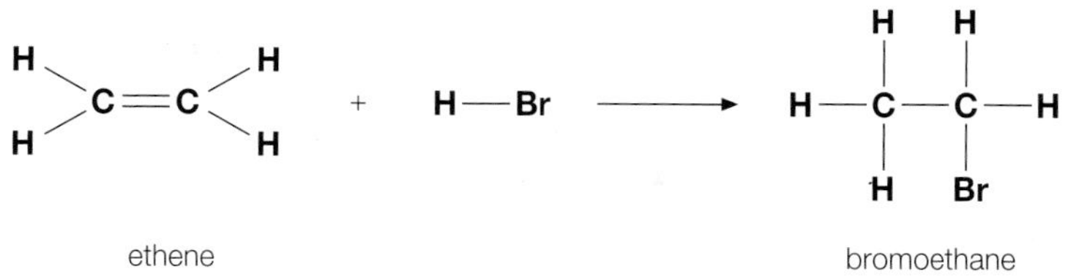

$$CH_2{=}CH_2 + H{-}Br \longrightarrow CH_3CH_2Br$$

ethene → bromoethane

Figure 12.12 ▲
The addition of hydrogen bromide to ethene forming bromoethane.

Addition of water (steam)

The addition of steam to alkenes in the presence of an acid catalyst can be used to form alcohols. This method is used to manufacture ethanol in the UK. A mixture of ethene and steam passes over an acid catalyst at 300 °C under pressure. The catalyst is phosphoric acid supported on an inert solid.

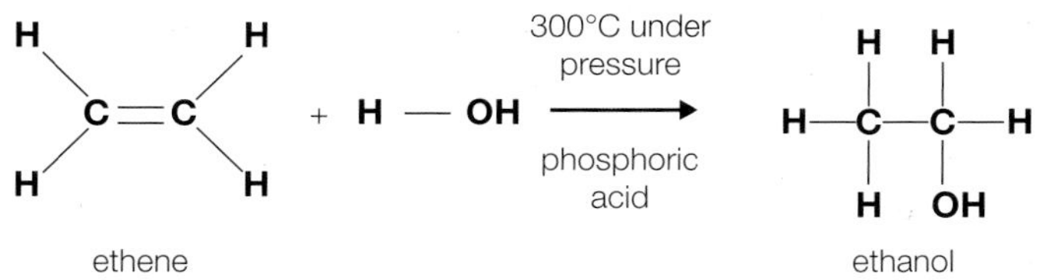

300°C under pressure, phosphoric acid

ethene ethanol

Figure 12.13 ◀
The addition of water (steam) to ethene producing ethanol.

On a smaller laboratory scale, ethene can be converted to ethanol by first absorbing ethene in cold concentrated sulfuric acid and then diluting the product with water. In the first stage of the reaction, sulfuric acid adds to ethene forming ethyl hydrogensulfate. In the second stage, water reacts with this to produce ethanol and sulfuric acid.

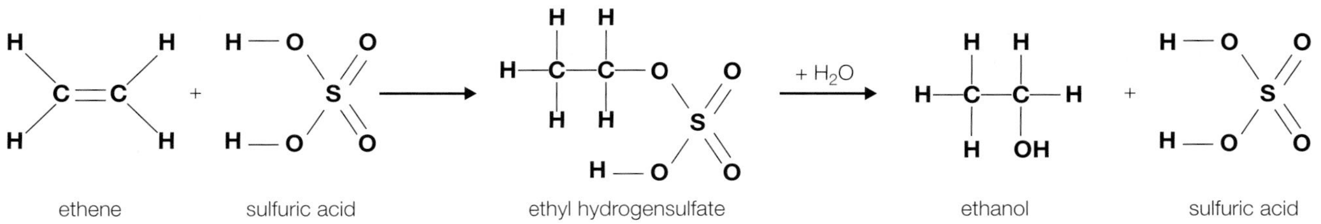

ethene sulfuric acid ethyl hydrogensulfate $+ H_2O$ ethanol sulfuric acid

Figure 12.14 ▲
The two-stage conversion of ethene to ethanol.

Test yourself

6 Write the structures and names of the products and the conditions for the reactions when propene reacts with:
  a) hydrogen
  b) chlorine
  c) concentrated sulfuric acid and then water.

7 Catalytic hydrogenation is sometimes used in the manufacture of spreads, such as 'Flora', from vegetable oils.
  a) What is meant by the term 'catalytic hydrogenation'?
  b) Explain the terms 'saturated' and 'unsaturated' as applied to organic compounds like those found in low-fat spreads and vegetable oils.
  c) Why are unsaturated fats such as olive oil and sunflower oil thought to be healthier foods than more saturated fats such as cream?

8 Why is sulfuric acid sometimes described as a catalyst when ethene reacts first with sulfuric acid and then water to form ethanol?

9 Name and draw the displayed formulae of the two possible products when hydrogen chloride reacts with propene.

12.4 Electrophilic addition to alkenes

Most of the reactions of alkenes described in Section 12.3 are electrophilic addition reactions. Electrophiles are electron-pair acceptors (Section 10.6). They attack the electron-rich region of the double bond in alkenes, particularly the exposed π bond. Electrophiles that add to alkenes include halogens, hydrogen halides and sulfuric acid.

The addition of hydrogen bromide to ethene

Hydrogen bromide molecules are polar (Section 6.8). The hydrogen atom, with its δ+ charge at one end of the molecule, acts as an electrophile (Figure 12.15).

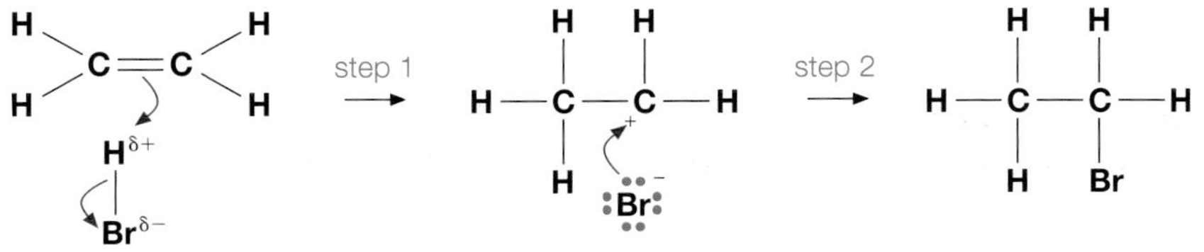

step 1 step 2

Figure 12.15 ◀
Electrophilic addition of hydrogen bromide to ethene. The reaction takes place in two steps.

In the first step of the reaction between hydrogen bromide and ethene, an HBr molecule approaches an ethene molecule. The δ+ hydrogen end of the HBr is attracted to the electron-dense double bond. As the HBr molecule gets even closer, heterolytic fission of the π bond occurs. The electrons in the π bond form a covalent bond to the hydrogen atom and, at the same time, heterolytic fission of the H–Br bond also occurs. Electrons in the H–Br bond are taken over by the bromine atom producing a Br^- ion.

The other product of step 1 is the highly reactive cation $CH_3CH_2^+$. This is so reactive that it reacts immediately with the Br^- ion in the second step of the reaction to form bromoethane, CH_3CH_2Br.

The addition of bromine to ethene

The addition of bromine to ethene also involves electrophilic addition. Bromine molecules are not polar, but they become polarised as they approach the electron-dense region of the double bond. Electrons in the double bond repel electrons in the bromine molecule and the δ+ end of the bromine molecule (nearer the double bond) becomes electrophilic (Figure 12.16). Heterolytic fission of the π bond and the Br–Br bond then occur simultaneously, and, in this case, the final product is 1,2-dibromoethane.

Figure 12.16 ▶ Electrophilic addition of bromine to ethene.

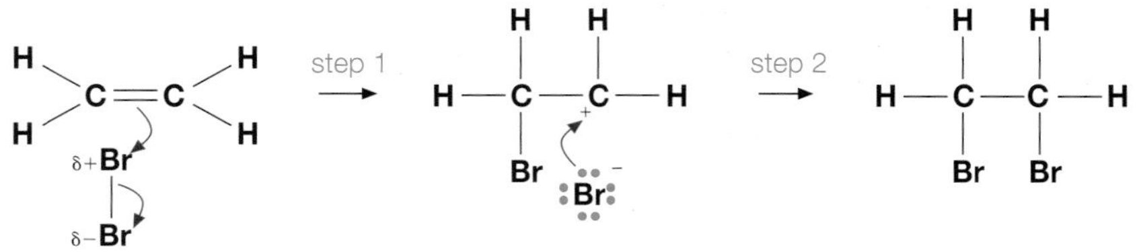

These mechanisms for electrophilic additions are not simply hypotheses or good ideas. They are supported by significant pieces of experimental evidence that help our understanding of the reactions at a molecular level.

Firstly, the rate of addition with hydrogen halides is in the order HCl > HBr > HI. This is, of course, the order of their strengths as acids, which supports the first step of the mechanism in which an H^+ ion adds to the alkene.

Secondly, if HBr addition is carried out in the presence of NaCl, chloroethane, CH_3CH_2Cl, is obtained in addition to the expected product CH_3CH_2Br. Similarly, if Br_2 addition is carried out in the presence of NaCl, 1-bromo-2-chloroethane, CH_2BrCH_2Cl, is obtained in addition to CH_2BrCH_2Br. This supports the second step of the mechanism in which the highly reactive cations $CH_3CH_2^+$ and $CH_2BrCH_2^+$ can be attacked by either Br^- or Cl^-.

Addition to unsymmetrical alkenes

When a molecule of the type HX adds to an unsymmetrical alkene such as propene, there are two possible products. When heterolytic fissions of the π bond and the H–X bond occur, two possible cation intermediates may form depending on which carbon atom in the double bond forms a covalent bond to hydrogen in the H–X molecule (Figure 12.17, stage 1).

Figure 12.17 ▶ Possible products when a compound of the type HX adds to propene.

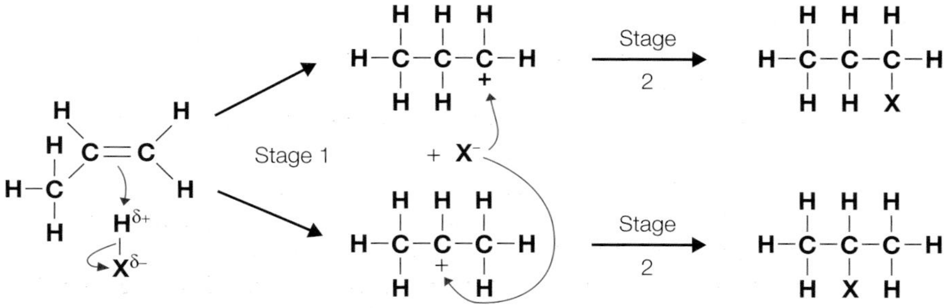

In the second stage of the reaction, either of the cation intermediates can react with the remaining X^- anion to form one or other of the possible products. These two possible products are isomers.

Definition

Intermediates are atoms, molecules, ions or free radicals that do not appear in the overall equation for a reaction because they are formed during one step of the reaction and then used up in the next.

Test yourself

10 Write the names and draw the structures of the possible products when:
   a) HBr adds to Z-but-2-ene
   b) H_2O adds to pent-2-ene.
11 For the reaction of hydrogen chloride with but-1-ene:
   a) draw the structures of the two possible cation intermediates
   b) draw the structures and name the two possible products.

12.5 Addition polymers from alkenes

If the conditions are right, molecules of ethene will undergo addition reactions with each other to form polyethene or, more correctly, poly(ethene). Two different kinds of poly(ethene) are manufactured: low-density poly(ethene) and high-density poly(ethene).

Low-density poly(ethene) is manufactured by heating ethene at high pressures and high temperatures with special substances called initiators. These initiators are often peroxides, which break apart to form free radicals that initiate (start) the reaction (Figure 12.18). The poly(ethene) produced has very long chains with lots of branches. The branches prevent the molecules from packing closely and this results in a low-density material.

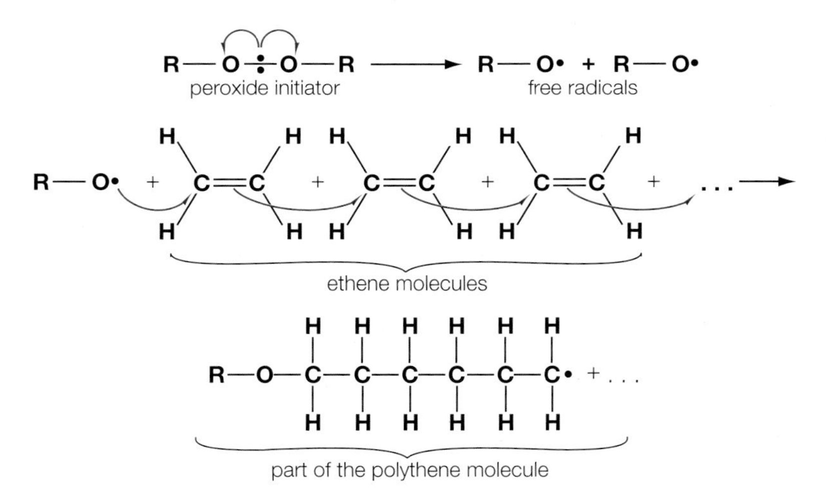

Figure 12.18 ◄
Molecules of ethene can undergo addition reactions with each other to form poly(ethene).

Definition

Addition polymerisation is an addition reaction in which small molecules, called monomers, join together forming a giant molecule called a polymer.

High-density poly(ethene) is manufactured at relatively low pressure and low temperature with special Ziegler-Natta catalysts. These are made from titanium(IV) chloride and aluminium alkyls, such as triethylaluminium, $Al(C_2H_5)_3$, in a hydrocarbon solvent. This method produces extra long chains with very little branching, so the chains pack more closely.

Processes like those that produce high-density and low-density poly(ethene) are called addition polymerisations. During addition polymerisation, small molecules known as monomers, like ethene, add to each other to form a giant molecule called a polymer.

Poly(ethene) is by far the most important polymer at present. After poly(ethene), the two most useful polymers are poly(propene) and poly(chloroethene) or pvc. These are also manufactured by addition polymerisation. Poly(ethene), poly(propene) and pvc are soft, flexible and slightly elastic. Because of this, they are often called plastics.

The monomer, polymer repeat unit, properties and uses of poly(ethene), poly(propene), pvc and poly(tetrafluoroethene) are shown in Table 12.1.

Figure 12.19 ▲
Clingfilm is just a thin film of pvc (polyvinylchloride). Its correct name is poly(chloroethene).

Monomer	Polymer repeat unit	Properties	Uses
Ethene $H_2C{=}CH_2$	Poly(ethene) Polythene $\left[-CH_2-CH_2-\right]_n$	Light Flexible Easily moulded Transparent Good insulator Resistant to water, acids and alkalis	Plastic bags and bottles Beakers Insulation for cables
Propene $H_2C{=}CH(CH_3)$	Poly(propene) Polypropene $\left[-CH_2-CH(CH_3)-\right]_n$	Tough Easily moulded Easily coloured Very resistant to water, acids and alkalis	Fibre for ropes and carpets Crates Toys
Chloroethene $H_2C{=}CHCl$	Polyvinylchloride (pvc) Poly(chloroethene) $\left[-CH_2-CHCl-\right]_n$	Tough Rigid or flexible Very resistant to water, acids and alkalis	Clingfilm Guttering and window frames Insulation for cables Waterproof clothing Flooring
Tetrafluoroethene $F_2C{=}CF_2$	Poly(tetrafluoroethene) (ptfe) $\left[-CF_2-CF_2-\right]_n$	Tough Rigid Low friction solid Resistant to water, acids and alkalis	Coating for non-stick pans and skis Low friction surfaces for bridges to move as metals expand and contract

Table 12.1 ▲
The monomer, polymer structure, properties and uses of the most important polymers.

Figure 12.20 ▲
This little boy is wearing a pvc waterproof jacket.

Test yourself

12 $CF_2{=}CF_2$ is the monomer for the polymer ptfe.
- a) What is the systematic name for $CF_2{=}CF_2$?
- b) Draw a section of the polymer ptfe composed from three monomer units.
- c) Explain why the polymer is usually called ptfe.

13 a) What conditions are used to manufacture low-density poly(ethene)?
- b) Why is poly(ethene) so useful?
- c) What is the major disadvantage of plastics?

14 Poly(styrene) is made from the following monomer.

$H_2C{=}CH(C_6H_5)$

- a) What is the common name for the monomer?
- b) The systematic name of the monomer is phenylethene. Write a systematic name for the polymer.
- c) Draw a section of the polymer composed of three monomer units.

15 Name and draw the structures of the monomers that would produce the polymers shown in the following repeat units.

a) $\left[-CH(CH_3)-CH(CH_3)-\right]_n$; b) $\left[-CH_2-C(CH_3)_2-\right]_n$; c) $\left[-CH(CH_2CH_3)-CH_2-\right]_n$?

12.6 Managing and disposing of plastic wastes

There are many benefits from the processing of alkenes to produce polymers and plastics. In particular, we have a whole range of new materials with different properties from which to make hundreds, if not thousands, of different products. Some of these materials and their uses are outlined in Table 12.1. However, there are some drawbacks to polymers and plastics, especially in the management and disposal of their waste.

Plastic waste, like all waste, causes pollution problems if it is not carefully managed. More waste is buried in landfill sites around the world than is managed and processed by other methods such as recycling, composting or energy recovery. About 7% of household waste is plastics, but another 30% or so is organic material, such as green garden refuse, wood and food waste (Figure 12.21). In recent years, there has been increased political and government pressure plus an increased social desire to reduce the landfill disposal of waste and increase its recycling and use for energy production.

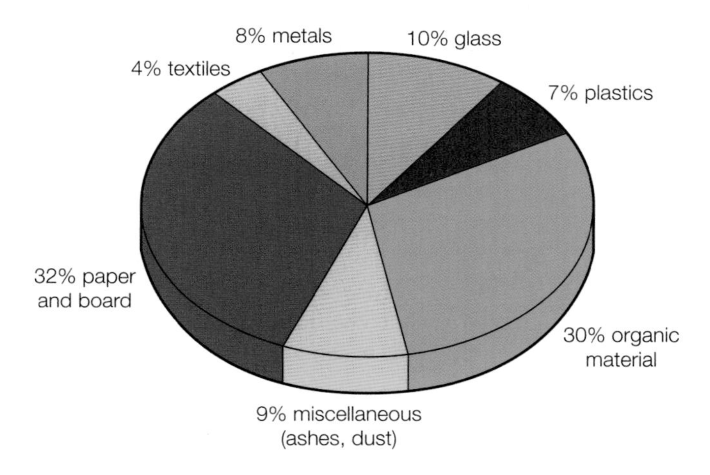

Figure 12.21 ◀ The typical composition of household waste in Europe.

Recycling

Plastics are much more difficult to recycle than metals such as iron and steel. One problem arises from the low density of plastic waste. It can take as many as 20 000 plastic bottles to make up a tonne of waste. Even so, the typical European throws away about 36 kg of plastic waste each year.

A second problem is that there are many types of plastic, so they have to be sorted. This is difficult, although the industry has introduced codes for labelling plastic products to help consumers separate them for recycling.

Recycling rates have increased in recent years as automatic machines for sorting plastic bottles have been developed. Separation methods include the use of infrared detectors, chemical scanning for chlorine in PVC and flotation to separate plastics according to their densities.

Recovery of plastic waste becomes easier when designers plan plastic products with recycling in mind. Engineers in the motor industry, for example, are putting much effort into designing components that can be recycled at the end of the life of a car.

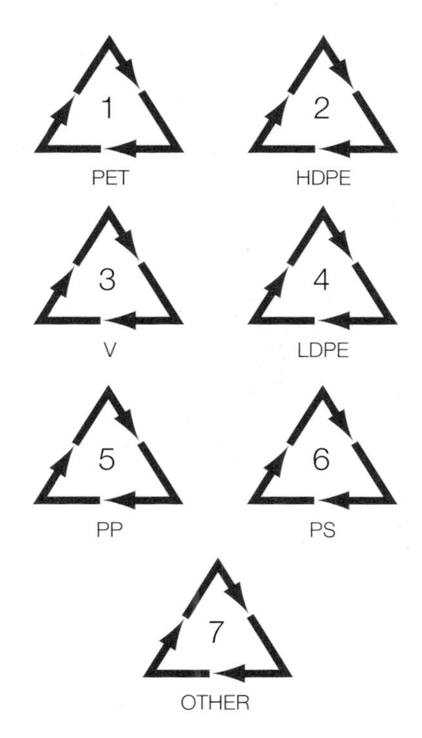

Figure 12.22 ▲ These recycling symbols are being used increasingly on plastic products to facilitate the separation of different polymers for recycling. PET stands for polyethene terephthalate (polyester), HDPE stands for high-density polyethene and PS stands for polystyrene.

Feedstock for cracking

An alternative approach to recycling is to use the plastic waste as feedstock for cracking in order to break the polymer molecules and regenerate small, monomer molecules. These monomer molecules can then be reused in the production of plastics and other chemicals.

Energy from waste

Another option being used more frequently is the combustion of plastic waste and domestic refuse to produce energy and generate electricity. The energy produced on combustion of various materials, including plastic waste, is shown in Table 12.2. Just dumping rubbish in landfill sites is a waste of resources. Burning all the plastic and domestic waste from homes in Europe could generate at least 5% of European energy needs.

Material	Energy produced on combustion/MJ kg^{-1}
Oil	40
Mixed plastic waste	25–40
Domestic refuse	10
Fuel made from refuse	16

Table 12.22 ◀ Energy produced on combustion of various materials.

Definition

Dioxins are a family of stable compounds that persist in the environment. Some dioxins are toxic and can cause skin disease, cancer and birth defects.

However, there has been strong opposition to the burning of waste. Old incinerators were inefficient producing clouds of soot and smoke. They operated at too low a temperature, and most were not designed to generate electricity. Incomplete combustion also produced carbon monoxide which is toxic.

Another worry is that if the waste contains halogenated plastics like pvc, an incinerator will give off corrosive or toxic chemicals such as hydrogen chloride and dioxins. However, chemists and chemical engineers have developed high temperature incinerators that work at 800–1250 °C. These help to achieve complete combustion and eliminate the emission of toxic chemicals. Modern combustion plants also have elaborate gas cleaning systems to control the emission of dust particles and harmful gases such as hydrogen chloride.

Although there are problems, burning waste has its benefits. Incineration greatly reduces the volume of waste to be dumped in landfill sites. By using the energy produced on combustion, it can also replace and save our reserves of non-renewable fossil fuels.

Figure 12.23 ▲ Scientists monitoring the effluent from a landfill site. Decomposition and fermentation of various materials take place at landfill sites. When rain water runs off the site, it will contain soluble substances from these processes and from the original waste. Scientists must monitor this effluent to be sure it does not contain any toxic products that might affect the environment, wildlife and, particularly, underground water supplies.

Landfill

Most plastic waste still ends up in landfill sites and the major problem of plastic waste becomes very apparent. It is non-biodegradable – unlike paper, wood and food waste – and cannot be broken down by micro-organisms and other decomposers.

In recent years, however, chemists have begun to develop biodegradable and compostable polymers and these are slowly replacing some of the uses of non-biodegradable plastics. The biodegradable polymers have been produced from maize, wheat and isoprene (2-methylbuta-1,3-diene), which is the natural monomer for rubber.

Test yourself

16 a) Suggest two problems of recycling plastics.
b) What techniques have been developed to help in recycling plastics?
c) Look at Figure 12.22. What do you think the labels V, LDPE and PP stand for?

17 Why are recycling rates much higher for metals than for plastics?

18 a) Suggest two problems created by the combustion of plastic waste for energy production.
b) In what ways have chemists minimised damage to the environment from the incineration of plastic waste?
c) What environmental problems does incineration exacerbate, even when it is clean?

19 What are local councils and the government now doing to reduce all landfill waste?

20 How are chemists helping to reduce the problems caused by non-biodegradable plastic waste?

Activity

A brighter future for plastics

Biodegradable plastics are now being synthesised and produced commercially as replacements for non-biodegradable plastics. One of the most important of these new biodegradable plastics is polylactic acid (PLA), which is produced from starch.

Starch is a natural polymer produced during photosynthesis. Cereal plants, such as maize and wheat, and tubers, such as potatoes, contain starch in large proportions. Starch can be processed directly into a bioplastic, but because this is soluble in water, articles made from starch swell and deform when exposed to moisture. The problem can be overcome by modifying the starch into a different polymer.

First, the starch is harvested from maize, wheat or potatoes. Micro-organisms then transform it into the monomer lactic acid. Finally, the lactic acid is chemically treated so that the molecules of lactic acid link into long chains or polymers, which bond together forming PLA.

Polylactic acid can be formed into containers and packaging for food and consumer goods as diverse as plant pots and disposable nappies. The biodegradable and compostable nature of PLA means that it will break down into harmless natural products under normal outdoor conditions. This gives it big political and environmental advantages over conventional plastic packaging, which uses an estimated 500 000 barrels of oil every day throughout the world. Its biodegradable qualities could also take the pressure off the world's mounting landfills, in which plastics take up an estimated 25% by volume. However, as PLA is significantly more expensive than conventional oil-based plastics, it has failed to gain widespread use, although it is becoming a more viable alternative as the price of crude oil continues to rise.

1 What do you understand by the following terms used in the introduction?

a) Bioplastic
b) Compostable

2 What limits the use of starch itself as a plastic?

3 The structural formula of lactic acid is:

```
       H
      /
  H  O    O
  |  |   //
H—C—C—C
  |  |   \
  H  H    O—H
```

a) Write down the molecular formula of lactic acid.
b) Give the names of the functional groups in lactic acid.

4 Starch reacts with water when micro-organisms transform it into lactic acid. Assuming that the formula of starch can be written as $(C_6H_{10}O_5)_n$, write an equation for the conversion of starch to lactic acid.

5 The –OH group and the –COOH group can react as shown below.

$$\mathbf{R{-}OH + HOOC{-}R^1 \rightarrow R{-}O{-}\underset{\underset{\displaystyle O}{\|}}{C}{-}R^1 + H_2O}$$

Use this to explain how lactic acid molecules can link up to form long chains or polymers.

6 State three advantages of PLA over oil-based plastics.

7 a) What is the major disadvantage of PLA compared with oil-based plastics?
b) Why is this disadvantage becoming less important?

REVIEW QUESTIONS

Extension questions

1 The diagram below shows three important reactions of ethene.

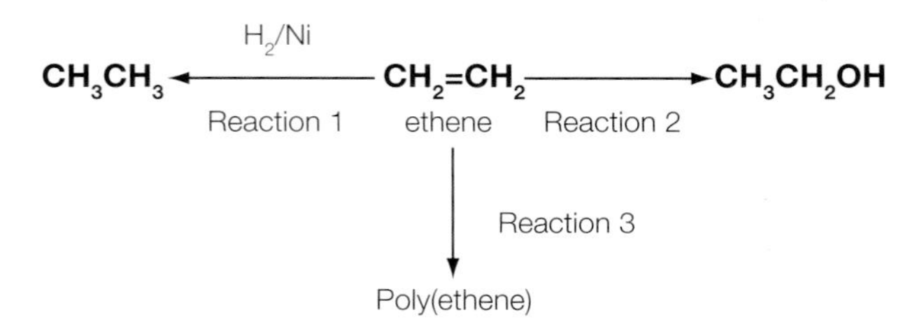

a) i) What conditions of temperature and pressure are used in Reaction 1? (2)

ii) Reaction 1 is used to convert unsaturated alkenes to saturated alkanes. What is meant by the terms unsaturated and saturated in this context? (3)

iii) Why are saturated and unsaturated chemicals important to dieticians and nutritionists? (2)

b) i) What chemicals are used to produce CH_3CH_2OH in Reaction 2? (2)

ii) A common name for CH_3CH_2OH is alcohol. What is its systematic name? (1)

iii) What is the major use of CH_3CH_2OH from Reaction 2? (1)

c) i) What conditions are used in Reaction 3 to convert ethene to high-density poly(ethene)? (2)

ii) Draw a section of the poly(ethene) structure that consists of three monomer units. (1)

d) State four properties of poly(ethene) that make it particularly suitable for making plastic bags. (2)

2 a) Propene and but-2-ene are used in the petrochemical industry to produce important polymers.

i) Explain the term 'polymer'. (3)

ii) Poly(propene) does not have a sharp melting point but softens over a wide temperature range. Why is this? (2)

iii) Draw a section of the polymer from but-2-ene showing two repeat units. (1)

b) But-2-ene can be converted to buta-1,3-diene by a process called dehydrogenation. This is used to make synthetic rubber.

i) Explain the term 'dehydrogenation'. (1)

ii) Draw the structure of buta-1,3-diene. (1)

iii) Write an equation for the dehydrogenation of but-2-ene to form buta-1,3-diene. (2)

3 The reaction between propene, $CH_3CH{=}CH_2$, and hydrogen bromide involves electrophilic addition to produce $CH_3CHBrCH_3$ as the major product.

a) Explain the term 'electrophilic addition'. (3)

b) What is the name of the compound $CH_3CHBrCH_3$? (1)

c) Draw displayed structures to show the mechanism of the reaction. (3)

d) Propene has one structural isomer.

i) Name and draw the structure of this isomer. (2)

ii) Why do you think the structural isomer is unstable, readily changing into propene? (1)

13 Alcohols

Ethanol is the best known member of the family of alcohols. It is the alcohol in beer, wine and spirits. Alcohols are useful solvents in the home, in laboratories and in industry. Understanding the properties of the –OH functional group in alcohols helps to make sense of the reactions of some important biological chemicals, especially carbohydrates such as sugars and starch. Alcohols are much more reactive than alkanes because the C–O and O–H bonds are polar.

13.1 Alcohol names and structures

Alcohols are compounds with the formula R–OH, where R represents an alkyl group. The hydroxy group (–OH) is the functional group that gives alcohols their characteristic reactions.

The IUPAC rules name alcohols by changing the ending of the corresponding alkane to -ol. So ethane becomes ethanol.

Definition

The terms **primary**, **secondary** and **tertiary** are used with organic compounds to show the position of functional groups. A primary alcohol has the –OH group at the end of the chain. A secondary compound has the –OH group somewhere along the chain but not at a branch in the chain or at the ends. A tertiary alcohol has the –OH group at a branch in the chain.

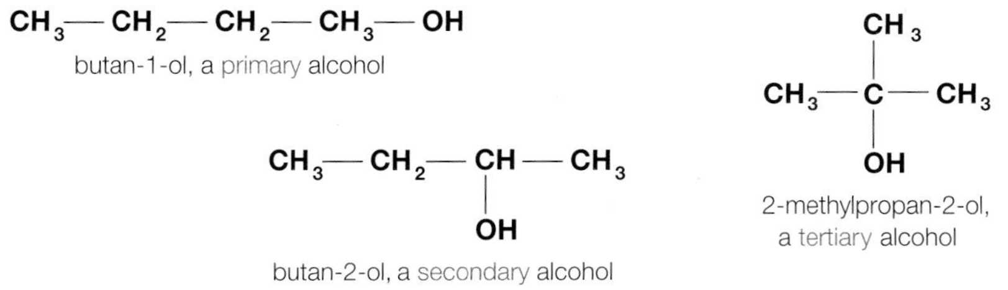

$CH_3—CH_2—CH_2—CH_3—OH$

butan-1-ol, a primary alcohol

$CH_3—CH_2—CH(OH)—CH_3$

butan-2-ol, a secondary alcohol

$(CH_3)_3C—OH$

2-methylpropan-2-ol, a tertiary alcohol

Figure 13.1 ◄ Names and structures of alcohols.

Test yourself

1 Draw the structures of the following alcohols and state whether they are primary, secondary or tertiary compounds.
 a) ethanol
 b) propan-1-ol
 c) propan-2-ol
 d) 2-methylbutan-2-ol
2 Draw the skeletal formula and name one isomer with the formula $C_5H_{11}OH$ that is:
 a) a primary alcohol
 b) a secondary alcohol
 c) a tertiary alcohol.

13.2 Physical properties of alcohols

The simplest alcohols – methanol and ethanol – are liquids at room temperature because of hydrogen bonding between –OH groups (see Section 6.14). For the same reason, alcohols are much less volatile than hydrocarbons with roughly the same molar mass and alcohols with relatively short hydrocarbon chains mix freely with water.

Test yourself

Data

3 Use the data tables on your CD-ROM to show that alcohols are less volatile than alkanes with similar molar masses.
4 a) Draw a diagram to show the hydrogen bonding between a methanol molecule and a water molecule.
 b) Explain why hydrogen bonding accounts for the fact that, at room temperature, methanol is a liquid that mixes freely with water, while ethane is a gas that is insoluble in water.

13.3 Manufacture and uses of ethanol

Ethanol can be made by fermentation of chemicals from plant crops or synthesis from petrochemicals. Fermentation is much the most important method: less than 5% of ethanol is made from oil and almost all of it comes from fermentation.

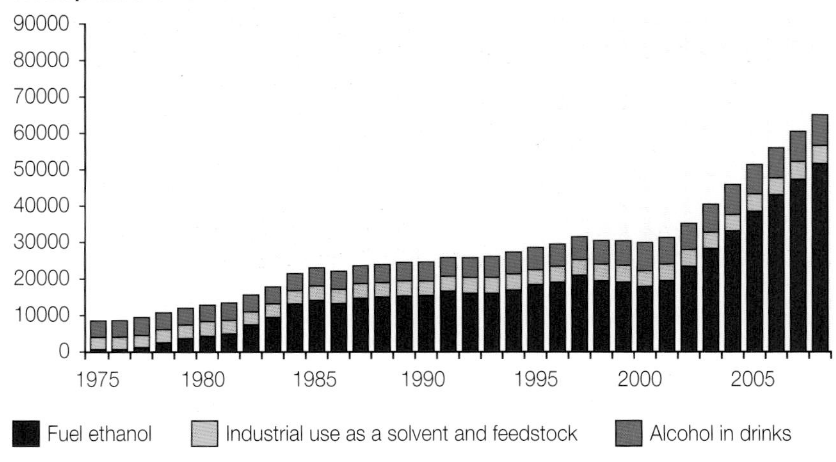

Figure 13.2 ▶ Worldwide production and uses of ethanol.

Traditionally, the fermentation of starch and sugars was used to make alcoholic drinks such as beer, wine and spirits. Until the late 1970s, this was the main use of alcohol. The growing concerns about oil supplies and climate change mean that the main use of ethanol is now as a biofuel.

Test yourself

5 Look at the chart in Figure 13.2.
   a) Estimate the proportion of ethanol used as a fuel in:
      i) 1980
      ii) 2008.
   b) Suggest reasons for
      i) the growth in ethanol production from 1975 to 2000
      ii) the steeper growth in ethanol production since the year 2000.

In industry, ethanol is a very useful solvent. High taxes are paid on the ethanol in drinks, and to avoid these, ethanol supplied for other purposes has to be mixed with chemicals that make it poisonous and give it a repellent taste. Methylated spirit is ethanol mixed with about 5% methanol, which makes it undrinkable. Industrial methylated spirit is widely used as a solvent. Surgical spirit is industrial methylated spirit with other additives, including castor oil. The methylated spirit ('meths') available from hardware stores as a solvent and fuel is coloured blue by a dye.

Ethanol from fermentation

Ethanol in alcoholic drinks is made by fermenting sugars from starchy crops such as barley or potatoes. This is a two-stage process. The starch first has to be broken down to simple sugars. Fermentation of sugars with yeast then produces ethanol.

Fermentation converts sugars, such as glucose, into ethanol and carbon dioxide. Enzymes from yeast catalyse the reactions. Ethanol is the product of the respiration of yeast in the absence of air, which is known as anaerobic respiration

$$\underset{\text{glucose}}{C_6H_{12}O_6(aq)} \xrightarrow{\text{enzymes}} \underset{\text{ethanol}}{2C_2H_5OH(l)} + 2CO_2(g)$$

Figure 13.3 ◄
Worker checking pipes in a distillery in Uttar Pradesh, India. The plant produces ethanol by fermenting sugar from sugar cane. Worldwide, there is rapid growth in the use of ethanol as a fuel.

Yeast ferments sugars faster in the temperature range 25–37 °C. The reaction is too slow below 25 °C, while temperatures above 37 °C start to alter the enzymes in yeast and denature them so that they are no longer effective. Yeast produces solutions with up to 14% ethanol. At higher concentrations, ethanol is toxic and kills the yeast.

Industrial feedstocks for making bioethanol include crops, such as sugar beet, sugar cane and maize, and waste materials, such as straw, wood chips and household waste. Cellulose is the main carbohydrate in waste. It is much more difficult to break down to sugars than starch, so acids or enzymes can be used to make glucose from the cellulose in wastes.

Definition

Fermentation converts sugars to ethanol and carbon dioxide. Fermentation is catalysed by enzymes from yeast and is an example of anaerobic respiration.

Test yourself

6 Identify and explain the product of fermentation that is important to:
 a) brewers
 b) bakers.

Ethanol from ethene

The petrochemical industry makes ethanol by hydrating ethane with steam in the presence of an acid catalyst. A mixture of ethanol and steam passes over the catalyst at 300 °C under pressure. The catalyst is phosphoric acid supported on an inert solid.

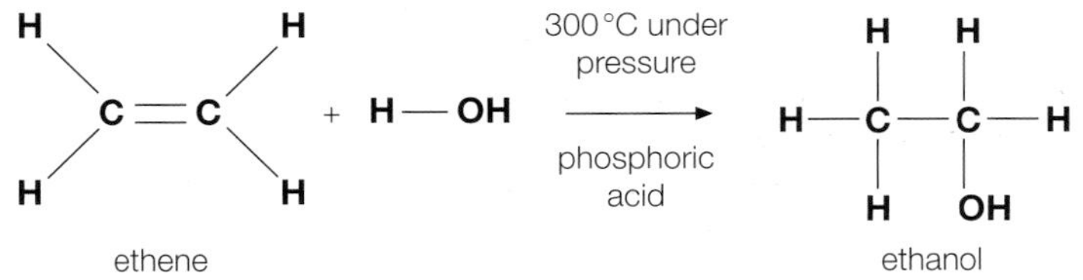

$$H_2C{=}CH_2 + H{-}OH \xrightarrow[\text{phosphoric acid}]{300\,°C\text{ under pressure}} CH_3CH_2OH$$

ethene ethanol

Figure 13.4 ◄
Addition of water to ethene to make ethanol.

The yield is high, but some side reactions produce chemicals such as methanol, ethanal and poly(ethene). The process produces a mixture of ethanol and water. The ethanol is concentrated by fractional distillation.

Test yourself

7 a) What is the theoretical atom economy (Section 10.7) for the synthesis of ethanol from ethene?
 b) Why is the ideal atom economy not achieved in practice?

8 Suggest a reason why ethanol made from ethene is never used to make alcoholic drinks.

Figure 13.5 ▲
Methanol burns cleanly in engines and is also much less flammable than petrol. It is the only fuel used in race cars that compete in the Indianapolis 500-mile race.

13.4 Manufacture and uses of methanol

Methanol is a liquid alcohol that is becoming more important as a fuel or petrol additive and as a raw material for the chemical industry. The main use of methanol is as a feedstock for the plastics industry. Millions of tonnes per year are needed to produce a variety of polymers. This means that the chemical industry manufactures methanol on a very large scale.

Methanol can be made directly from a mixture of carbon monoxide and hydrogen. This gas mixture is called synthesis gas and is available from the reaction of methane and steam in the presence of a catalyst.

Methanol is also one of several organic oxygen compounds that can be added to fuel to improve performance and cut the level of carbon monoxide and unburned fuel in exhaust gases (see Section 18.7). However, it is much more usual to add ethanol or one of another class of oxygen compounds called ethers. About a quarter of all of the methanol produced is converted to the ether MTBE, which is widely used as a petrol additive.

Use of MTBE is banned, or being phased out, in some states of the United States because it has leaked from storage tanks and contaminated underground water supplies. It is not particularly toxic, but it is not biodegradable and has a strong taste. The taste and smell of MTBE can be detected at concentrations around 5–10 mg dm^{-3} – much lower than the concentration of several milligrams per litre at which it becomes toxic.

Test yourself

9 Write equations for the reactions of:
a) methane and steam to make synthesis gas
b) synthesis gas to make methanol.

10 The initials MTBE are based on the traditional name for the compound 2-methoxy-2-methylpropane. Write the displayed formula for MTBE.

11 Suggest why adding alcohols and ethers such as MTBE to petrol reduces the concentration of carbon monoxide in vehicle exhausts.

13.5 Chemical properties of alcohols

Alcohols are much more reactive than alkanes because the C–O and O–H bonds in the molecules are polar (Section 5.2).

Combustion

Alcohols burn in air with a clean colourless flame. This makes them attractive as fuels and explains why methanol and ethanol are both common fuels and fuel additives.

$$2CH_3OH(l) + 3O_2(g) \rightarrow 2CO_2(g) + 4H_2O(l)$$

Oxidation by potassium dichromate(VI)

An acidifed solution of potassium dichromate(VI) is orange. It turns green on warming with alcohols such as ethanol. Potassium dichromate(VI) oxidises primary and secondary alcohols during such a reaction, but it does not react with tertiary alcohols.

Oxidation of a primary alcohol takes place in two steps to produce an aldehyde and then a carboxylic acid. During the reaction, the potassium dichromate(VI) turns from the orange colour of $Cr_2O_7^{2-}(aq)$ to the green colour of $Cr^{3+}(aq)$.

```
     H   H   H                                       H   H
     |   |   |          Cr2O7 2-/H+(aq)              |   |      //O
H ── C ─ C ─ C ── OH + [O] ─────────────▶  H ── C ─ C ─ C          + H2O
     |   |   |               warm                    |   |      \
     H   H   H                                       H   H       H

                                                     propanal
                                                    (aldehyde)
```

Figure 13.6 ▶
Oxidation of propan-1-ol to propanal by acidified $Cr_2O_7^{2-}(aq)$. The symbol [O] is a shorthand way of balancing the equation, in which [O] represents oxygen atoms from the oxidising agent.

Oxidation is completed when a primary alcohol is heated with an excess of acidified potassium dichromate(VI) in the apparatus shown in Figure 13.8. This converts the alcohol first to an aldehyde and then to a carboxylic acid.

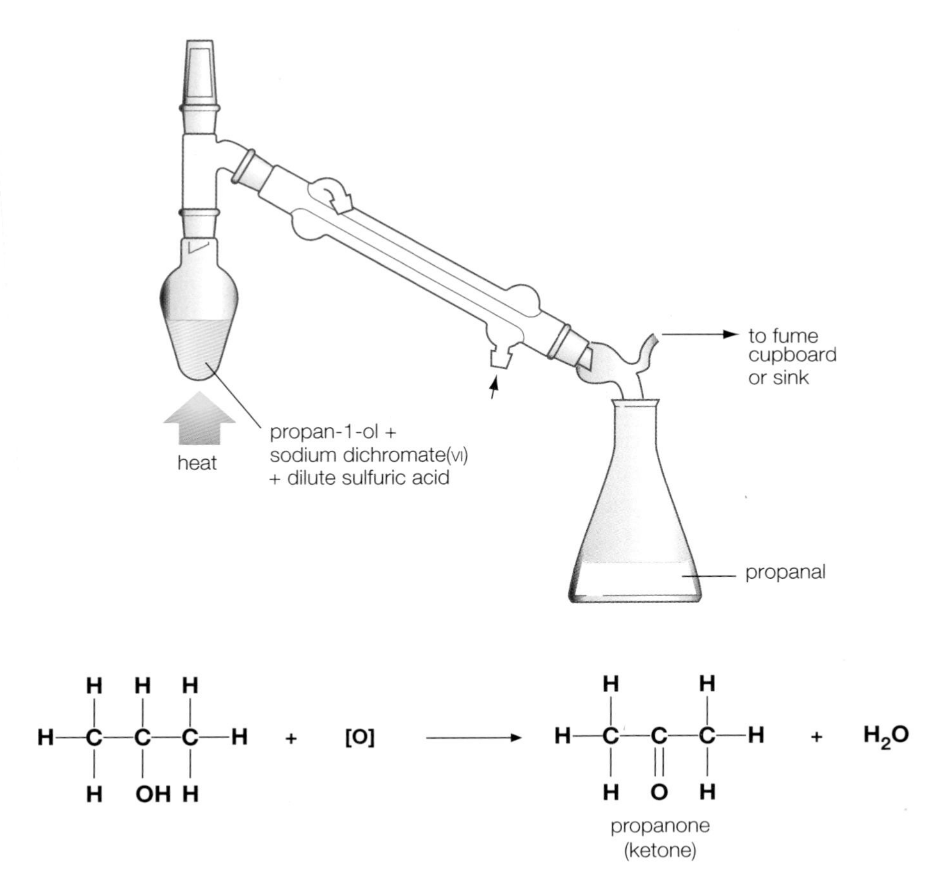

Figure 13.7 ◀
Apparatus used to oxidise a primary alcohol to an aldehyde. The aldehyde distils off as it forms, which prevents further oxidation of the aldehyde.

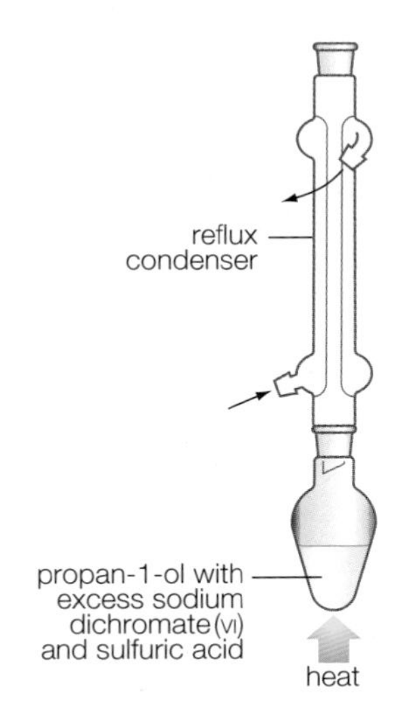

Figure 13.8 ▲
Apparatus used to oxidise a primary alcohol to a carboxylic acid. The reflux condenser ensures that any volatile aldehyde condenses and flows back into the flask, where excess oxidising agent ensures complete conversion.

$$\mathrm{CH_3CH(OH)CH_3 + [O] \longrightarrow CH_3COCH_3 + H_2O}$$

propanone (ketone)

Figure 13.9 ▲
Oxidation of propan-2-ol produces the ketone propanone.

Oxidation of a secondary alcohol produces a ketone.

These oxidation reactions help to distinguish between primary, secondary and tertiary alcohols.

Definition

A **reflux condenser** is fitted to a flask to prevent vapour escaping while a liquid is being heated. Vapour from the boiling reaction mixture condenses and flows back into the flask.

Practical guidance

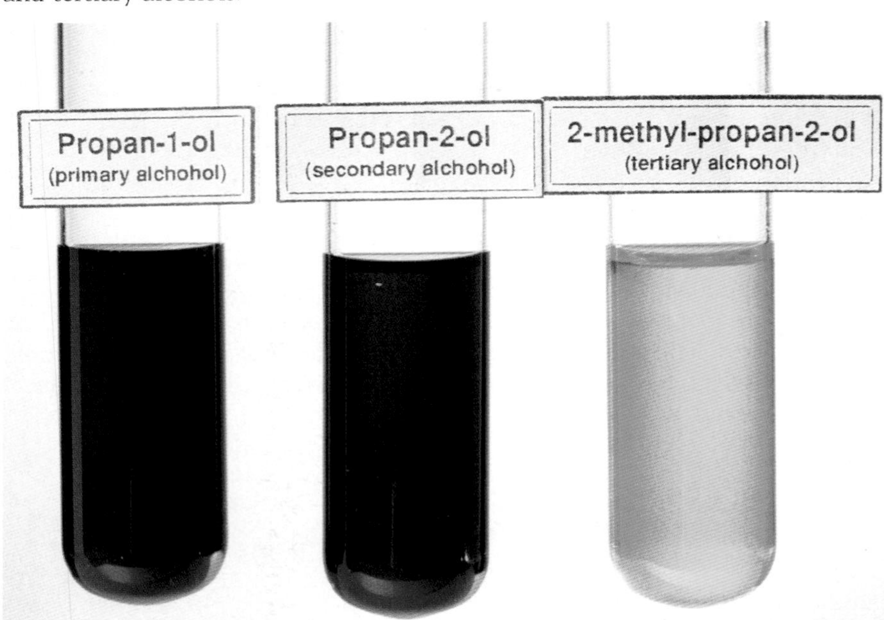

Figure 13.10 ◀
Result of warming three alcohols with an acidic solution of potassium dichromate(VI). Dichromate(VI) is reduced to green chromium(III) ions when there is a reaction. Tertiary alcohols do not react with potassium dichromate(VI).

Infrared spectroscopy can be used to detect the functional groups in organic molecules. It is an analytical tool that can be used to show the change in functional groups when alcohols are oxidised (see Section 15.1).

Test yourself

12 Use oxidation numbers to show that chromium is reduced when $Cr_2O_7^{2-}$(aq) ions turn into Cr^{3+}(aq) ions.
13 Write an equation for the oxidation of propanal to propanoic acid.
14 Predict the products, if any, of oxidising the following alcohols with acidified dichromate(VI).
 a) Butan-1-ol (product distilled off immediately)
 b) Butan-1-ol (reagents heated under reflux for some time)
 c) Butan-2-ol
 d) 2-methylbutan-2-ol

Esterification

Alcohols react with carboxylic acids to form esters. The reaction happens on gentle heating in the presence of a little concentrated sulfuric acid to act as a catalyst. The reaction is reversible.

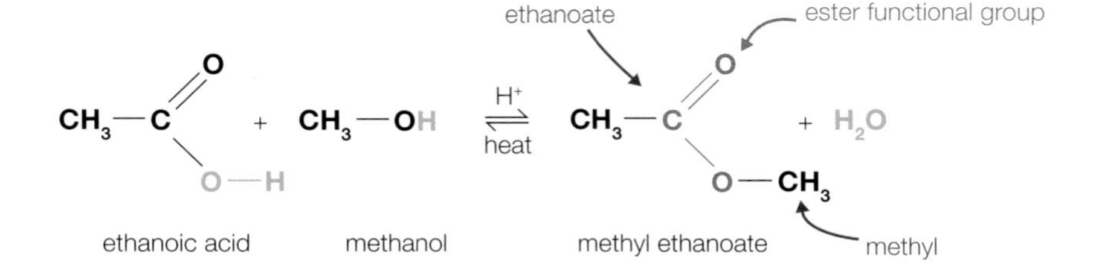

Figure 13.11 ▶ Formation of an ester from a carboxylic acid and an alcohol.

Definition

An **ester** is a compound that forms when a carboxylic acid reacts with an alcohol. The formula of an ester is R–CO–OR', where R and R' are alkyl groups.

Esters are compounds with a fruity smell that contribute to the flavour of bananas, pineapples and many other fruits. The smell of esters contributes to the scent of perfumes. Fats and vegetable oils are esters.

Test yourself

15 Write out the structural formulae of the following esters.
 a) Ethyl ethanoate
 b) Ethyl propanoate
 c) Propyl ethanoate
16 Identify the alkenes formed when the following alcohols are heated with H_3PO_4.
 a) Propan-2-ol
 b) 2-methylpropan-2-ol

Definition

An **elimination reaction** splits off a simple molecule from within a larger molecule to form a double bond. Simple molecules that may be eliminated include water and hydrogen halides.

Elimination of water

Eliminating water from an alcohol produces an alkene. As this change involves splitting off water molecules, its alternative name is dehydration. The procedure involves heating the alcohol with concentrated sulfuric acid or phosphoric acid. The advantage of phosphoric acid is that it is a non-oxidising acid and thus leads to fewer side reactions. The acid acts as a catalyst for the elimination reaction.

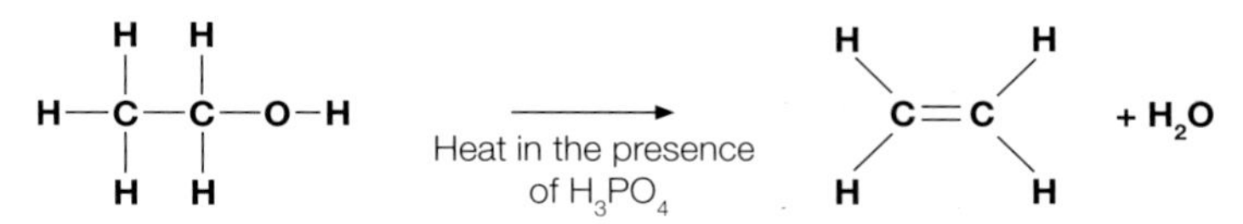

Figure 13.12 ▶ Elimination of water from ethanol to form ethene.

Elimination generally happens more readily from a tertiary alcohol than from a primary alcohol.

Activity

Eliminating water from cyclohexanol

Figure 13.13 shows the steps involved in an elimination reaction to make, separate and purify an organic product from cyclohexanol. The volume of cyclohexanol (density 0.96 g cm^{-3}) in the flask is 10.4 cm^3. The yield of pure product is 6.9 g.

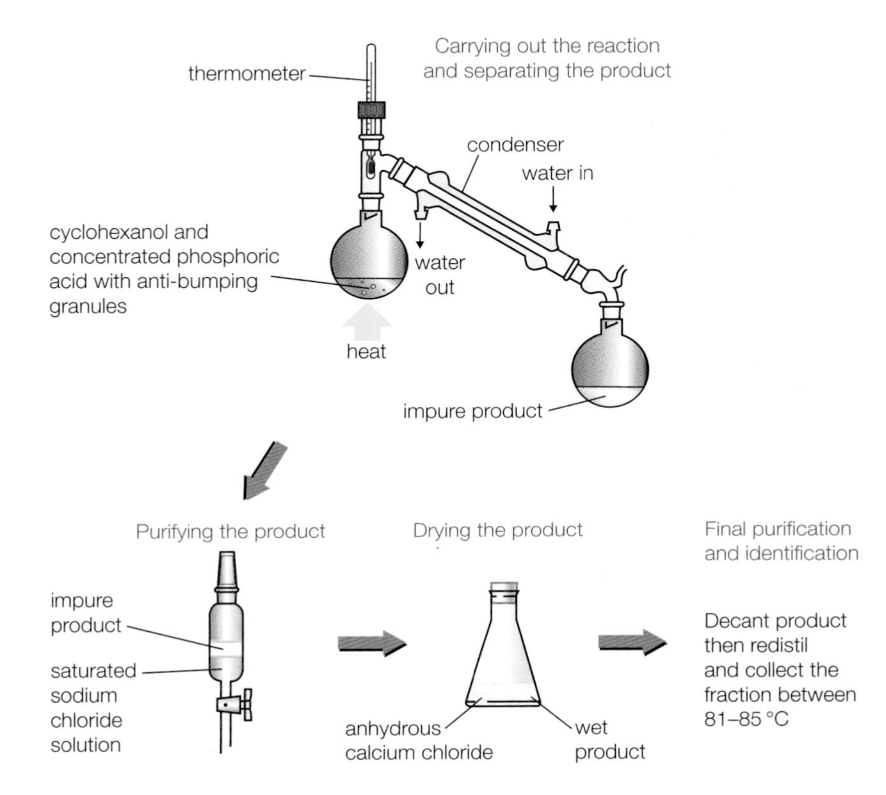

Figure 13.13 ◀ Eliminating water from cyclohexanol.

1 Give the name and skeletal formula of the organic product of the reaction.

2 Explain why the product separates from the reaction mixture and why the liquid collected is impure.

3 a) What types of impurities are removed by shaking the product with water in a tap funnel?

b) Suggest a reason for saturating the water in the tap funnel with sodium chloride.

4 What is the purpose of adding anhydrous calcium chloride to the impure product and allowing the mixture to stand for a few minutes?

5 Why is a final distillation necessary?

6 Write an equation for the reaction.

7 a) Calculate the theoretical yield of product.

b) Calculate the percentage yield.

c) Suggest reasons why the percentage yield is less than 100%.

Practical guidance

REVIEW QUESTIONS

Extension questions

1 Ethanol is used as an industrial solvent, as a fuel and as a raw material for making other organic compounds. Worldwide, most ethanol is made by fermentation, but much of the industrial ethanol in the UK is made from ethene.

a) What properties of ethanol mean that it is widely used as a solvent? **(3)**

b) i) Outline the production of ethanol by fermentation. Name the raw materials, and state the conditions for the final step of making ethanol.

ii) Write an equation for the fermentation reaction that produces ethanol.

iii) Explain why fermentation is becoming increasingly important as a way of making ethanol. **(6)**

c) i) State the conditions for the production of ethanol from ethane.

ii) Write an equation for the reaction.

iii) Explain why ethanol for alcoholic drinks is made by fermentation and not from ethene. **(6)**

2 Consider the following reaction scheme.

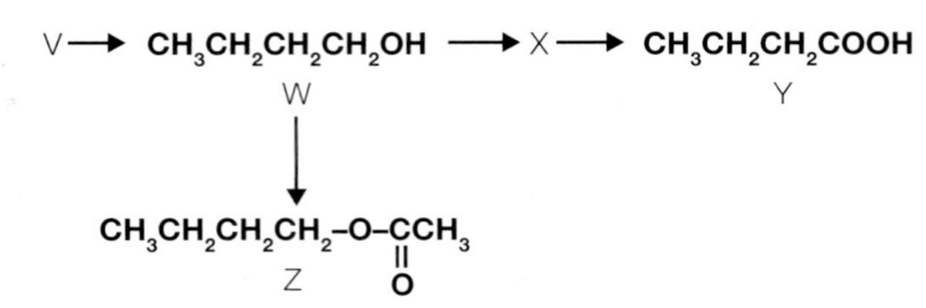

a) State the reagents and conditions needed to convert compound W directly to compound Y. **(3)**

b) Draw the structure of X and state how the conditions you have given in part a) could be modified to produce compound X instead of compound Y. **(2)**

c) Compound W forms compound V when heated with concentrated phosphoric acid.

i) Write an equation for the reaction.

ii) Describe a simple test that could be used to distinguish V from W. **(3)**

d) Give the reagents and conditions for converting compound W into compound Z. **(2)**

14 Halogenoalkanes

Halogenoalkanes are important to organic chemists in research and industry. The reason for this is that many halogenoalkanes are reactive compounds that can be converted to other more valuable products. This makes them useful as intermediates for converting one chemical to another. Organic halogen compounds are also useful in their own right, but the many benefits have to be weighed against the risks. There are growing restrictions on the uses of many halogenoalkanes because of concerns about their hazards to health, their persistence in the environment and their effect on the ozone layer.

Figure 14.1 ◄
The halogenoalkane bromomethane is still used in parts of the United States to sterilise soils and protect strawberry crops from pests despite the harm it can do to the ozone layer. Halogenoalkanes have a wide range of other uses: as solvents, refrigerants and fire extinguishers.

14.1 Halogenoalkanes

In the structure of a halogenoalkane, one or more of the hydrogen atoms in an alkane molecule is replaced with a halogen atom.

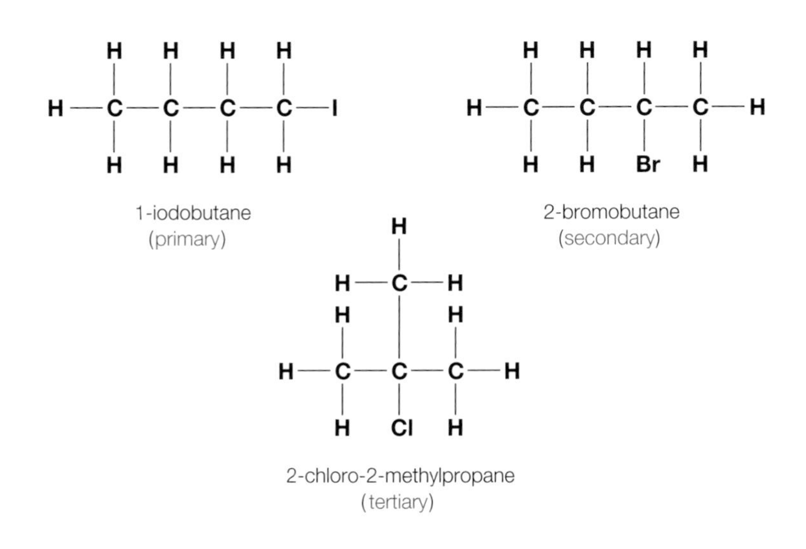

Figure 14.2 ▲
Names and structures of three halogenoalkanes.

Test yourself

1 Draw the structures of the following compounds and identify the primary, secondary and tertiary compounds.
   a) 1-iodopropane
   b) 2-chloro-2-methylbutane
   c) 3-bromopentane
2 Name and give the structure of a halogenoalkane that forms when:
   a) ethane reacts with chlorine
   b) propene reacts with hydrogen bromide.

14.2 Physical properties of halogenoalkanes

Chloromethane, bromomethane and chloroethane are gases at room temperature. Most other halogenoalkanes are colourless liquids that do not mix with water.

Test yourself

3 Which of the following molecules are polar and which are non-polar: $CHCl_3$, CH_2Cl_2, $CHCl_3$ and CCl_4?
4 Look up the boiling points of the following compounds on the data sheet and suggest an explanation for the trend in values: 1-chlorobutane, 1-bromobutane and 1-iodobutane.
5 Compare the boiling points of the isomers 1-bromobutane, 2-bromobutane and 2-bromo-2-methylpropane. Suggest an explanation for the differences in boiling points of the primary, secondary and tertiary compounds.

14.3 Hydrolysis of halogenoalkanes

The most important reactions of halogenoalkanes are substitution reactions. These reactions replace the halogen atom with another atom or group of atoms.

Substitution by reaction with water

Definition

Hydrolysis is a reaction in which a compound splits apart in a reaction that involves water. Hydrolysis reactions are often catalysed by acids or alkalis.

The substitution reactions of halogenoalkanes with water are examples of hydrolysis. They split the alkyl group from the halogen atom. The word 'hydrolysis' comes from two other words: 'hydro', which is related to water, and 'lysis', which means splitting. So the term hydrolysis is used to describe any reaction in which water causes another molecule to split apart. Hydrolysis reactions are often catalysed by acids or alkalis. They are also often substitution reactions in which the chemical attack is by water molecules (or hydroxide ions).

Cold water slowly hydrolyses halogenoalkanes and replaces the halogen atoms with –OH groups to form alcohols.

$$CH_3CH_2CH_2I(l) + H_2O(l) \rightarrow CH_3CH_2CH_2OH(l) + H^+(aq) + I^-(aq)$$

The rate of hydrolysis of different halogenoalkanes can be compared by carrying out the reaction in the presence of silver ions. The halogen atoms in halogenoalkanes are covalently bonded to carbon and give no precipitate of a silver halide. Hydrolysis releases halide ions, which immediately precipitate as the silver halide (see Section 9.4). Halogenoalkanes do not mix with water or aqueous solutions, so the reaction is carried out in the presence of ethanol, which can dissolve the halogenoalkane and mix with aqueous silver nitrate.

Test yourself

6 Explain the use of the terms 'substitution' and 'hydrolysis' to describe the reaction of 2-iodopropane with water.
7 Write equations for the two reactions that take place when 2-bromobutane reacts with water in the presence of silver ions.

Substitution by reaction with hydroxide ions

Replacement of the halogen atom of a halogenoalkane by an –OH group is much quicker with an aqueous solution of an alkali such as sodium or potassium hydroxide. Heating increases the rate of reaction even further.

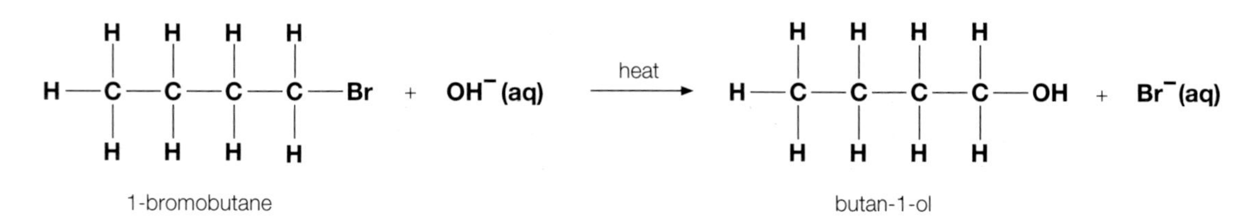

Figure 14.3 ▲
Reaction of a halogenoalkane with an alkali on heating.

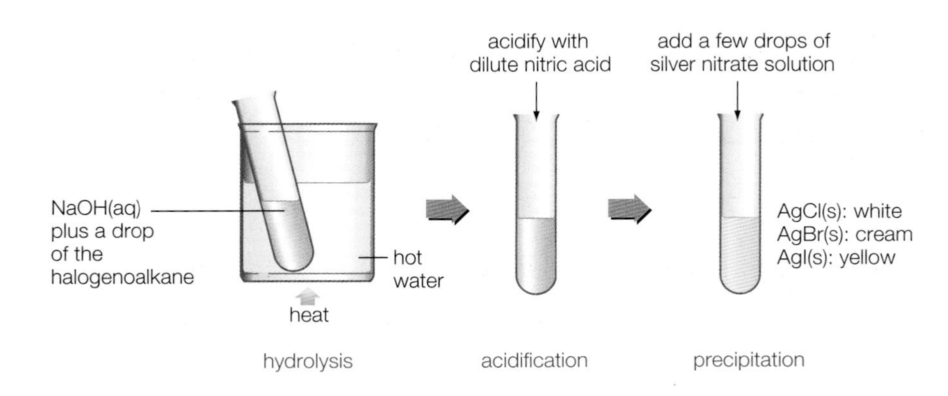

Figure 14.4 ◄
Hydrolysis of halogenoalkanes makes it possible to distinguish between chloro-, bromo- and iodo-compounds. Heating the compound with an alkali releases halide ions. Acidifying with nitric acid and then adding silver nitrate produces a precipitate of the silver halide.

The rate of reaction of halogenoalkanes with water and alkalis is in the order: RI > RBr > RCl, where R represents an alkyl group. The reaction rates do not correlate with bond polarity in the halogenoalkane. Chlorine is the most electronegative and iodine the least electronegative, so the C–Cl bond is the most polar and the C–I bond the least polar. Bond polarity is therefore not the key factor that determines the rates of reaction.

The reaction rates do, however, correlate with the strength of the bonds. The C–I bond is the longest and weakest (as measured by the mean bond enthalpy). The C–Cl bond is the shortest and strongest. This suggests that the C–I bond breaks more readily than the C–Cl bond.

Test yourself

8 Refer to Figure 14.4.
- a) Why is hydrolysis necessary before testing with silver nitrate?
- b) Why must nitric acid be added before the silver nitrate solution?
- c) Write the equations for the three reactions that take place when detecting bromide ions in 1-bromobutane by this method.

14.4 Nucleophilic substitution in halogenoalkanes

The carbon–halogen bond in a halogenoalkane is polar, with the carbon atom at the δ+ end of the dipole. This makes the carbon atom open to attack by molecules or ions that can provide a pair of electrons to form a new covalent bond.

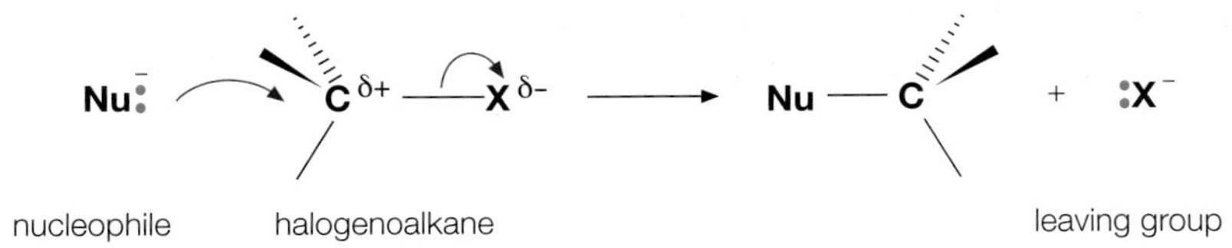

$$Nu{:}^{-} \quad C^{\delta+}—X^{\delta-} \longrightarrow Nu—C \quad + \quad {:}X^{-}$$

nucleophile — halogenoalkane — leaving group

Figure 14.5 ◄
Generalised description of the attack of a nucleophile on a halogenoalkane, which leads to a substitution reaction.

Nucleophiles that react with halogenoalkanes include water molecules and hydroxide ions (see Section 10.6). When a hydroxide ion ($:OH^-$) attacks a halogenoalkane, it forms a new bond with the carbon atom as the C–halogen bond breaks to set free a halide ion.

Figure 14.6 ▶
Nucleophile attacking a polar bond, which leads to heterolytic fission.

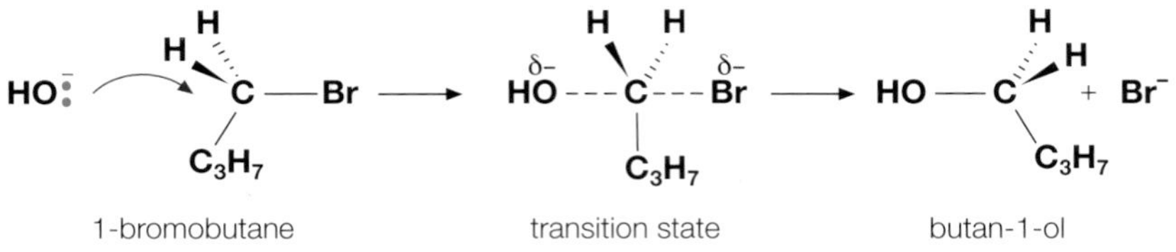

$$HO^{-}: + H_2C(C_3H_7)^{\delta+}{-}Br^{\delta-} \longrightarrow [HO^{\delta-}\cdots CH_2(C_3H_7)\cdots Br^{\delta-}] \longrightarrow HO{-}CH_2C_3H_7 + Br^{-}$$

1-bromobutane — transition state — butan-1-ol

Tutorial

Test yourself

9 a) Show the bond forming and bond breaking when a hydroxide ion reacts with 1-iodopropane. Use 'curly arrows' to show the movement of pairs of electrons and identify relevant dipoles.
b) Use this example to explain why the hydroxide ion is described as a 'nucleophile'.

14.5 Uses of halogenoalkanes and their impacts

Intermediates for synthesis

Halogenoalkanes are important intermediates in the making of new chemicals. This is illustrated by the processes used to make the monomers of the plastics pvc and ptfe (see Section 12.5).

The manufacture of pvc is a three-stage process:

- Ethene reacts with chlorine to make 1,2-dichloroethane.
- 1,2-dichloroethane is heated strongly to break it down into chloroethene and hydrogen chloride.
- Chloroethene is suspended as droplets in water and heated in the presence of an initiator.

The reaction of ethene with chlorine in the first stage is exothermic. The reaction mixture has to be cooled to keep the temperature below 60 °C to stop further reaction of the product with chlorine.

Test yourself

10 Where does the ethene that is used to make pvc come from?
11 Write equations for the three reactions used to make pvc from ethene.
12 What type of reaction is taking place at each of the three stages?
13 Suggest how 1,2-dichloroethane might react further with chlorine if the temperature of the reaction mixture in the first stage is allowed to rise above 60 °C.

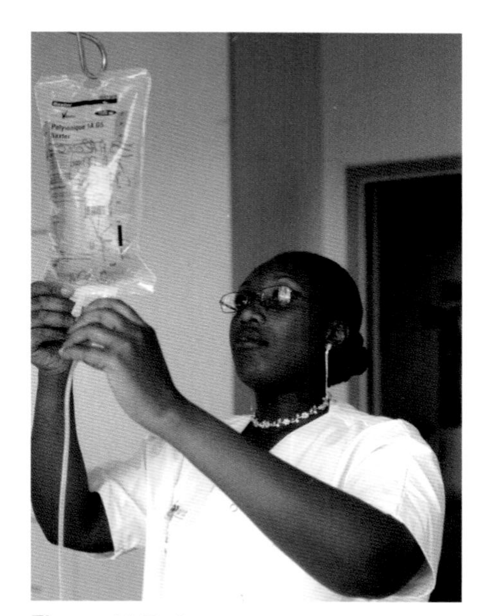

Figure 14.7 ▲
Plasticised pvc is widely used in medicine to make blood bags and flexible sterile tubing.

The monomer for making ptfe is made from methane in a three-stage process. Methane first reacts with chlorine in the presence of a catalyst to make trichloromethane. Trichloromethane is then mixed with hydrogen fluoride in the presence of another catalyst to make chlorodifluoromethane. Finally, the chlorodifluoromethane is heated strongly so that it cracks to make tetrafluoroethene and hydrogen chloride. Tetrafluoroethene is a highly unstable gas, so it has to be cooled rapidly as it forms to prevent an explosive decomposition.

The monomer is fed under pressure into water that contains an initiator at about 70 °C. The polymer forms as a suspension of granules, which can be ground to a fine powder ready for moulding.

Test yourself

14 Write equations for the three reactions involved in making tetrafluoroethene.
15 Suggest a reason why tetrafluoroethene is normally made when and where it is needed for polymerisation.

Chlorofluorocarbons

One particular class of unreactive halogenoalkanes has become notorious in recent years because of the damaging effect the compounds have on the ozone layer. These are the chlorofluorocarbons (CFCs), which were developed in the early twentieth century as safer alternatives to toxic refrigerants such as ammonia. As CFCs are also powerful greenhouse gases, their use is now known to be doubly damaging (see Section 18.3).

Chlorofluorocarbons are compounds that contain only the elements chlorine, fluorine and carbon, such as CCl_3F, CCl_2F_2 and CCl_2FCClF_2. They contain no hydrogen. They have some very desirable properties: they are unreactive, they do not burn and they are not toxic. It is also possible to make CFCs with different boiling points to suit different applications.

The manufacture of CFCs and their general use was banned by the Montreal Protocol in 1987 because of the effects on the ozone layer (Section 18.6). At the same time, governments also agreed to ban the production of bromofluoroalkanes, which were then in common use as halon fire extinguishers in homes, on aircraft and on ships, as well as in a wide range of electronic equipment.

These bans set chemists the challenge of finding replacement chemicals with similar properties to CFCs but without the harmful impacts on the environment. The chemists' first step was to try adding hydrogen to make hydrochlorofluorocarbons (HCFCs). These are much less stable in the lower atmospheres, so they break down before reaching the ozone layer. Other compounds that can be used are the hydrofluorocarbons (HFCs), which survive for an even shorter time in the lower atmosphere than HCFCs. The HFCs have no effect on the ozone layer, but they are greenhouse gases. Hydrocarbons and carbon dioxide can also be used instead of CFCs for some purposes.

Figure 14.8 ▲
Goretex magnified 100 times under an electron microscope. This is a false colour image that shows PTFE (yellow) sandwiched between woven layers of nylon threads (pink). The tiny pores in the PTFE layer are 20 000 times smaller than a drop of water but 700 times larger than a water molecule. Goretex keeps rain out but allows perspiration to escape.

Refrigeration and air conditioning

Refrigerators and air conditioners contain a working fluid that circulates around a closed system. The compound chosen as the refrigerant must have the right boiling point: high enough for it to liquefy when compressed but low enough to vaporise easily.

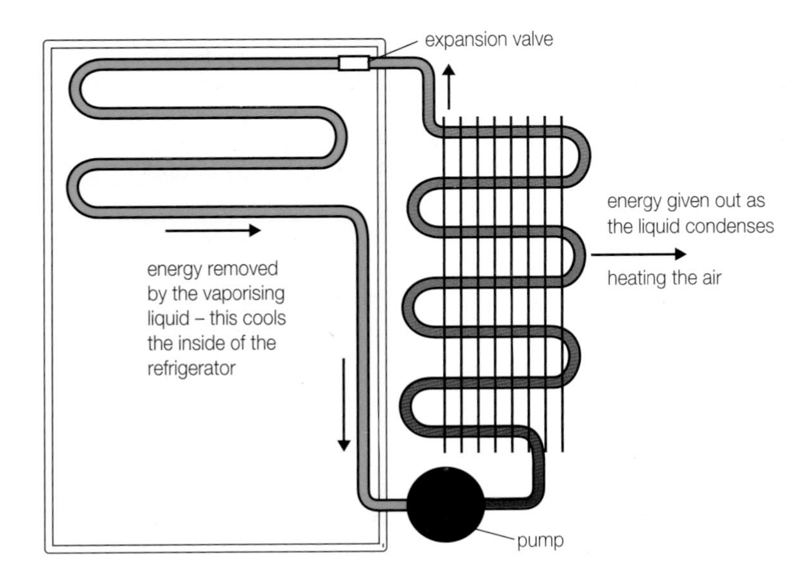

Figure 14.9 ▲
Diagram of a refrigerator showing the circulation of the refrigerant.

Until the discovery of CFCs in the 1930s, the main refrigerant in use was ammonia. Ammonia is corrosive, toxic and flammable. It also has a very strong smell. The American engineer Thomas Midgeley seemed to have achieved a remarkable breakthrough when he produced CFCs that were non-toxic and unreactive and had the right properties.

Figure 14.10 ▶
Many old refrigerators and freezers have CFCs in the coolant pipes and insulating foam. The CFCs have to be recovered when they are scrapped.

Aerosol propellants

Aerosol cans contain a propellant mixed with its contents. When the valve on the can opens, the propellant evaporates and forces the contents out of the can in a fine spray. CFCs were widely adopted for this purpose before their harmful effects on the environment were understood.

Blowing agents

A blowing agent is a chemical used to create the gas bubbles in insulating foam and expanded plastics. CFCs were also used for this purpose on a large scale. They were mixed with the ingredients of the plastic as it was made. The energy from the polymerisation reaction heated up the CFC so that it vaporised and produced the bubbles. Inevitably, some of the CFC escaped into the air during the process. More is lost when the foam is finally disposed of.

Test yourself

16 Explain why a refrigerant cools its surrounding as it evaporates.

17 Why were CFCs so widely adopted by the makers of refrigerators after they were developed in the1930s to replace ammonia?

18 What properties of CFCs made them attractive to use as the propellants in aerosol cans?

19 Insulating foams can make a big contribution to reducing the emission of greenhouse gases. Suggest why.

Activity

Alternatives to CFCs

Formula	Type of compound	Boiling point/°C	Flammable	Ozone depleting potential	Global warming potential over 100 years
CCl_3F	CFC-11	24	No	1.0	4600
CCl_2F_2	CFC-12	−30	No	1.0	7300
$CHClF_2$	HCFC-22	−41	No	0.05	1700
CF_3CCl_2H	HCFC-123	+29	No	0.02	100
CH_2F_2	HFC-32	−51.6	No	0	650
CF_3CH_2F	HFC-134a	−26.6	No	0	1300
CH_4	Alkane	−161	Yes	0	23
C_3H_8	R-290	−42	Yes	0	
CO_2	R-744	Liquid under pressure	No	0	1
NH_3	R-717	−33	Yes	0	0

Table 13.1 ▲
Alternatives to CFCs.

1 What is the chemical name for HFC-134a?

2 Draw *'dot-and-cross'* diagrams for these two compounds:
 a) CFC-12
 b) ammonia.

3 Suggest reasons why CFC-12 and ammonia have similar boiling points.

4 Suggest reasons why ammonia is much more chemically reactive than CFC-12.

5 What properties made CFC-12 suitable as a replacement for ammonia as a refrigerant?

6 Refrigerators and air-conditioning units keep the working fluid in a sealed system. Even so, the refrigerants continue to cause environmental problems. Suggest reasons why.

7 Suggest reasons why ammonia is now being reintroduced as a refrigerant.

8 The HCFCs and HFCs have been introduced as replacements for CFCs. They are now being phased out. Suggest why.

9 Carbon dioxide from burning fuels is an environmental problem, yet when it is used as a blowing agent or propellant, it is classified as having minimal global warming potential. How do you account for this?

10 Suggest disadvantages of using propane as an aerosol propellant for spray paints.

REVIEW QUESTIONS

Extension questions

1 A few drops of a halogenoalkane were added to 2 cm^3 of ethanol in a test tube, and 5 cm^3 of aqueous silver nitrate were then added. Finally, the test tube was placed in a water bath for a few minutes. A cream precipitate formed. This precipitate was soluble in concentrated ammonia.

a) Why was ethanol used in this experiment? **(1)**

b) Why was the test tube containing the mixture warmed in a hot water bath? **(1)**

c) What was the formula of the precipitate? Explain your answer. **(2)**

d) What can you conclude about the halogenoalkane? **(1)**

2 Classify the following conversions as: addition, elimination, substitution, oxidation, reduction, hydrolysis or polymerisation reactions. (Note that a reaction may belong to more than one category.)

a) Butan-2-ol to but-2-ene **(1)**

b) Butane to 1-bromobutane **(1)**

c) But-1-ene to 1,2-dichlorobutane **(1)**

d) Butanal to butanoic acid **(1)**

e) 1-bromobutane to butan-1-ol **(1)**

f) Buta-1,3-diene to synthetic rubber **(1)**

3 'Technologies based on science can provide us with many things that make our lives more comfortable and safer. Some technologies, however, have unintended and undesirable impacts. These need to be weighed against the benefits.' Illustrate and discuss this statement with examples based on the uses of CFCs. **(6)**

15 Instrumental analysis

In modern chemistry, the use of instruments for analysis is more important than analysis with chemical tests. Two of the techniques that chemists use frequently are infrared spectroscopy and mass spectrometry.

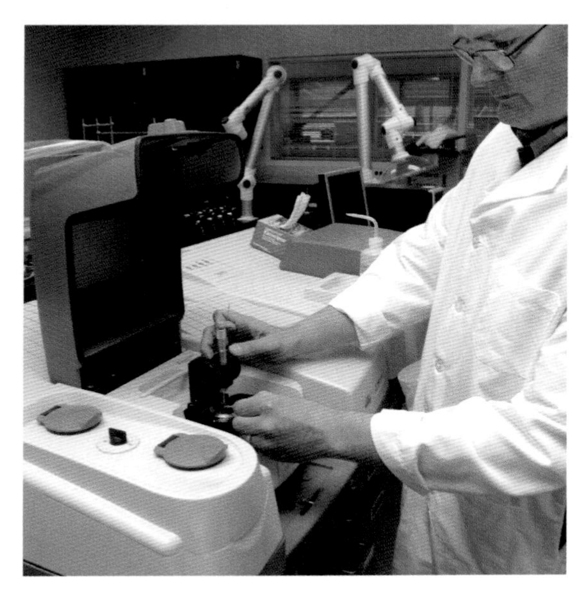

Figure 15.1 ◀
Scientists using an infrared spectrometer. The instrument covers a range of infrared wavelengths, and a detector records how strongly the sample absorbs at each wavelength. Wherever the sample absorbs, there is a dip in the intensity of the radiation transmitted, which shows up as a dip in the plot of the spectrum.

15.1 Infrared spectroscopy

Spectroscopy is a term that covers a range of practical techniques for studying the composition, structure and bonding of compounds. Spectroscopic techniques are now the essential 'eyes' of chemistry.

The range of techniques available covers many parts of the electromagnetic spectrum, including the infrared, visible and ultraviolet regions. The instruments used are either called spectroscopes (which emphasises the use of techniques for making observations) or spectrometers (which emphasises the importance of measurements).

Infrared spectroscopy is used to identify functional groups in organic molecules. Infrared radiation from a glowing lamp or fire makes us feel warm. This is because the frequencies of infrared radiation correspond to the natural frequencies of vibrating atoms in molecules. Our skin warms up as the molecules absorb infrared radiation and vibrate faster.

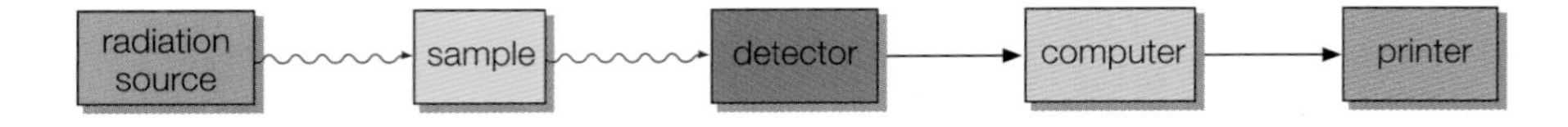

Figure 15.2 ◀
Essential features of a modern single-beam spectrometer.

Most compounds absorb infrared radiation. The wavelengths of the radiation they absorb correspond to the natural frequencies at which vibrating bonds in the molecules bend and stretch. However, it is only molecules that change their polarity as they vibrate that interact with infrared radiation.

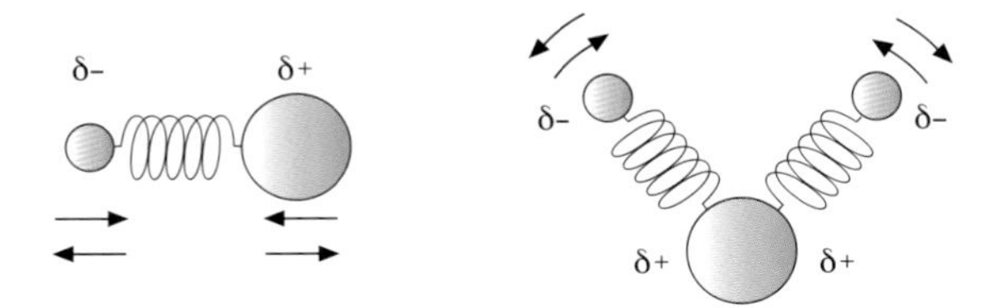

Figure 15.3 ◀
Bond vibrations give rise to absorptions in the infrared region. Vibrations of molecules that cause a fluctuating polarity interact with electromagnetic waves.

Definitions

An **absorption spectrum** is a plot that shows how strongly a chemical absorbs radiation over a range of frequencies.

Transmittance on the vertical axis of infrared spectra measures the percentage of radiation that passes through the sample. The troughs appear at the wavenumbers at which the compound absorbs strongly.

Bonds vibrate in particular ways and absorb radiation at specific wavelengths. This means that it is possible to look at an infrared spectrum and identify functional groups.

Spectroscopists have found that it is possible to correlate absorptions in the region 4000–1500 cm^{-1} with the stretching or bending vibrations of particular bonds. As a result, the infrared spectrum gives valuable clues to the presence of functional groups in organic molecules. The important correlations between different bonds and observed absorptions are shown in Figure 15.4. Hydrogen bonding broadens the absorption peaks of –OH groups in alcohols and even more so in carboxylic acids.

Note

Most organic molecules contain C–H bonds. As a result, most organic compounds have a peak in their infrared spectrum at around 3000 cm^{-1}.

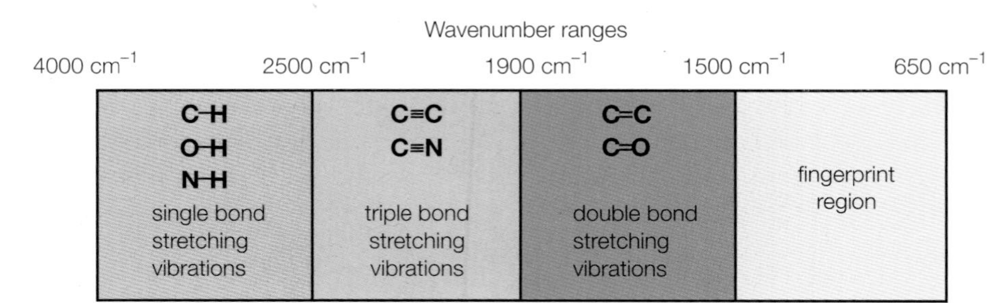

Figure 15.4 ▲
Main regions of the infrared spectrum and important correlations between bonds and observed absorptions.

Molecules with several atoms can vibrate in many ways because the vibrations of one bond affect others close to it. The complex pattern of vibrations can be used as a 'fingerprint' to be matched against the recorded infrared spectrum in a database.

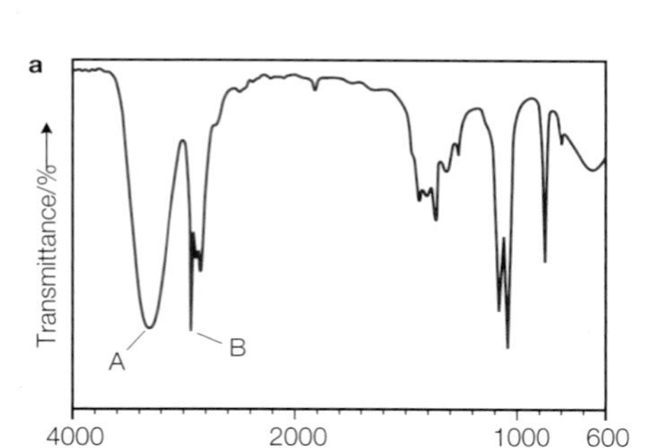

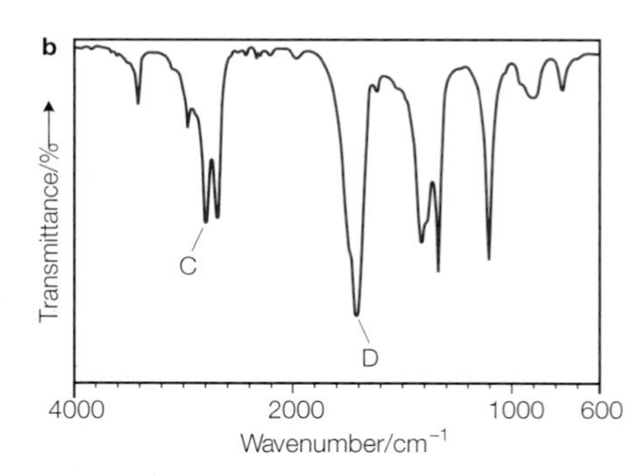

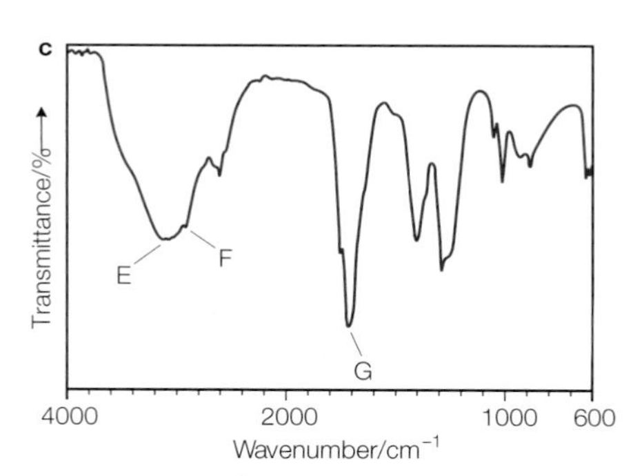

Figure 15.5 ▲
Infrared spectra for three organic compounds. Note that the quantity plotted on the vertical axis is transmittance, which means that the line dips at the wavenumbers at which the molecules absorb radiation. Chemists often refer to these dips in the line as 'peaks', because they indicate high levels of absorption.

Infrared spectroscopy is an analytical tool that can be used to monitor the progress of an organic synthesis. Comparing the spectrum of the final product with the known spectrum in a database can be used to check that the product is pure.

Definition

Infrared **wavenumbers** range from 400 cm^{-1} to 4500 cm^{-1}. Spectroscopists find the numbers more convenient than wavelengths. The wavenumber is the number of waves in 1 cm.

Test yourself

1 Why do the vibrations of O–H, C–O and C=O bonds show up strongly in infrared spectra, while C–C vibrations do not.

2 Figure 15.5 shows the infrared spectra of ethanol, ethanal and ethanoic acid.
   a) Which vibrations give rise to the peaks marked with the letters A–G?
   b) Which spectrum belongs to which compound?
   c) Why do all three spectra have a peak at around 3000 cm^{-1}?
   d) Why do two of the spectra have broad peaks at wavenumbers between 3000 and 3500?

3 Suggest reasons why it is better to use infrared spectroscopy to check the purity of a liquid product from a synthesis than to measure its boiling temperature.

Activity

Detecting alcohol

Infrared spectroscopy is one of the methods used to analyse breath samples from drivers suspected of drinking.

Alcohol from drinks is absorbed into the bloodstream through the walls of the stomach and intestines. The alcohol circulates with the blood through all parts of the body, including the lungs. It is slowly removed by the liver at a rate of about one unit an hour.

Alcohol moves from the blood into the breath in the lungs. Analysis of the blood and breath of a range of people after drinking has established that, on average, the ratio of the concentration of the alcohol in the blood to the breath is around 2300:1. This makes it possible to estimate someone's blood alcohol level by measuring the concentration of alcohol in their breath. The current legal limits for drivers in the United Kingdom correspond to 35 mg of alcohol per 100 cm^3 breath or 80 mg of alcohol per 100 cm^3 of blood.

Figure 15.6 ▲
Police use infrared instruments to test the breath of motorists in order to provide evidence of drink driving.

The police use test instruments that contain fuel cells for their roadside tests. In these devices, any alcohol in a driver's breath acts as fuel to produce an electrical voltage from the cell. The voltage is automatically converted by the instrument to a measure of the concentration of alcohol in the blood. These instruments are not used as evidence in court, but they allow the police to decide whether to take a driver to the police station for a further test. For many years, the accurate breath tests at police stations relied solely on infrared instruments. The newer machines combine infrared and fuel cell technologies.

The infrared spectrum of ethanol has strong peaks corresponding to the O–H and C–H bond vibrations (see Figure 15.5). The absorption at 2950 cm^{-1} corresponds to the C–H bond vibration and is used for analysis. The test instrument passes infrared radiation at 2950 cm^{-1} through a standard cell containing a sample of the driver's breath. The strength of the absorption is a measure of the alcohol concentration. A built-in computer processes the signal from the instrument, and the machine prints out the result.

1 A unit of alcohol in the UK is 10 cm^3 ethanol. Estimate the number of units of alcohol in:
   a) half a pint (284 cm^3) of lager that contains 5% by volume of alcohol
   b) a small glass (125 cm^3) of wine that contains 14% by volume of alcohol.

2 Show that the values of 80 mg of alcohol per 100 cm^3 of blood and 35 mg of alcohol per 100 cm^3 breath correspond to a concentration ratio of 2300:1.

3 Suggest reasons why:
   a) prosecution does not follow a breath test unless the breath alcohol concentration is at least 40 mg per 100 cm^3
   b) drivers are given the option of having a blood or urine test if the value is between 40 and 50 mg per 100 cm^3 but not if it is higher than this range.

4 Suggest reasons why the C–H peak in the infrared spectrum and not the O–H peak is used to analyse alcohol in breath.

5 Suggest a procedure for checking the accuracy of a breath test machine in a police station.

15.2 Mass spectrometry and the elements

Mass spectrometry provides an incredibly sensitive method of analysis not only for chemical research but also in areas such as space research, medical research, monitoring pollutants in the environment and the detection of illegal drugs in sport.

A mass spectrometer separates atoms and molecules according to their mass and also shows the relative numbers of the different atoms and molecules present (see Section 1.6). All mass spectrometers have the components shown in Figure 15.7.

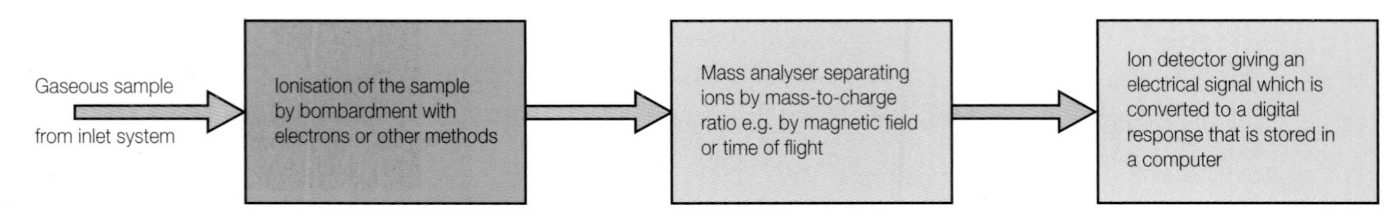

Figure 15.7 ▲
Schematic diagram showing the key features of a mass spectrometer.

There are five main types of mass spectrometer, which differ in the method used to separate ions with different mass-to-charge ratios. The type illustrated in Figure 1.22 (see Section 1.6) uses an electric field to accelerate ions into a magnetic field, which then deflects the ions onto the detector. A second type accelerates the ions and then separates them by their flight time through a field-free region. The so-called transmission quadrupole instrument is now the most common, because it is very reliable, compact and easy to use. It varies the fields in the instrument in a subtle way that allows ions with a particular mass-to-charge ratio to pass through to the detector at one time.

Inside a mass spectrometer is a high vacuum. This allows ionised atoms and molecules from the chemical being tested to be studied without interference from atoms and molecules in the air.

Definition

The **mass-to-charge ratio** (m/e) is the ratio of the relative mass of an ion to its charge, where e is the number of charges (1, 2 and so on). In all the examples that you will meet, e = 1.

Activity

Mass spectrometry in space research

In December 1978, space probes landed on Venus for the first time. Weight limitations meant that the low-resolution mass spectrometers on board could give measurements of relative masses to only one decimal place. A molecule of relative molecular mass 64.0 was identified, but the analysis could not show whether it was SO_2 or S_2.

Since then, mass spectrometers with higher resolutions have been developed. One reason for this is that space scientists are keen to explore whether or not there is, or ever has been, life on Mars. Scientists want to explore the north pole of the planet because evidence from surveys by orbiting spacecraft suggests that the pole is rich in ice just below the surface.

Beagle was a British-led effort to land a spacecraft on Mars in 2003 to look for signs of life. Like half of all missions to Mars, the project failed. In this case, it failed because the spacecraft crashed on landing. However, the mass spectrometers developed for the mission have been developed further and have been used in later projects, such as the Phoenix lander that was launched in 2007.

Instruments on spacecraft sent to Mars include a combination of high-temperature ovens and a mass spectrometer. These are designed to study ice and soil samples. After the spacecraft lands, a robotic arm digs a trench. It then scoops up samples and drops them into a hopper that feeds them into ovens no bigger than a ballpoint pen. As the oven temperatures increase up to 1000 °C, any gases can be mixed with oxygen and then carried into the mass spectrometer by a stream of gas. The mass spectrometer can detect and measure the amount of carbon dioxide formed by organic material burning or from decomposing minerals or released from gases trapped in rocks.

Figure 15.8 ▲
Artist's impression of the Phoenix lander on Mars just as it is beginning to dig a trench and gather samples for analysis by mass spectrometry.

Data from the mass spectrometer also allow scientists to determine ratios of various isotopes of hydrogen, oxygen, carbon and nitrogen. The ability to measure the ratios of carbon isotopes is key to the search for life, because photosynthesis is known to bring about a slight separation of these isotopes. During photosynthesis on Earth, plants show a slight preference for carbon-12 rather than carbon-13. This means that the proportion of the carbon-13 isotope is slightly lower than average in the chemicals in plants.

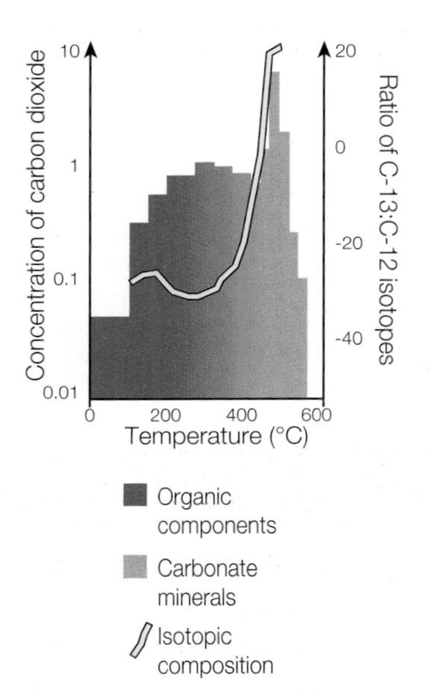

Figure 15.9 ▲
Results from a test with a mass spectrometer designed for exploration of Mars. The vertical scale on the left applies to the bar chart and shows the concentration of carbon dioxide in the gas collected from the sample. The scale on the right gives a measure of the carbon-13:carbon-12 ratio and applies to the yellow line. The test was carried out on a sample that contained organic and inorganic carbon.

1 Why could the mass spectrometers used in 1978 not show whether the molecule found on Venus was SO_2 or S_2? (S = 32, O = 16)

2 High-resolution mass spectrometers can measure relative masses to three, four or even five decimal places (S = 31.972, O = 15.995). Explain how a high-resolution mass spectrometer could determine whether a molecule is SO_2 or S_2.

3 What assumptions about the possibilities for life on Mars are built into the design of the methods of analysis for missions to Mars?

4 Suggest reasons for landing spacecraft to sample and test the soil at the north pole of Mars.

5 Suggest why it is necessary to dig below the surface to look for signs of life on Mars.

6 Refer to Figure 15.9 and the answer the following questions.
- **a)** Suggest a reason why the carbon from the organic component in the sample is detected at a lower temperature than carbon from carbonate minerals.
- **b)** Why does the ratio of carbon-13:carbon-12 increase sharply as the temperature rises above 400 °C?
- **c)** What would the plot look like from an analysis of a sample from Mars that contained carbonate minerals but no carbon from living things?

7 What data from the mass spectrometer results would provide evidence that there is, or has been, life on Mars?

15.3 Mass spectra of organic compounds

Mass spectrometry can also help to determine the relative molecular masses and molecular structures of organic compounds. In this way, it can be used to identify unknown compounds. The technique is extremely sensitive and requires very small samples, which can be as small as one nanogram (10^{-9} g).

Figure 15.10 ▶
The inside of a mass spectrometer. The sample is vaporised and ionised at the bottom left. The ions are accelerated and steered round the U-shaped tube by magnetic fields to the detector on the right.

A beam of high-energy electrons bombards the molecules of the sample. This turns them into ions by knocking out one or more electrons.

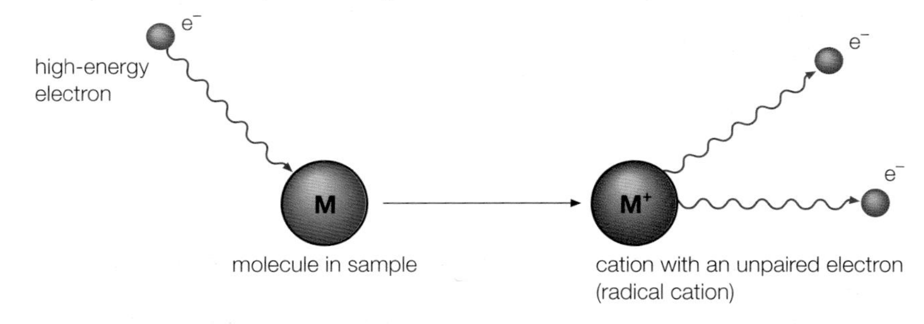

Figure 15.11 ▶
High-energy electrons ionising a molecule in a mass spectrometer. Knocking out one electron leaves a positive ion, hence the symbol M^+.

Bombarding molecules with high-energy electrons not only ionises them but usually splits them into fragments. As a result, the mass spectrum consists of a 'fragmentation pattern'.

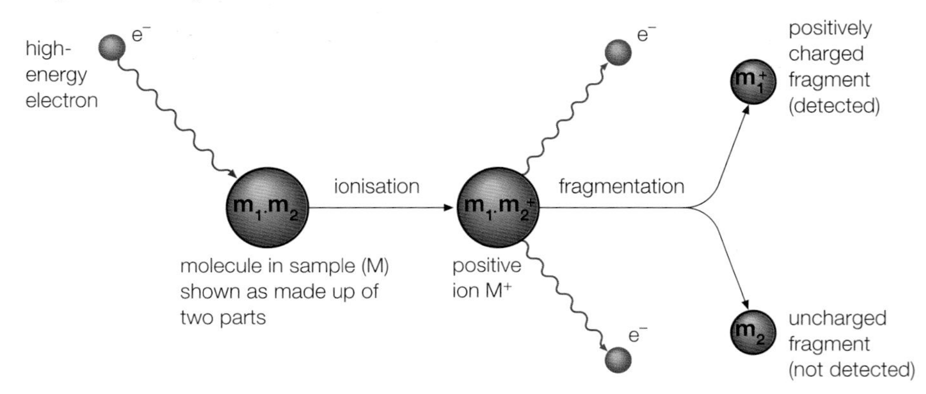

Figure 15.12 ▶
Ionisation and fragmentation of a single molecule $m_1.m_2$, which fragments into two parts: m_1^+ and m_2. Only charged species show up in the mass spectrum, because electric and magnetic fields have no effect on neutral fragments. In this case, therefore, the instrument detects only m_1^+.

Molecules break up more readily at weak bonds or at bonds that give rise to more stable fragments. It turns out that positive ions with the charge on a secondary or tertiary carbon atom are more stable than ions with the charge on a primary carbon atom. Species such as CH_3CO^+ are also more stable because of the presence of the C=O bond.

After ionisation and fragmentation, the charged species are separated to produce the mass spectrum. The extent of the deflection depends on the ratio of the mass of the fragment to its charge. When analysing molecular compounds, the peak of the ion with the highest mass is usually the whole molecule ionised. The mass of this 'parent ion', M, is therefore the relative molecular mass of the compound.

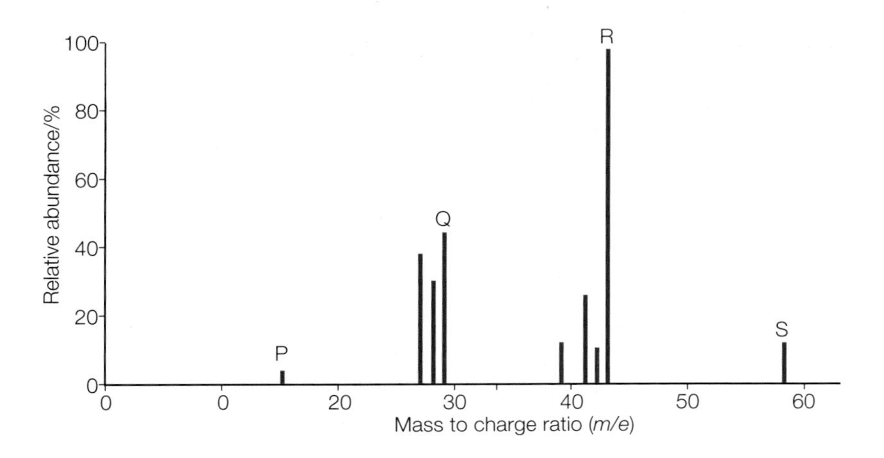

Figure 15.13 ◀ Mass spectrum of butane C_4H_{10} – the pattern of fragments is characteristic of this compound.

Chemists study mass spectra in order to gain insights into the structure of new molecules. They identify the fragments from their relative masses and then piece together likely structures with the help of evidence from other methods of analysis, such as infrared spectroscopy.

Chemists have now built up a very large database of mass spectra of known compounds for use in analysis. Chemists regard the spectra in the database as a set of 'fingerprints' for identifying chemicals during analysis. The computer of a mass spectrometer is programmed to search the database to find a good match between the spectrum of a compound being analysed and a spectrum in the database.

Test yourself

4 Give two reasons why it is important to have a high vacuum inside a mass spectrometer.

5 a) i) Which peak in the mass spectrum of butane in Figure 15.13 corresponds to the parent ion?

ii) What is the relative mass of this ion?

b) Suggest the identity of the fragments labelled P, Q and R.

c) Suggest a reason why the peak at *m/e* = 15 is relatively weak.

d) Use symbols to show one way in which the parent ion of butane could fragment.

REVIEW QUESTIONS

Extension questions

1 With the help of the data sheet, sketch and label diagrams to predict the pattern of the main peaks in the infrared spectra of the following compounds between wavenumbers 1500 cm^{-1} to 3500 cm^{-1}.

a) Cyclohexane (2)

b) 1-bromopropane (2)

c) Propanone (2)

2 Oxidation of butan-1-ol gives a product with the infrared spectrum shown in Figure 15.14. Use the data sheet to interpret the spectrum. State the reagents and conditions that were used to carry out the oxidation of the alcohol and give your reasons. (4)

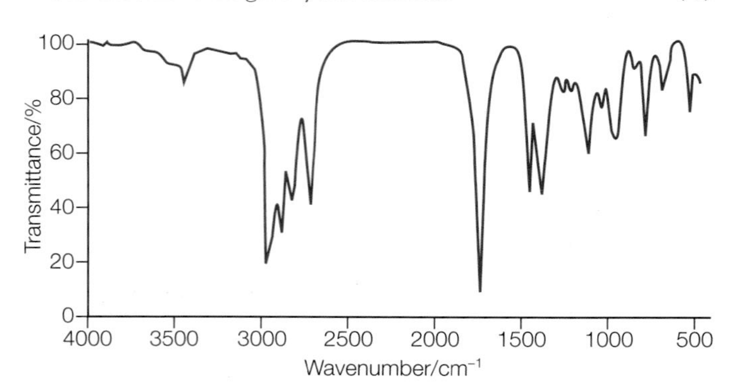

Figure 15.14 ▲

Data

Data

Practical guidance

3 Figure 15.15 shows the mass spectrum of ethanol

a) Match the numbered peaks with the formulae of these positive ions formed: $C_2H_5^+$, CH_2OH^+, $C_2H_5O^+$, $C_2H_5OH^+$ and $C_2H_3^+$. (3)

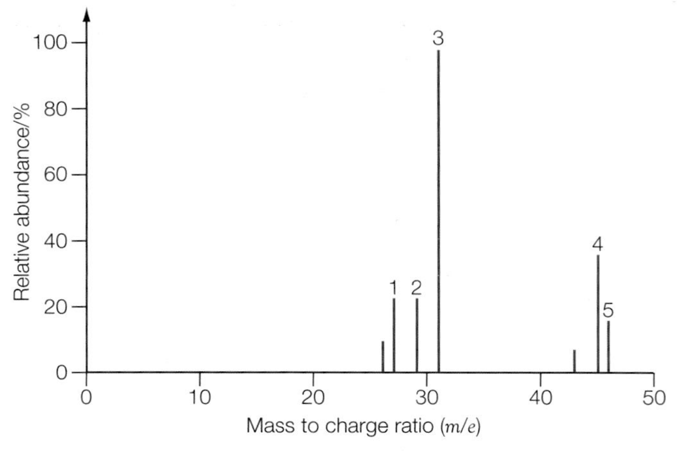

Figure 15.15 ▲

b) Write an equation to represent the formation of the molecular ions. (2)

c) i) Write an equation to show how the molecular ion fragments to give CH_2OH^+.

ii) Why does the other chemical species formed during this fragmentation process not show up in the mass spectrum? (2)

16 Enthalpy changes and energetics

In the natural environment, energy changes from the Sun create differences in temperature which stir the winds and vaporise water. The Sun also provides the energy for photosynthesis in plants, which produces the concentrated sources of chemical energy in foods. In many mechanical systems, the energy changes from burning fuels create the high temperatures needed to keep machines turning and vehicles moving. There is a close connection between these energy changes and chemical reactions.

16.1 Energy changes

The study of energy changes during chemical reactions is called thermochemistry. Thermochemistry is important because it helps chemists to measure the energy changes in chemical reactions and explain the stability of compounds. With the help of thermochemistry, chemists can also decide whether or not reactions are likely to occur.

The term 'system' is important in thermochemistry and it has a precise meaning. It describes just the chemical or the mixture of chemicals being studied. Everything around the system is called the surroundings (Figure 16.1). The surroundings include the apparatus, the air in the laboratory – in theory everything else in the Universe.

In a closed system like that in Figure 16.1, the system cannot exchange matter with its surroundings because the flask is closed with a bung. It can, however, exchange energy, such as heat, with the surroundings. If the bung is removed, the system is described as 'open'. An open system can exchange both energy and matter with its surroundings.

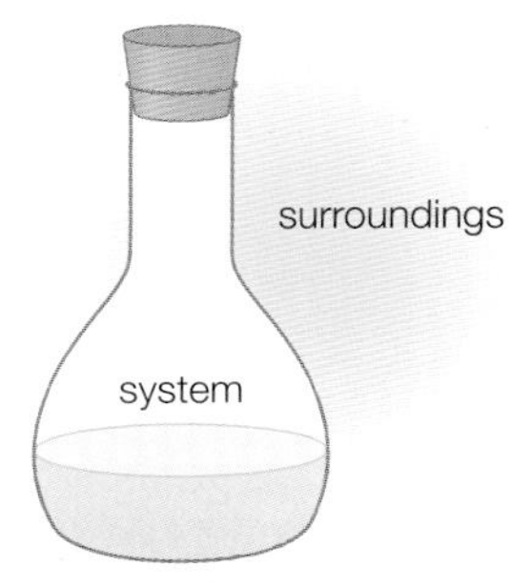

Figure 16.1 ▲
A system and its surroundings.

Definition

An **enthalpy change**, ΔH, is the overall energy exchanged with the surroundings when a change happens at constant pressure and the final temperature is the same as the starting temperature.

16.2 Enthalpy changes

Whenever a change occurs in a system, there is almost always an energy change involving transfer of energy between the system and its surroundings.

The energy transferred between a system and its surroundings is described as an enthalpy change when the change happens at constant pressure. The symbol for an enthalpy change is ΔH and its units are kJ mol^{-1}.

Note

Scientists use the capital Greek letter 'delta', Δ, for a change or difference in a physical quantity. So, ΔH means change in enthalpy and ΔT means change in temperature.

Exothermic and endothermic changes

Exothermic changes give out energy to the surroundings. In an **ex**othermic change, energy leaves the system, just as people leave a building by the **ex**it.

Burning is an obvious exothermic chemical reaction. Respiration is another exothermic reaction in which foods are oxidised to provide energy for living things to grow and move. Hot packs used in self-warming drinks and in treating painful rheumatic conditions also involve exothermic reactions (Figure 16.2).

Figure 16.2 ◄
A self-warming can of coffee – pressing the bulb on the bottom of the can starts an exothermic reaction in a sealed compartment. The energy released heats the coffee.

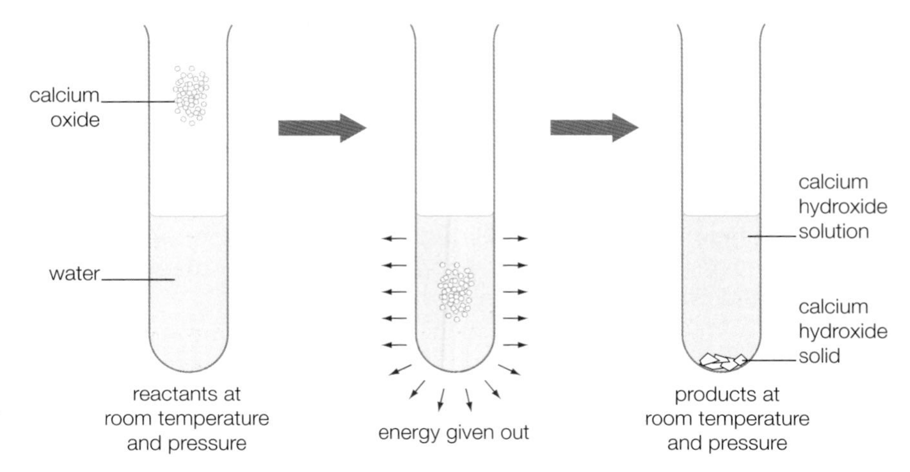

Figure 16.3 ▶
The exothermic reaction that takes place in some hot packs.

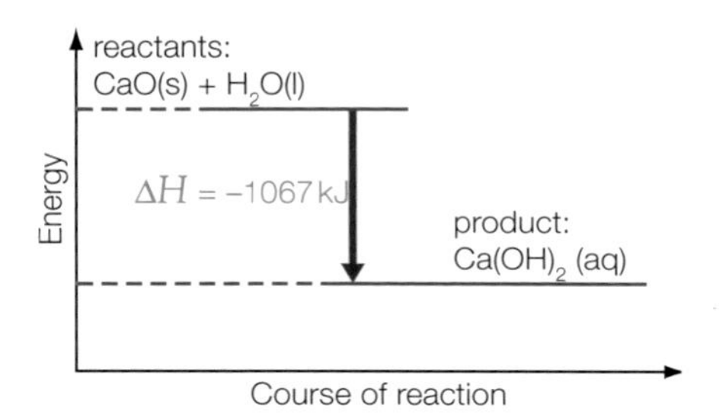

Figure 16.4 ▲
An enthalpy profile diagram for the reaction of calcium oxide with water.

Figure 16.3 shows what happens in the exothermic reaction between calcium oxide and water. This reaction is used in hot packs such as the self-warming can of coffee in Figure 16.2.

When one mole of solid calcium oxide reacts with water to form calcium hydroxide solution, 1067 kJ of heat are given out. The system has lost energy (heat) to the surroundings. This loss of energy from the system means that ΔH is negative and we can write:

$$CaO(s) + H_2O(l) \rightarrow Ca(OH)_2(aq) \qquad \Delta H = -1067\ \text{kJ mol}^{-1}$$

The energy changes in chemical reactions can be summarised in enthalpy profile diagrams.

Figure 16.4 shows the enthalpy profile diagram for the reaction of calcium oxide with water. Energy is lost to the surroundings as heat and therefore the products are at a lower energy level than the reactants. For this and all other exothermic reactions, ΔH is negative.

Endothermic changes take in energy from the surroundings. They are the opposite of exothermic changes. Melting and vaporisation are endothermic changes of state. Photosynthesis is an endothermic chemical change. During photosynthesis, plants take in energy from the Sun in order to convert carbon dioxide and water to glucose. Figure 16.5 illustrates the use of an endothermic reaction in a cold pack.

An enthalpy profile diagram shows that the system has more energy after an endothermic reaction than it had at the start. So, for endothermic reactions, the enthalpy change, ΔH, is positive and the products are at a higher energy level than the reactants (Figure 16.6).

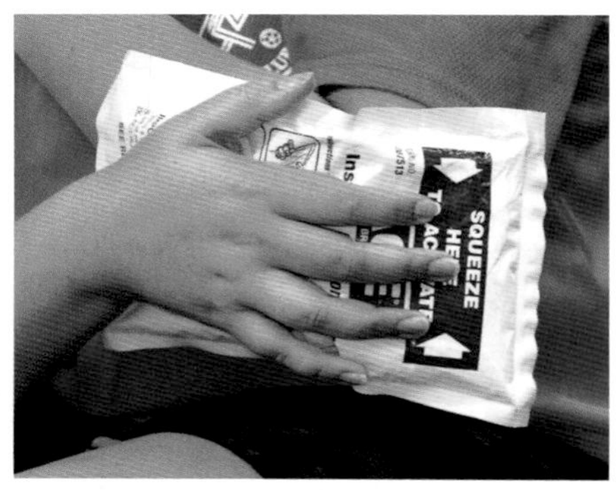

Figure 16.5 ▲
Twist a cold pack and it gets cold enough to reduce the pain of a sports injury. When chemicals in the cold pack react, they take in energy as heat and the pack gets cold. This, in turn, cools the sprained or bruised area and helps to reduce painful swelling.

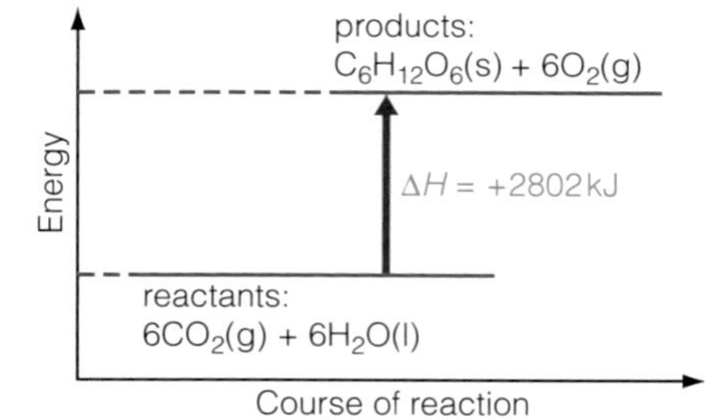

Figure 16.6 ▲
An enthalpy profile diagram for photosynthesis.

Note

Enthalpy profile diagrams are often called enthalpy level diagrams.

Test yourself

1 Which of the following changes are exothermic and which are endothermic?
   a) melting ice   b) burning wood   c) condensing steam
   d) metabolising sugar   e) subliming iodine.
2 When 1 mole of carbon burns completely, 394 kJ of heat is given out.
   a) Write an equation for the reaction including state symbols and show the value of the enthalpy change.
   b) Draw an enthalpy profile diagram for the reaction including the enthalpy change.
3 When 0.2 g of methane, CH_4 (natural gas), was burned completely, it produced 11 000 J.
   a) Write an equation for the reaction when methane burns completely.
   b) Calculate the molar mass of methane.
   c) How much heat is produced when 1 mole of methane burns completely?
   d) Draw an enthalpy profile diagram for the reaction showing the value of the enthalpy change.

Enthalpy profile diagrams and activation energy

During a chemical reaction, bonds are first broken and others are then formed. This means that energy is needed to break bonds in the reactants and start the process for both endothermic and exothermic reactions. In many cases, the reacting atoms and molecules do not have sufficient energy in themselves for the necessary bonds to be broken. The reaction will only go if energy is provided for the system, usually in the form of heat.

The minimum energy needed by the reactants if they are to react is called the activation energy of the reaction. The symbol for an activation energy is E_A and its units are $kJ\ mol^{-1}$.

The fact that a minimum amount of energy is needed to initiate a reaction is well illustrated by fuels and explosives. These chemicals usually need a small input of energy to get the reaction started, even though their reactions are highly exothermic overall (Figure 16.7).

Definition

The **activation energy** of a reaction is the minimum energy needed by the amounts of reactants shown in the equation if they are to react.

The activation energy is a kind of energy barrier for a reaction to occur.

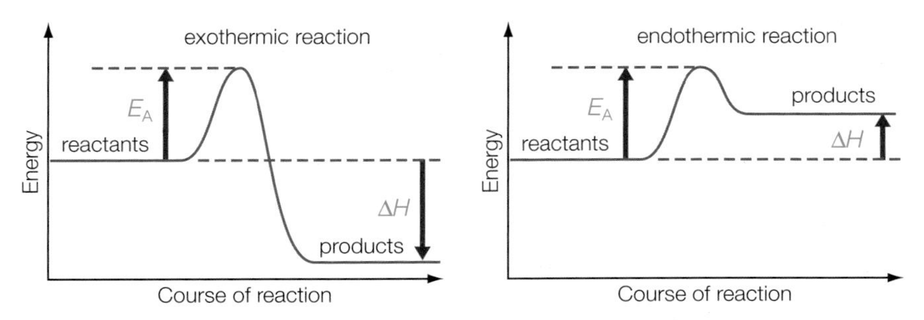

Figure 16.7 ▲
Energy profile diagrams showing the activation energy for both exothermic and endothermic reactions.

16.3 Measuring enthalpy changes

The heat given out or taken in during many chemical reactions can be measured fairly conveniently and this allows us to determine enthalpy changes.

Enthalpy changes from burning fuels

Figure 16.8 shows the simple apparatus that could be used to measure the heat given out when a liquid fuel like meths burns. Wear eye protection if you try this experiment and remember that liquid fuels are highly flammable.

In the experiment we will assume that *all* the energy produced from the burning meths heats up the water. The results from one experiment are shown in Table 16.1.

- From the mass of water in the can and its temperature rise, we can work out the energy produced.
- From the loss in mass of the liquid burner, we can find the mass of meths which has burned.
- We can then calculate the energy produced when 1 g of the fuel burns.

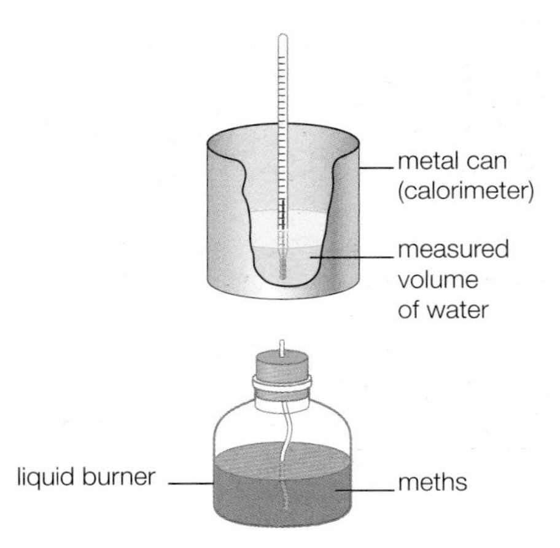

Figure 16.8 ▲
Measuring the enthalpy change when meths is burned.

Mass of burner + meths at start of experiment	= 271.8 g
Mass of burner + meths at end of experiment	= 271.3 g
∴ Mass of meths burned	= 0.5 g
Volume of water in can	= 250 cm^3
∴ Mass of water in can	= 250 g
Rise in temperature of water	= 10 °C = 10 K

Table 16.1 ▲

The specific heat capacity of water is $4.2\ J\ g^{-1}\ K^{-1}$. This means that:

4.2 J will raise the temperature of 1 g of water by 1 K (1 °C)
∴ $m \times 4.2$ J will raise the temperature of m g of water by 1 K (1 °C)
and $m \times 4.2 \times \Delta T$ J will raise the temperature of m g of water by ΔT K (ΔT °C)

In general, if $\mathbf{m}$ g of a substance with a specific heat capacity of $\mathbf{c}$ $J\ g^{-1}\ K^{-1}$ changes in temperature by $\Delta \mathbf{T}$ K,

$$\textbf{Energy change/J} = \mathbf{m}/\text{g} \times \mathbf{c}/\text{J g}^{-1}\text{K}^{-1} \times \Delta \mathbf{T}/\text{K}$$

> **Definition**
>
> The **specific heat capacity** of a material, c, is the energy needed to raise the temperature of 1 g of the material by 1 K. For water $c = 4.2\,J\,g^{-1}\,K^{-1}$.

Using this equation with the results of our experiment:

$$0.5\,g \text{ of burning meths transfers } 250\,g \times 4.2\,J\,g^{-1}\,K^{-1} \times 10\,K = 10\,500\,J \text{ to the water}$$

$$\therefore 1\,g \text{ of meths produces } 21\,000\,J = 21\,kJ$$

Meths is mainly ethanol, C_2H_5OH. If we assume that meths is pure ethanol ($M_r = 46$), then:

$$\text{the heat produced when 1 mole of ethanol burns} = 46 \times 21\,kJ = 966\,kJ$$

So we can write:

$$C_2H_5OH(l) + 3O_2(g) \rightarrow 2CO_2(g) + 3H_2O(l) \qquad \Delta H = -966\,kJ\,mol^{-1}$$

The energy transferred when one mole of a fuel burns completely is usually called the enthalpy change of combustion of the fuel and given the symbol ΔH_c. Therefore:

$$\Delta H_c[C_2H_5OH(l)] = -966\,kJ\,mol^{-1}$$

Assumptions and errors in thermochemical experiments

In the experiment just described, we assumed that all the heat (energy) from the flame heats the water. In practice a proportion of the energy heats the metal can and the surrounding air. In addition, the flame is affected by draughts and sometimes the fuel burns incompletely, leaving soot on the bottom of the metal can. Our initial assumption is clearly flawed and our result is certainly inaccurate. These major sources of error (loss of heat, flame disturbance and incomplete combustion) all reduce the energy transferred to the water. This leads to a result that is lower than the true value.

Accurate values for enthalpy changes of combustion are obtained using a bomb calorimeter (Figure 16.9). The apparatus is specially designed to ensure that the sample burns completely and that heat losses are avoided. A measured amount of the sample burns in excess oxygen under pressure. There must be enough oxygen to ensure that all carbon in the compound is fully oxidised to carbon dioxide and no carbon monoxide or soot are produced.

Heat losses are eliminated altogether by coupling the thermochemical experiment with an electrical calibration. After the chemical reaction has been carried out, the experiment is repeated, but this time an electrical heating coil replaces the reactants. The current in the coil is continually adjusted to give the same temperature rise in the same time as the chemical reaction. By recording the current during the time of this electrical

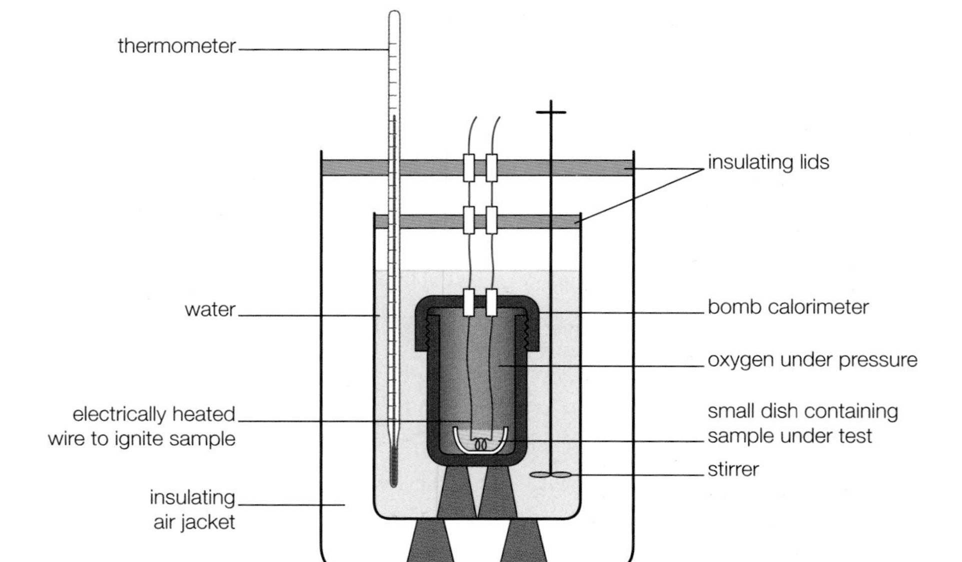

Figure 16.9 ▶ A bomb calorimeter.

calibration, it is possible to calculate very accurately the electrical energy supplied. This is the same as the energy transferred in the chemical reaction.

Enthalpy changes of combustion are important in the fuel and food industries (Figure 16.10). Scientists use bomb calorimeters like that in Figure 16.9 to determine the energy values of fuels and foods. The prices of fuels are closely related to their energy values and dieticians give advice related to their knowledge of energy-providing foods.

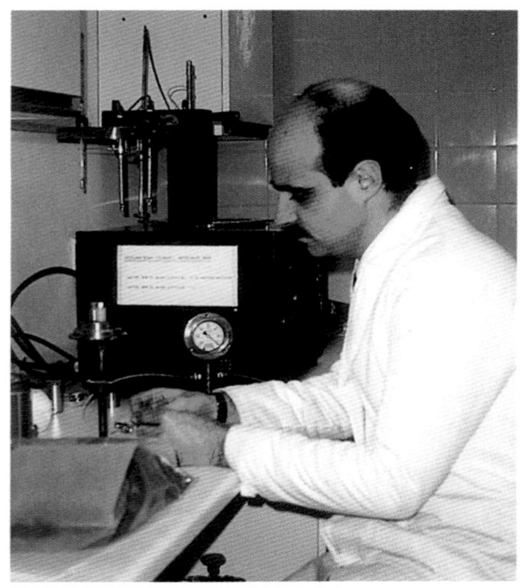

Figure 16.10 ▲
Scientists use bomb calorimeters like the one in this photo to measure the energy produced by different fuels and foods.

Enthalpy changes in solution

Enthalpy changes for reactions in solution can be measured using insulated plastic containers such as polystyrene cups (Figure 16.11). Polystyrene is an excellent insulator and it has a negligible specific heat capacity.

If the reaction is exothermic, the energy released cannot escape to the surroundings so it heats up the solution. If the reaction is endothermic, no energy can enter from the surroundings so the solution cools. If the solutions are dilute, it is sufficiently accurate to calculate the enthalpy changes by assuming that the solutions have the same density and specific heat capacity as water.

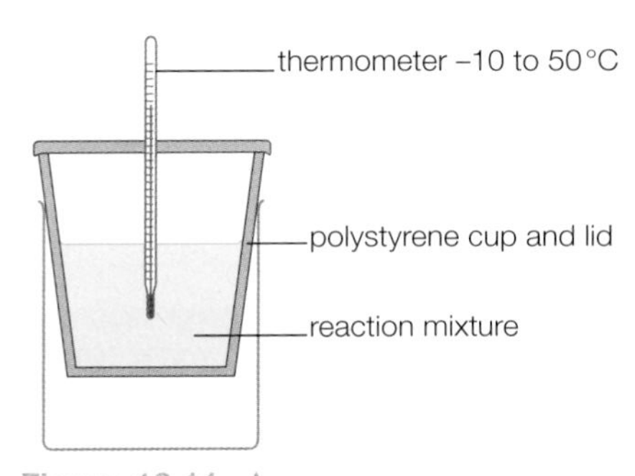

Figure 16.11 ▲
Measuring the enthalpy change of a reaction in solution.

Worked example

When 4.0 g of ammonium nitrate, NH_4NO_3, dissolves in 100 cm^3 of water, the temperature falls by 3.0 °C. Calculate the enthalpy change per mole when NH_4NO_3 dissolves in water under these conditions.

Answer

$$\text{Amount of } NH_4NO_3 \text{ used} = \frac{\text{mass}}{M_r(NH_4NO_3)} = \frac{4\,\text{g}}{80\,\text{g mol}^{-1}} = 0.05\,\text{mol}$$

Assuming that the solution has the same density and specific heat capacity as water:

Energy taken in from the solution

$$= \text{mass} \times \text{specific heat capacity} \times \text{temperature change}$$
$$= 100\,\text{g} \times 4.2\,\text{J kg}^{-1}\text{K}^{-1} \times 3\,\text{K}$$
$$= 1260\,\text{J}$$

$$\therefore \text{Energy taken in per mole of } NH_4NO_3 = \frac{1260\,\text{J}}{0.05\,\text{mol}} = 25\,200\,\text{J mol}^{-1}$$
$$= 25\,\text{kJ mol}^{-1}$$

The reaction is endothermic, so the enthalpy change for the system is positive.

$$NH_4NO_3(s) \xrightarrow{(aq)} NH_4NO_3(aq) \qquad \Delta H = +25\,\text{kJ mol}^{-1}$$

Test yourself

4 Burning butane, C_4H_{10}, from a Camping Gaz® container raised the temperature of 200 g water from 18 °C to 28 °C. The Gaz container was weighed before and after and the loss in mass was 0.29 g. Estimate the molar enthalpy change of combustion of butane.

5 On adding 25 cm^3 of 1.0 mol dm^{-3} nitric acid to 25 cm^3 1.0 mol dm^{-3} potassium hydroxide in a plastic cup, the temperature rise is 6.5 °C.
  a) Write an equation for the reaction.
  b) Calculate the enthalpy change for the neutralisation reaction per mole of nitric acid.

6 On adding excess powdered zinc to 25.0 cm^3 of 0.2 mol dm^{-3} copper(II) sulfate solution, the temperature rises by 9.5 °C.
  a) Write an equation for the reaction.
  b) Calculate the enthalpy change of the reaction for the molar amounts in the equation.

Activity

Measuring and evaluating the enthalpy change for the reaction of zinc with copper(II) sulfate solution

Two students decided to measure the enthalpy change for the reaction between zinc and copper(II) sulfate solution.

$$Zn(s) + CuSO_4(aq) \rightarrow ZnSO_4(aq) + Cu(s)$$

The method they used is shown in Figure 16.12 and their results are shown in Table 16.2. After adding the zinc, it took a little while for the temperature to reach a peak and then the mixture began to cool.

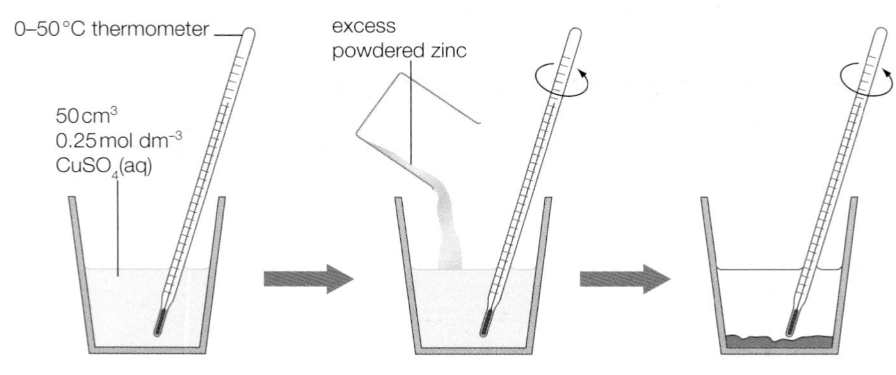

Figure 16.12 Measuring the enthalpy change for the reaction of zinc with copper(II) sulfate solution.

Table 16.2

Time/min	Temperature /°C	Time/min	Temperature /°C	Time/min	Temperature /°C
0	24.1	3.5	34.2	6.5	33.7
0.5	24.0	4.0	34.8	7.0	33.6
1.0	24.1	4.5	35.0	7.5	33.5
1.5	24.1	5.0	34.6	8.0	33.4
2.0	24.2	5.5	34.2	8.5	33.2
2.5	24.1	6.0	33.9	9.0	33.1
3.0	–				

1 Plot a graph of temperature (vertically) against time (horizontally) using the results in Table 16.2.

2 Extrapolate the graph backwards from 9 minutes to 3 minutes, as shown in Figure 16.13, in order to estimate the maximum temperature. This assumes that all the zinc reacted at once and there was no loss of heat to the surroundings.

a) What is the estimated maximum temperature at 3 minutes?
b) What is the temperature rise, ΔT, for the reaction?

3 Calculate the energy given out during the reaction using the equation:

Energy transferred = mass × specific heat capacity × temperature change

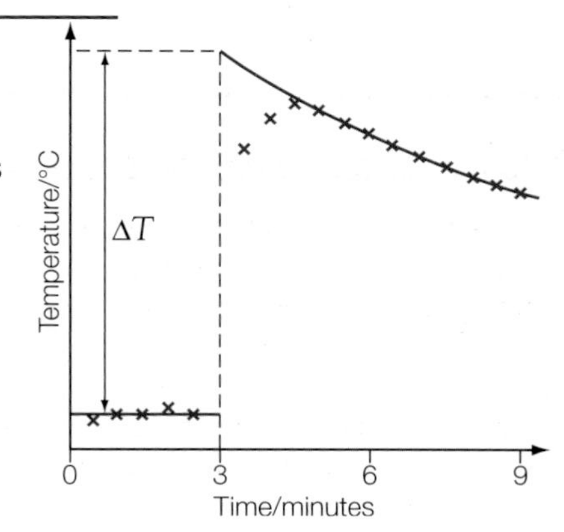

Figure 16.13 Estimating the maximum temperature of the mixture when zinc reacts with copper(II) sulfate solution.

Assume that:

- all the heat is transferred to the solution in the polystyrene cup
- the density of the solution is 1 g cm^{-3}
- the specific heat capacity of the solution is 4.2 J g^{-1} K^{-1}.

4 How many moles of each chemical reacted?

a) $CuSO_4$ **b)** Zn

5 What is the enthalpy change of the reaction, $\Delta H_{reaction}$, for the amounts of Zn and $CuSO_4$ in the equation?
(State the value of $\Delta H_{reaction}$ in kJ mol^{-1} with the correct sign.)

Practical guidance

6 Copy and complete Table 16.3 for the various measurements in the experiment.

Measurement	Value	Uncertainty	Percentage uncertainty
Concentration of copper(II) sulfate solution			
Volume of copper(II) sulfate solution measured from a 100 cm^3 measuring cylinder			
Temperature rise, ΔT, estimated from the difference in two readings taken with a 0–50 °C thermometer			

Table 16.3 ◀ The values, uncertainties and percentage uncertainties of measurements in the experiment.

7 What is the total percentage uncertainty in the experiment?

8 What is the total uncertainty in the value you have calculated for the enthalpy change?

9 Write a value for the enthalpy change showing the uncertainty as a ± amount.

10 What are the main sources of error in the measurements and procedure for the experiment?

11 Look critically at the procedures in the experiment and suggest improvements to minimise errors and increase the reliability of the result.

16.4 Standard enthalpy changes

Our studies in this topic have already shown that it is only energy changes and enthalpy changes that can be measured and not absolute levels of energy or enthalpy in a system. It is, however, very useful to compare enthalpy changes.

In order to compare different enthalpy changes accurately, it is important that the conditions under which measurements and experiments are carried out are the same in all cases. These conditions for the accurate comparison of enthalpy changes and other thermochemistry measurements are called standard conditions.

Standard conditions are:

- a pressure of 1 atmosphere (10^5 Pa = 100 kPa)
- a temperature of 298 K (25 °C)
- substances in their standard (most stable) state at 1 atm pressure and 298 K
- solutions with a concentration of 1 mol dm^{-3}.

Definition

The **standard enthalpy change of a reaction**, $\Delta H^{\ominus}_{reaction}$, is the energy transferred when the molar quantities of reactants as stated in the equation react under standard conditions.

Any enthalpy change measured under these conditions is described as a standard enthalpy change and given the symbol $\Delta H^{\ominus}_{298}$ or simply $\Delta H^{\ominus}$, pronounced 'delta *H* standard'.

In thermochemistry, it is important to specify the states of the substances and therefore to include state symbols in equations. So, $\Delta H^{\ominus}$ for the reaction:

$$2H_2(g) + O_2(g) \rightarrow 2H_2O(l)$$

must relate to hydrogen gas, oxygen gas and liquid water (not steam). The states of elements and compounds must also be the most stable at 298 K and 1 atmosphere. Thus, $\Delta H^{\ominus}$ measurements involving carbon should use graphite, which is energetically more stable than diamond.

Worked example

When $50\ cm^3$ of $2.0\ mol\ dm^{-3}$ hydrochloric acid is mixed with $50\ cm^3$ of $2.0\ mol\ dm^{-3}$ sodium hydroxide in a bomb calorimeter at 25 °C and 1 atmosphere pressure, the temperature rises by 13.7 °C. What is the enthalpy change of the reaction, $\Delta H_{reaction}$?

Answer

The equation for the reaction is:

$$HCl(aq) + NaOH(aq) \rightarrow NaCl(aq) + H_2O(l)$$

Amount of HCl used = amount of NaOH used

$$= \frac{50\ dm^3}{1000} \times 2.0\ mol\ dm^{-3} = 0.1\ mol$$

Energy given out and used to heat the solution

$$= 100\ g \times 4.2\ J\ g^{-1}\ K^{-1} \times 13.7\ K = 5754\ J$$

$$\therefore \text{Energy given out per mole of acid} = \frac{5754\ J}{0.1\ mol} = 57\,540\ J\ mol^{-1}$$

$$\therefore \Delta H_{reaction} = -57.5\ kJ\ mol^{-1}$$

Standard enthalpy changes of combustion

For combustion reactions, there is a special name for the enthalpy change of reaction – it is called the enthalpy change of combustion. The standard enthalpy change of combustion of an element or compound, $\Delta H^{\ominus}_c$, is the enthalpy change when one mole of the substance burns completely in oxygen using standard conditions. The substance and the products of burning must be in their stable (standard) states. For a carbon compound, complete combustion means that all the carbon burns to carbon dioxide and that there is no soot or carbon monoxide. If the substance contains hydrogen, the water formed must end up as liquid and not as a gas.

Values of enthalpies of combustion are much easier to measure than many other enthalpy changes. They can be calculated from measurements taken with a bomb calorimeter (Section 16.3).

Chemists use two ways to summarise standard enthalpy changes of combustion. One way is to write the equation with the enthalpy change alongside it. So, for the standard enthalpy change of combustion of carbon, they write:

$$C_{(graphite)} + O_2(g) \rightarrow CO_2(g) \qquad \Delta H^{\ominus}_c = -393.5\ kJ\ mol^{-1}$$

The other way is to use a shorthand form. For the standard enthalpy change of combustion of methane, this is written as:

$$\Delta H^{\ominus}_c[CH_4(g)] = -890\ kJ\ mol^{-1}$$

Remember that all combustion reactions are exothermic, so $\Delta H^{\ominus}_c$ values are always negative.

Definition

The **standard enthalpy change of combustion** of a substance, $\Delta H^{\ominus}_c$, is the enthalpy change when one mole of the substance completely burns in oxygen under standard conditions with the reactants and products in their standard states.

Standard enthalpy changes of formation

Another important enthalpy change for any compound is its enthalpy change of formation. The standard enthalpy change of formation of a compound, $\Delta H_f^\ominus$, is the enthalpy change when one mole of the compound forms from its elements. The elements and the compound formed must be in their stable standard states. The more stable state of an element is chosen where there are allotropes (different forms in the same state) such as graphite and diamond.

As with standard enthalpies of combustion, there are two ways of representing standard enthalpy changes of formation.

One way is to write the equation with the enthalpy change alongside it. For the standard enthalpy change of formation of water this is:

$$H_2(g) + \tfrac{1}{2}O_2(g) \rightarrow H_2O(l) \qquad \Delta H_f^\ominus = -286\,\text{kJ mol}^{-1}$$

The other way is to use shorthand. For the standard enthalpy change of formation of ethanol this is:

$$\Delta H_f^\ominus [C_2H_5OH(l)] = -277\,\text{kJ mol}^{-1}$$

Like all thermochemical quantities, the precise definition of the standard enthalpy change of formation is important. Books of data tabulate values for standard enthalpies of formation. These tables are very useful because they make it possible to calculate the enthalpy changes for many reactions (Section 16.5).

Unfortunately, it is difficult to measure some enthalpy changes of formation directly. For example, it is impossible to convert the elements carbon, hydrogen and oxygen straight to ethanol under any conditions. Because of this, chemists have had to find an indirect method of measuring the standard enthalpy change of formation of ethanol and many other compounds (Section 16.5).

One important consequence of the definition of standard enthalpy changes of formation is that, for an element, $\Delta H_f^\ominus = 0\,\text{kJ mol}^{-1}$. This is because there is no change in substance or state when an element forms from itself, and therefore no enthalpy change. In other words, the standard enthalpy change of formation of an element is zero. So:

$$\Delta H_f^\ominus [Cu(s)] = 0 \quad \text{and} \quad \Delta H_f^\ominus [O_2(g)] = 0$$

> **Definition**
>
> The **standard enthalpy change of formation** of a compound, $\Delta H_f^\ominus$, is the enthalpy change when one mole of the compound forms from its elements under standard conditions with the elements and the compound in their standard (stable) states.

16.5 Hess's Law and the indirect determination of enthalpy changes

Practical guidance

The enthalpy change of a reaction is the same whether the reaction happens in one step or in a series of steps. As long as the reactants and products are the same, the overall enthalpy change is the same whether the reactants are converted straight to products or through two or more intermediates. This is Hess's Law. In Figure 16.14 the enthalpy change for route 1 and the overall enthalpy change for route 2 are the same.

Hess's Law is an example of a mathematical model. This is shown by the precise quantitative relationship between ΔH_1, ΔH_2, ΔH_3 and ΔH_4 in Figure 16.14. Using Hess's Law it is possible to bring together data and calculate enthalpy changes which cannot be measured directly by experiment. So, Hess's Law can be used to calculate:

- standard enthalpy changes of reaction from standard enthalpy changes of combustion
- standard enthalpy changes of reaction from standard enthalpy changes of formation.

> **Definition**
>
> **Hess's Law** says that the enthalpy change in converting reactants to products is the same regardless of the route taken, provided the initial and final conditions are the same.

Hess's Law is a chemical version of the law of conservation of energy. Suppose the enthalpy change for route 1 in Figure 16.14 was more exothermic than the total enthalpy change for route 2. It would be possible to go round the cycle in Figure 16.14 from A to D direct and back to A via C and B, ending up with the same starting chemical but with a net release of energy. This would contravene the law of conservation of energy.

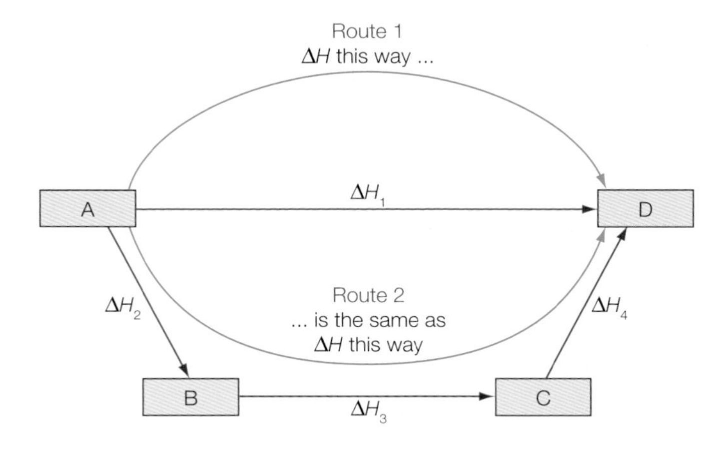

Figure 16.14 ▶ A diagram to illustrate Hess's Law:

$\Delta H_1 = \Delta H_2 + \Delta H_3 + \Delta H_4$

Note

Reversing the direction of a reaction reverses the sign of ΔH.

Enthalpies of reaction from enthalpies of combustion

Figure 16.15 shows the form of an enthalpy cycle which we can use to calculate enthalpy changes of reaction from enthalpy changes of combustion.

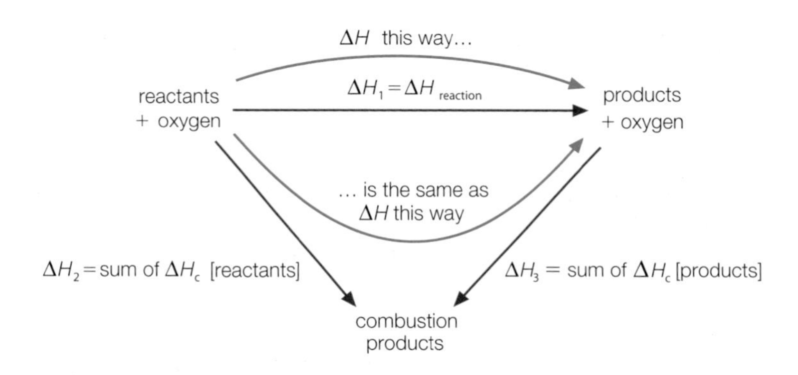

Figure 16.15 ▶ An enthalpy cycle for calculating standard enthalpies of reaction from standard enthalpies of combustion:

$\Delta H_1 = \Delta H_2 - \Delta H_3$

Worked example

Catalytic cracking (Section 11.3) can now be used to convert ethane to ethene and hydrogen. Calculate the standard enthalpy change of this reaction given the following standard enthalpies of combustion:

ethane, $\Delta H_c^{\ominus}[C_2H_6(g)] = -1560\,kJ\,mol^{-1}$
ethene, $\Delta H_c^{\ominus}[C_2H_4(g)] = -1411\,kJ\,mol^{-1}$
hydrogen, $\Delta H_c^{\ominus}[H_2(g)] = -286\,kJ\,mol^{-1}$

Notes on the method
Draw up an enthalpy cycle using the model in Figure 16.15. Use Hess's Law to produce an equation linking the relevant enthalpy changes – pay careful attention to the signs. Put the value and sign for a quantity in brackets when adding or subtracting enthalpy values.

Answer
An enthalpy cycle linking the cracking of ethane with its combustion and the combustion of the products of cracking (ethene and hydrogen) is shown in Figure 16.16. Sometimes enthalpy cycles like the one in Figure 16.16 are called Hess cycles.

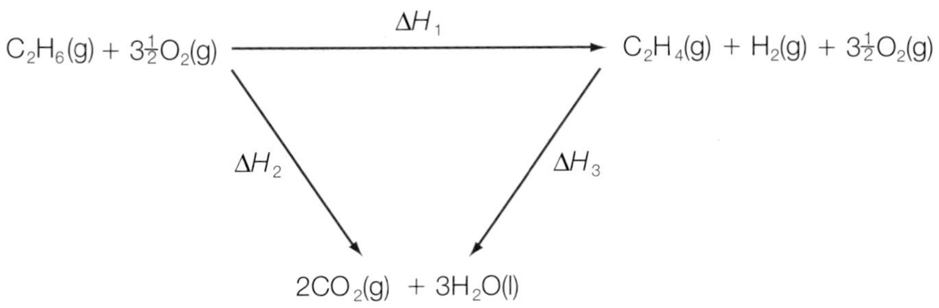

Figure 16.16

$$C_2H_6(g) + 3\tfrac{1}{2}O_2(g) \xrightarrow{\Delta H_1} C_2H_4(g) + H_2(g) + 3\tfrac{1}{2}O_2(g)$$

ΔH_2: $C_2H_6(g) + 3\tfrac{1}{2}O_2(g) \rightarrow 2CO_2(g) + 3H_2O(l)$

ΔH_3: $C_2H_4(g) + H_2(g) + 3\tfrac{1}{2}O_2(g) \rightarrow 2CO_2(g) + 3H_2O(l)$

According to Hess's Law, $\Delta H_1 = \Delta H_2 - \Delta H_3$

$\Delta H_1 = \Delta H^{\ominus}_{reaction}$

$\Delta H_2 = \Delta H^{\ominus}_c[C_2H_6(g)] = -1560\ kJ\ mol^{-1}$

$\Delta H_3 = \Delta H^{\ominus}_c[C_2H_4(g)] + \Delta H^{\ominus}_c[H_2(g)]$

$= (-1411)\ kJ\ mol^{-1} + (-286)\ kJ\ mol^{-1}$

$= -1697\ kJ\ mol^{-1}$

Hence: $\Delta H^{\ominus}_{reaction} = (-1560)\ kJ\ mol^{-1} - (-1697)\ kJ\ mol^{-1}$

$= -1560 + 1697 = +137\ kJ\ mol^{-1}$

Probably the most useful application of this type of calculation is in determining the enthalpies of formation of combustible compounds. This is illustrated in the tutorial entitled 'Calculating enthalpy changes of reaction from enthalpy changes of combustion' using Hess cycles on your CD-ROM.

Tutorial

Test yourself

7 a) By writing a balanced equation, show that the standard enthalpy change of formation of carbon dioxide is the same as the standard enthalpy change of combustion of carbon (graphite).

b) Write equations for the standard enthalpy change of formation of:
   i) aluminium oxide, Al_2O_3
   ii) hydrogen chloride, HCl
   iii) propane, C_3H_8.

8 Use the values for standard enthalpies of combustion below to calculate the standard enthalpy change of reaction for the formation of methanol, CH_3OH.

$\Delta H^{\ominus}_c[CH_3OH(l)] = -726\ kJ\ mol^{-1}$

$\Delta H^{\ominus}_c[C_{graphite}] = -393\ kJ\ mol^{-1}$

$\Delta H^{\ominus}_c[H_2(g)] = -286\ kJ\ mol^{-1}$

9 Why is it useful to have standard enthalpies of combustion which can be used to calculate standard enthalpies of reaction and formation?

Enthalpies of reaction from enthalpies of formation

Data books contain tables of standard enthalpies of formation for both inorganic and organic compounds. The great value of this data is that it allows us to calculate the standard enthalpy change for any reaction involving the substances listed in the tables.

The standard enthalpy change of a reaction is the enthalpy change when the amounts shown in the chemical equation react. Like other standard quantities in thermochemistry, the standard enthalpy change of reaction is defined at 298 K and 1 atmosphere pressure with the reactants and products in their normal stable states. The concentration of any solution is 1 mol dm^{-3}.

Thanks to Hess's Law it is easy to calculate the standard enthalpy change of a reaction using tabulated values of standard enthalpy changes of formation (Figure 16.17).

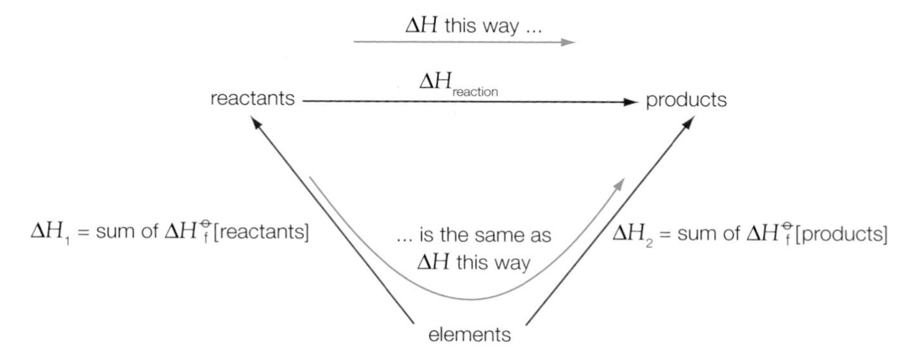

Figure 16.17 ▲
An enthalpy cycle for calculating standard enthalpies of reaction from standard enthalpies of formation.

According to Hess's Law:

$$\Delta H_{reaction} = -\Delta H_1 + \Delta H_2 = \Delta H_2 - \Delta H_1$$

Notice the negative sign in front of ΔH_1 because the enthalpy change from reactants to elements is the reverse of ΔH_1 in Figure 16.17.

So $\Delta H_{reaction}$ = sum of $\Delta H^{\ominus}_f$[products] – sum of $\Delta H^{\ominus}_f$[reactants]

Worked example

Calculate the enthalpy change for the reduction of iron(III) oxide by carbon monoxide.

$\Delta H^{\ominus}_f[Fe_2O_3] = -824\,kJ\,mol^{-1}$

$\Delta H^{\ominus}_f[CO] = -110\,kJ\,mol^{-1}$

$\Delta H^{\ominus}_f[CO_2] = -393\,kJ\,mol^{-1}$

Notes on the method

Write the balanced equation for the reaction and then draw an enthalpy cycle (Hess cycle) using the model in Figure 16.17.

Remember that, by definition, $\Delta H^{\ominus}_f$[element] = 0 kJ mol^{-1}.

Pay careful attention to the signs. Put the value and sign for a quantity in brackets when adding or subtracting enthalpy values.

Answers

$$Fe_2O_3(s) + 3CO(g) \rightarrow 2Fe(s) + 3CO_2(g)$$

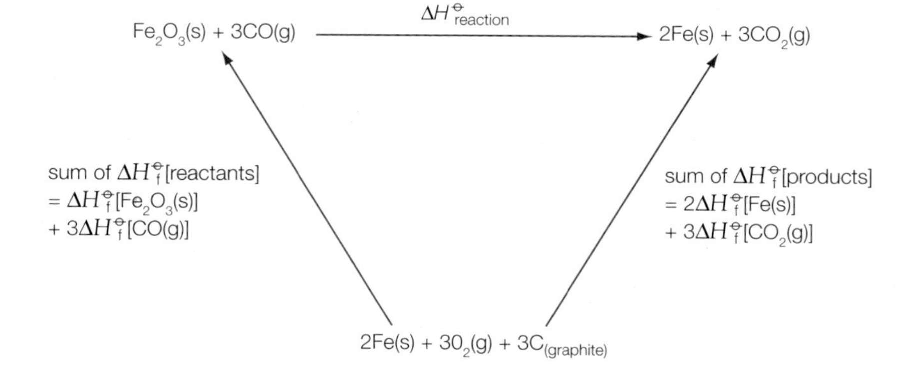

Figure 16.18 ▶
An enthalpy cycle (Hess cycle) for calculating the enthalpy change of reaction between iron(III) oxide and carbon monoxide.

Applying Hess's Law to Figure 16.18:

$$\Delta H^\ominus_{\text{reaction}} = -\text{ sum of } \Delta H^\ominus_f[\text{reactants}] + \text{sum of } \Delta H^\ominus_f[\text{products}]$$

$$= \text{sum of } \Delta H^\ominus_f[\text{products}] - \text{sum of } \Delta H^\ominus_f[\text{reactants}]$$

$$\therefore \Delta H^\ominus_{\text{reaction}} = 2 \times \Delta H^\ominus_f[\text{Fe}] + 3 \times \Delta H^\ominus_f[\text{CO}_2] - \Delta H^\ominus_f[\text{Fe}_2\text{O}_3] - 3 \times \Delta H^\ominus_f[\text{CO}]$$

$$= 0 + (3 \times -393) - (-824) - (3 \times -110)$$

$$= -1179 + 824 + 330$$

$$\Delta H^\ominus_{\text{reaction}} = -25\,\text{kJ mol}^{-1}$$

Tutorial

Test yourself

10 The standard enthalpy change of formation of sucrose (sugar), $C_{12}H_{22}O_{11}$, is $-2226\,\text{kJ mol}^{-1}$. Write the balanced equation for which the standard enthalpy change of reaction is $-2226\,\text{kJ mol}^{-1}$.

11 When calculating standard enthalpy changes for reactions involving water, why is it important to specify that the H_2O is present as water and not as steam?

12 Calculate the standard enthalpy of reaction of hydrazine, $N_2H_4(l)$, with oxygen, O_2, to form nitrogen, N_2, and water, H_2O.

$\Delta H^\ominus_f[N_2H_4(l)] = +51\,\text{kJ mol}^{-1}$

$\Delta H^\ominus_f[H_2O(l)] = -286\,\text{kJ mol}^{-1}$

16.6 Enthalpy changes and the direction of change

Strike a match – it catches fire and burns. Put a spark to petrol – it burns furiously. These are two exothermic reactions which, once started, tend to 'go'. They are examples of the many exothermic reactions which just keep going once they have started. In general, chemists expect that a reaction will go if it is exothermic.

This means that, in general, reactions which give out energy to their surroundings are the ones which happen. This ties in with the common experience that changes happen in the direction in which energy is spread around and dissipated in the surroundings. So, the sign of ΔH is a guide to the likely direction of change, but it is not a totally reliable guide for three main reasons.

- The direction of change may depend on the conditions of temperature and pressure. One example is the condensation of a liquid, such as steam. Steam condenses to water below 100 °C and heat is given out. This is an exothermic change.

  $$H_2O(g) \rightarrow H_2O(l) \qquad \Delta H = -44\,\text{kJ mol}^{-1}$$

  At temperatures above 100 °C the change goes in the opposite direction, and this process is endothermic.

- There are some examples of endothermic reactions which occur readily under normal conditions. So some reactions for which ΔH is positive can happen. One example of this is the reaction of citric acid solution with sodium hydrogencarbonate. The mixture fizzes vigorously and cools rapidly.

- Some highly exothermic reactions never occur because the rate of reaction is so slow and the mixture of reactants is effectively inert. For example, the change from diamond to graphite is exothermic, but diamonds don't suddenly turn into black flakes.

16.7 Enthalpy changes and bonding

During reactions, the bonds in reactants break and then new bonds form in the products. For example, when hydrogen reacts with oxygen:

$$2H_2(g) + O_2(g) \rightarrow 2H_2O(g)$$

Bonds in the H_2 and O_2 molecules first break to form H and O atoms (Figure 16.19). New bonds then form between the H and O atoms to produce water, H_2O.

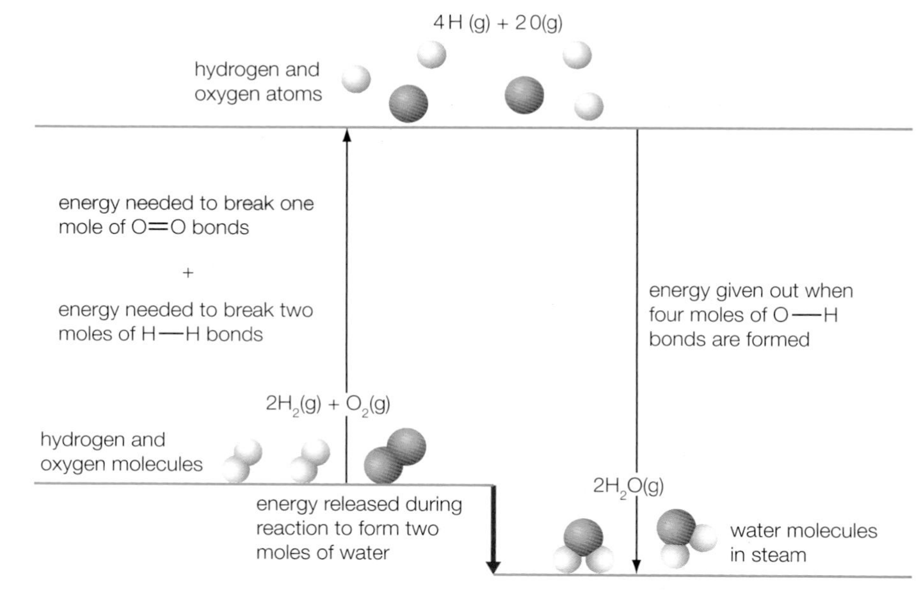

Figure 16.19 ▶ An energy level diagram for the reaction between hydrogen and oxygen.

Note

Bond breaking is endothermic.
Bond making is exothermic.

Obviously, energy is needed to break the bonds between atoms. So, energy must be released when the reverse occurs and a bond forms.

Chemical reactions involve bond breaking followed by bond making. This means that the enthalpy change of a reaction is the energy difference between bond-breaking and bond-making processes.

When hydrogen and oxygen react, more energy is released in making four new O–H bonds in the two H_2O molecules than in breaking the bonds in two H_2 molecules and one O_2 molecule. So, the overall reaction is exothermic (Figure 16.19).

A definite quantity of energy, known as the bond enthalpy or bond energy, is associated with each type of bond. This energy is absorbed when the bond is broken and evolved when the bond is formed. In measuring and using bond enthalpies, chemists distinguish between the terms 'bond enthalpy' and 'average bond enthalpy' (mean bond enthalpy).

Definition

The **average bond enthalpy** of a particular bond is an average of the energy required to break one mole of those bonds in molecules in the gaseous state.

- Bond enthalpies are precise values for specific bonds in compounds (e.g. for the C–Cl bond in CH_3Cl).
- Average bond enthalpies are average values for one kind of bond in different compounds (e.g. an average value for the C–Cl bond in all compounds).
- Average bond enthalpies take into account the fact that the bond enthalpy for a specific covalent bond varies slightly from one molecule to another.

The average (mean) bond enthalpies for various bonds are given in the data section of your CD-ROM. The symbol for bond enthalpy is E, so the C–H bond enthalpy is written as:

$$E(\text{C–H}) = 413\,\text{kJ mol}^{-1}$$

Using bond enthalpies

The most important use of average bond enthalpies is in estimating the approximate values of enthalpy changes in chemical reactions involving molecular substances with covalent bonds. These estimates are particularly helpful when experimental measurements cannot be made as in the following worked example.

Worked example

Use average bond enthalpies to estimate the enthalpy of formation of hydrazine, N_2H_4.

Notes on the method

Write out the equation for the reaction showing all the atoms and bonds in the molecules. This makes it easier to count the number of bonds broken and formed.

Answer

$$N{\equiv}N + 2H{-}H \longrightarrow H_2N{-}NH_2$$

Bonds broken	*kJ mol⁻¹*	*Bonds formed*	*kJ mol⁻¹*
1 N≡N	+945	1 N–N	–158
2 H–H	+(2 × 436)	4 N–H	–(4 × 391)

$$\therefore \Delta H_{reaction} = +945 + (2 \times 436) - 158 - (4 \times 391)$$

$$= +1817 - 1722 = +95\ \text{kJ mol}^{-1}$$

For many reactions, the values of ΔH estimated from average bond enthalpies agree closely with experimental values. However, there are limitations to the use of bond enthalpy data in this way, and significant differences between the values of ΔH estimated from bond enthalpies and those obtained by experiment do occur. These differences usually arise because of variations in the strength of one kind of bond in different molecules.

Test yourself

Data

13 a) Look back at Figure 16.19 and write out the equation $2H_2(g) + O_2(g) \rightarrow 2H_2O(g)$ showing all the bonds between atoms in the molecules.

b) Look up the average bond enthalpies for the bonds involved and calculate:

i) the energy needed to break one mole of O=O bonds plus two moles of H–H bonds

ii) the energy given out when four moles of O–H bonds are formed in two moles of water (steam) molecules

iii) the energy released during the reaction to form two moles of water (steam).

14 Look up the bond enthalpies for the H–H, Cl–Cl and H–Cl bonds.

a) Calculate the overall enthalpy change for the reaction:
$H_2(g) + Cl_2(g) \rightarrow 2HCl(g)$

b) Draw an energy level diagram for the reaction (similar to that in Figure 16.19).

15 Why are almost all spontaneous reactions exothermic?

16 a) Make a table to show the average bond enthalpies and bond lengths of the C–C, C=C and C≡C bonds.

b) What generalisations can you make based on your table?

17 Use average bond enthalpies to estimate the enthalpy change when ethene, $H_2C{=}CH_2(g)$, reacts with $H_2(g)$ to form ethane, $CH_3{-}CH_3(g)$.

18 a) Which are likely to give a more accurate answer to a calculation of the enthalpy change for a reaction – average bond enthalpies or enthalpies of formation?

b) Give a reason for your answer to part a).

19 Look carefully at the average bond enthalpies for hydrogen and the halogens (chlorine, bromine and iodine).

a) Write an equation for the reaction of hydrogen with chlorine.

b) Explain which bond (H–H or Cl–Cl) you think will break first in the reaction.

c) The mean bond enthalpy for fluorine, *E*(F–F), is 158 kJ mol^{-1}. How would you expect the reaction of fluorine with hydrogen to compare with the reaction of chlorine with hydrogen?

REVIEW QUESTIONS

1 When Epsom salts, $MgSO_4.7H_2O(s)$, are heated strongly, they decompose forming the anhydrous salt and water.

a) Write a balanced equation with state symbols for the decomposition of Epsom salts. (2)

b) Is the decomposition likely to be exothermic or endothermic? Explain your answer. (2)

c) Why can the enthalpy change for the decomposition of Epsom salts not be measured directly? (1)

d) Using the following values, calculate the standard enthalpy change for the decomposition. (5)

$\Delta H_f^\ominus[MgSO_4.7H_2O(s)] = -3389\ kJ\ mol^{-1}$

$\Delta H_f^\ominus[MgSO_4(s)] = -1285\ kJ\ mol^{-1}$

$\Delta H_f^\ominus[H_2O(g)] = -242\ kJ\ mol^{-1}$

2 Butane (Camping Gaz), C_4H_{10}, burns readily on a camp cooker. The equation for the reaction is:

$C_4H_{10}(g) + 6\frac{1}{2}O_2(g) \rightarrow 4CO_2(g) + 5H_2O(g)$

a) Rewrite the equation showing all the covalent bonds between atoms in the reactants and products. (4)

b) Make a table showing the bonds broken in the reactants and the bonds formed in the products during the reaction. (2)

c) Use the following average bond enthalpies to calculate the enthalpy change of the reaction. (4)

$E(C–C) = 347\ kJ\ mol^{-1}$ $E(C–H) = 413\ kJ\ mol^{-1}$

$E(O=O) = 498\ kJ\ mol^{-1}$ $E(C=O) = 740\ kJ\ mol^{-1}$

$E(H–O) = 464\ kJ\ mol^{-1}$

3 An excess of solid sodium hydrogencarbonate was added to 50 cm^3 of 1.0 $mol\ dm^{-3}$ ethanoic acid in an insulated polystyrene container under standard conditions. The temperature fell by 8.0 °C.

a) Copy and complete the following equation for the reaction.

$CH_3COOH(aq) + NaHCO_3(s)$

$\rightarrow$ ____ + ____ + ____ (3)

b) Why do you think the $NaHCO_3$ was added in small portions? (1)

c) Calculate the energy change during the reaction. (Assume that the specific heat capacity of the solution is $4.2\ J\ g^{-1}\ K^{-1}$ and its density is $1.0\ g\ cm^{-3}$. Ignore the mass of sodium hydrogencarbonate.) (2)

d) How many moles of ethanoic acid were used? (1)

e) Calculate the standard enthalpy change of the reaction. Show the correct sign and units. (3)

f) Explain the number of significant figures in your answer. (1)

4 A student suggested that ethane might react with bromine in two different ways in bright sunlight.

Reaction 1: $C_2H_6(g) + Br_2(l) \rightarrow C_2H_5Br(l) + HBr(g)$

Reaction 2: $C_2H_6(g) + Br_2(l) \rightarrow 2CH_3Br(g)$

a) Use the following average bond enthalpies to calculate the enthalpy changes for the two possible reactions.

$E(C–C) = 347\ kJ\ mol^{-1}$ $E(Br–Br) = 193\ kJ\ mol^{-1}$

$E(C–H) = 413\ kJ\ mol^{-1}$ $E(H–Br) = 366\ kJ\ mol^{-1}$

$E(C–Br) = 290\ kJ\ mol^{-1}$ (8)

b) Use your calculations to explain which of the reactions is more likely to occur. (2)

c) Suggest two reasons why your calculated enthalpy changes may not agree with the accurately determined experimental values. (2)

5 Tin is manufactured by heating tinstone, SnO_2, at high temperatures with coke (carbon).

There are two possible reactions for the process.

Reaction 1: $SnO_2(s) + C(s) \rightarrow Sn(s) + CO_2(g)$

Reaction 2: $SnO_2(s) + 2C(s) \rightarrow Sn(s) + 2CO(g)$

a) Calculate the standard enthalpy change for each of the possible reactions using the data below.

$\Delta H_f^\ominus[SnO_2(s)] = -581\ kJ\ mol^{-1}$

$\Delta H_f^\ominus[CO_2(g)] = -394\ kJ\ mol^{-1}$

$\Delta H_f^\ominus[CO(g)] = -110\ kJ\ mol^{-1}$ (6)

b) Use your calculations to explain which of the reactions would be most economic for industry. (2)

6 a) Draw a diagram of the apparatus that you could use to determine the enthalpy change of the reaction between powdered magnesium and excess copper(II) sulfate solution. (3)

b) When excess powdered magnesium was added to 50 cm^3 of 0.04 $mol\ dm^{-3}$ copper(II) sulfate solution, the temperature rose by 5.0 °C.

i) Write a balanced equation with state symbols for the reaction. (1)

ii) Calculate the energy transferred to the copper(II) sulfate solution. (2)

iii) What assumptions have you made in your calculation in part ii)? (2)

iv) Calculate the enthalpy change for the reaction shown in your equation in part i). (3)

17 Rates and equilibria

In the chemical industry, manufacturers aim to get the best possible yield in the shortest time. In this respect, the development of new catalysts to speed up reactions is one of the frontier aspects of modern chemistry. The aim is to make manufacturing processes more efficient so that they use less energy and produce little or no harmful waste. The need for greater efficiency in chemical processes is more pressing than ever as people become more aware of the harm that waste chemicals can do to our health and to the environment.

Definition

Chemical kinetics is the study of the rates of chemical reactions.

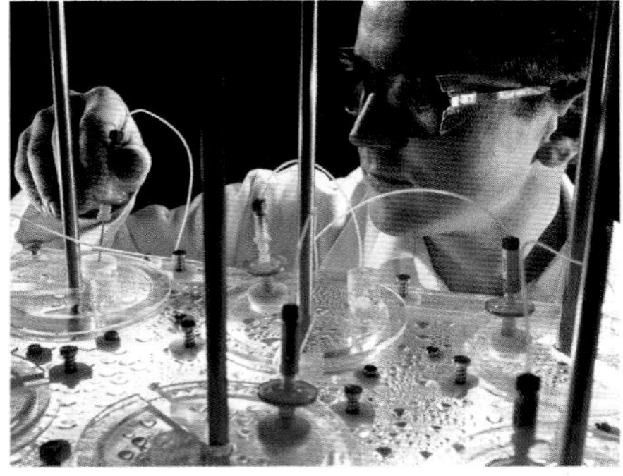

Figure 17.1 ▲
A pharmacy technician monitoring the rate at which a drug is released from a medicine tablet in conditions similar to those in the stomach.

17.1 Reaction rates

The study of rates of reaction is important because it helps chemists to control reactions both in the laboratory and on a large scale in industry.

Kinetics is also important in many other fields. The study of rates of reaction has helped environmental scientists to explain why CFCs and other chemicals are destroying the ozone layer in the upper atmosphere (see Section 18.6). Pharmacologists who study the chemistry of drugs must study the speed at which they change to other chemicals and break down in the human body. The pharmacists who formulate and supply medicines also need to know about the rate at which the chemicals slowly degrade in the bottle or pack. For many medicines, the shelf-life is the time for which they can be stored before the concentration of the active ingredient has decreased by 10%.

Chemical reactions happen at a variety of speeds. Ionic precipitation reactions are very fast. Explosions are even faster. However, the rusting of iron and other corrosion processes are slow and may continue for years.

Figure 17.2 ▶
Firefighters have to know how to slow down and stop burning. Water cools the burning materials as it evaporates, and the steam produced can help to keep out air.

Test yourself

1 How would you slow down or stop the following reactions?
  a) Iron corroding
  b) Toast burning
  c) Milk turning sour
2 How would you speed up the following reactions?
  a) Fermentation in dough to make bread rise
  b) Solid fuel burning in a stove
  c) Epoxy glues (adhesives) setting
  d) Conversion of chemicals in engine exhausts to harmless gases

17.2 Measuring reaction rates

Balanced chemical equations tell us nothing about how quickly the reactions occur. In order to get this information, chemists have to do experiments to measure the rates of reactions under various conditions.

The amounts of the reactants and products change during any chemical reaction. Products form as reactants disappear. The rates at which these changes happen give a measure of the rate of reaction.

The rate of the reaction between magnesium and hydrochloric acid:

$$Mg(s) + 2HCl(aq) \rightarrow MgCl_2(aq) + H_2(g)$$

can be measured by:

- the rate of loss of magnesium
- the rate of loss of hydrochloric acid
- the rate of formation of magnesium chloride
- the rate of formation of hydrogen.

In this example, it is probably easiest to measure the rate of formation of hydrogen by collecting the gas and recording its volume with time (Figure 17.3).

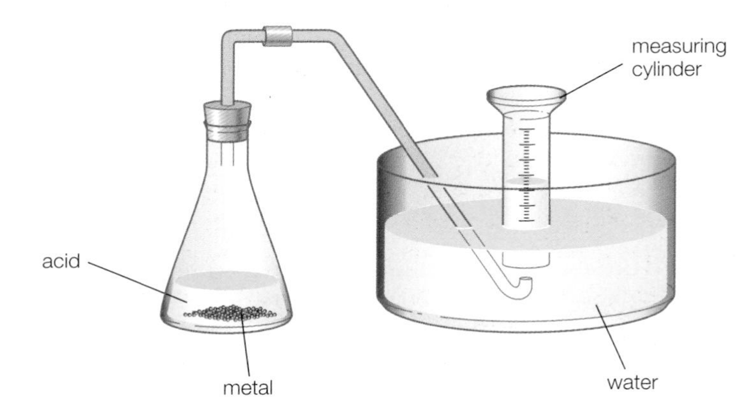

Figure 17.3 Collecting and measuring the gas produced when magnesium reacts with acid. A gas syringe can be used instead of a measuring cylinder full of water.

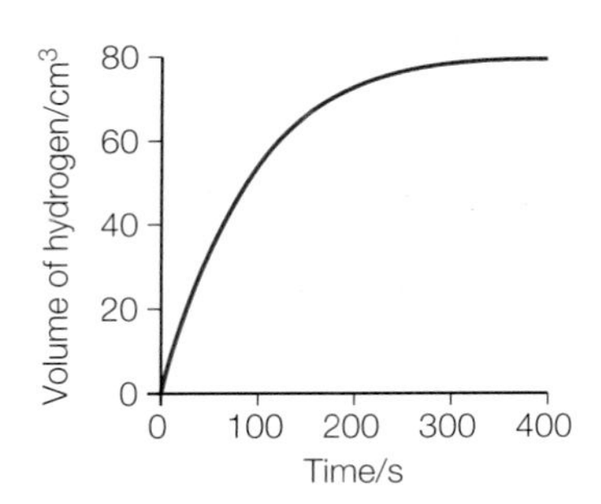

Figure 17.4 Volume of hydrogen plotted against time for the reaction of magnesium with hydrochloric acid.

Chemists design their rate experiments to measure a property that changes with the amount or concentration of a reactant or product:

$$\text{rate of reaction} = \frac{\text{change recorded in the property}}{\text{time for the change}}$$

In most chemical reactions, the rate changes with time. The graph in Figure 17.4 is a plot of the results from a study of the reaction of magnesium with dilute hydrochloric acid. The graph is steepest at the start, when the reaction is at its fastest. As the reaction continues, it slows down until it finally stops. This happens because one of the reactants is being used up until none of it is left.

The gradient at any point of a graph that shows the amount or concentration plotted against time measures the rate of reaction.

Test yourself

3 In an experiment to study the reaction of magnesium with dilute hydrochloric acid, 48 cm^3 of hydrogen forms in 10 s at room temperature. Calculate:

a) the average rate of formation of hydrogen in cm^3 s^{-1}

b) the average rate of formation of hydrogen in mol s^{-1} (see Section 2.4)

c) the rates of appearance or disappearance of the other product and the reactants in mol s^{-1}.

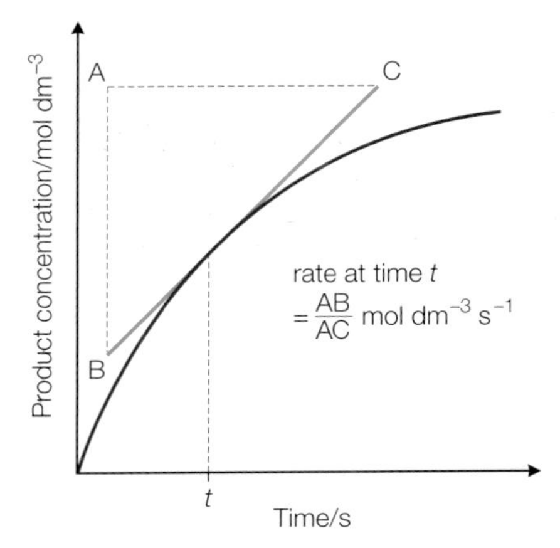

Figure 17.5 Graph showing the concentration of a product plotted against time. The gradient at any point measures the rate of reaction at that time.

A useful way of studying the effect of changing the conditions on the rate of a reaction is to find a way of measuring the rate just after mixing the reactants. Figure 17.6 is a graph for two different sets of conditions. One of the reactants was more concentrated and produced line A. Near the start, it took t_A seconds to produce x mol of product. When the same reactant was less concentrated, the results gave line B. This time, near the start, it took t_B seconds to produce x mol of product. The reaction was slower when the concentration was lower, so it took longer to produce x mol of product.

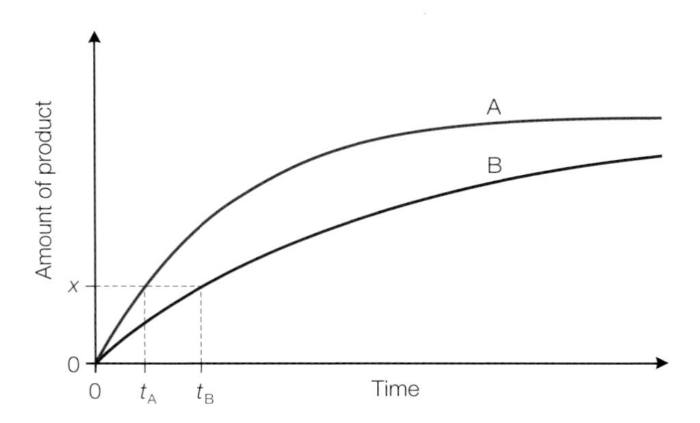

Figure 17.6 ▶ Formation of the same amount (x mol) of product starting with different concentrations of one of the reactants.

$$\text{Average rate of formation of product on line A} = \frac{x}{t_A}$$

$$\text{Average rate of formation of product on line B} = \frac{x}{t_B}$$

If x is kept the same, it follows that the average rate near the start is proportional to $\frac{1}{t}$.

This means that it is possible to arrive at a measure of the initial rate of a reaction by measuring how long the reaction takes to produce a small fixed amount of product or use up a small fixed amount of reactant.

17.3 What factors affect reaction rates?

Concentration

In general, the higher the concentration of the reactants, the faster the reaction. For gas reactions, a change in pressure has the same effect as changing the concentration because a higher pressure compresses a mixture of gases and increases their concentration. In a mixture of reacting gases, therefore, the higher the pressure, the faster the reaction.

Activity

Investigation of the effect of concentration on the rate of a reaction

Figure 17.7 illustrates an investigation of the effect of concentration on the rate at which thiosulfate ions in solution react with hydrogen ions to form a precipitate of sulfur.

$S_2O_3^{2-}(aq) + 2H^+(aq) \rightarrow S(s) + SO_2(aq) + H_2O(l)$

The observer records the time taken for the sulfur precipitate to obscure the cross on the paper under the flask. In this example, the quantity x in Figure 17.6 is the amount of sulfur needed to hide the cross on the paper. This is the same each time. The rate of reaction is therefore proportional to $\frac{1}{t}$.

The results of the investigation are shown in Table 17.1.

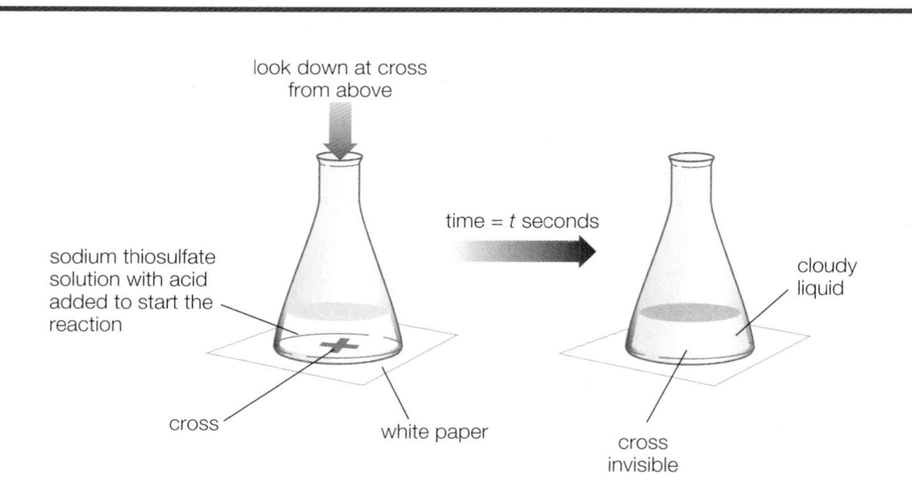

Figure 17.7 ◄ Investigating the effect of the concentration of thiosulfate ions on the rate of reaction in acid solution. The hydrogen ion concentration is the same in each experiment.

Table 17.1 ◄ Results of the investigation in Figure 17.7.

Experiment	Concentration of thiosulfate ions / mol dm^{-3}	Time, t, for the cross to be obscured/s	Rate of reaction, $\frac{1}{t}$ /s^{-1}
1	0.15	43	0.023
2	0.12	55	
3	0.09	66	0.015
4	0.06	105	0.0095
5	0.03	243	0.0041

1 How would you prepare 50 cm^3 of a solution of sodium thiosulfate solution with a concentration of 0.12 mol dm^{-3} from a solution with a concentration of 0.15 mol dm^{-3}?

2 Calculate the value for the rate of reaction when the concentration of thiosulfate ions is 0.12 mol dm^{-3}.

3 Plot a graph to show how the rate of reaction varies with the concentration of thiosulfate ions.

4 What is the relationship between reaction rate and concentration of thiosulfate for this reaction according to the graph?

Surface area of solids

Breaking a solid into smaller pieces increases the surface area in contact with a liquid or gas. This speeds up any reaction happening at the surface of the solid. This effect also applies to reactions between liquids that do not mix. Shaking breaks up one liquid into droplets that are then dispersed in the other liquid, thereby increasing the surface area for reaction.

Temperature

Increasing the temperature is a very effective way of increasing the rate of a reaction. In general, a 10 °C increase in temperature roughly doubles the rate of reaction (see Figure 17.8 on page 202).

Bunsen burners, hot plates and heating mantles are common in laboratories because it is often convenient to speed up reactions by heating the reactants. For the same reason, many industrial processes are carried out at high temperatures.

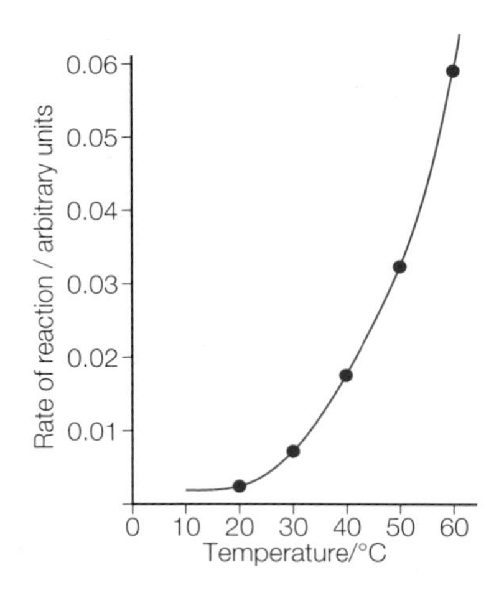

Figure 17.8 ▲
Effect of temperature on the rate of decomposition of thiosulfate ions to form sulfur.

Test yourself

4 Figure 17.10 illustrates the effect of changing conditions on the reaction of zinc metal with sulfuric acid. The red line shows the volume of hydrogen plotted against time using an excess of zinc turnings and 50 cm^3 of 2.0 mol dm^{-3} sulfuric acid at 20 °C.

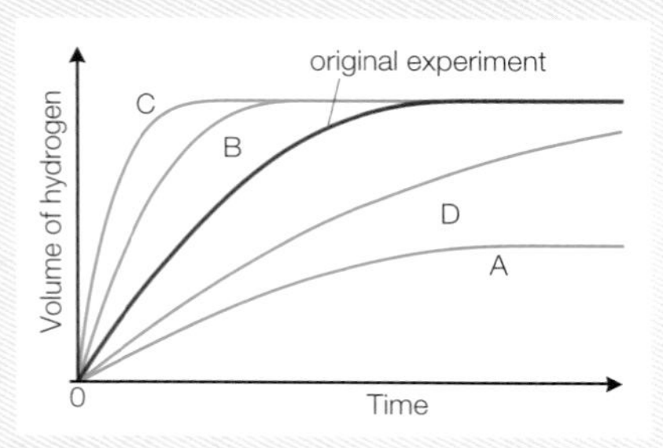

Figure 17.10 ▲

a) Write a balanced equation for the reaction.
b) Draw the apparatus that could be used to obtain the results to plot the graph.
c) Identify which line of the graph shows the effect of carrying out the same reaction under the same conditions with the following changes:
   i) adding a few drops of copper(II) sulfate solution to act as a catalyst
   ii) increasing the temperature to 30 °C
   iii) using the same mass of zinc but in larger pieces
   iv) using 50 cm^3 of 1.0 mol dm^{-3} sulfuric acid.

Catalysts

Catalysts have an astonishing ability to speed up the rates of chemical reactions without themselves changing permanently. Very small quantities of active catalysts can speed reactions to produce many times their own weight of chemicals.

Catalysts work by removing or lowering the barriers that prevent a reaction. They bring reactants together in a way that makes a reaction more likely.

Catalysts can also be extraordinarily selective. A catalyst may increase the rate of only one very specific reaction. Enzymes, the catalysts in living cells, are especially selective. Enzymes have the great advantage that they are effective at temperatures and pressures close to normal.

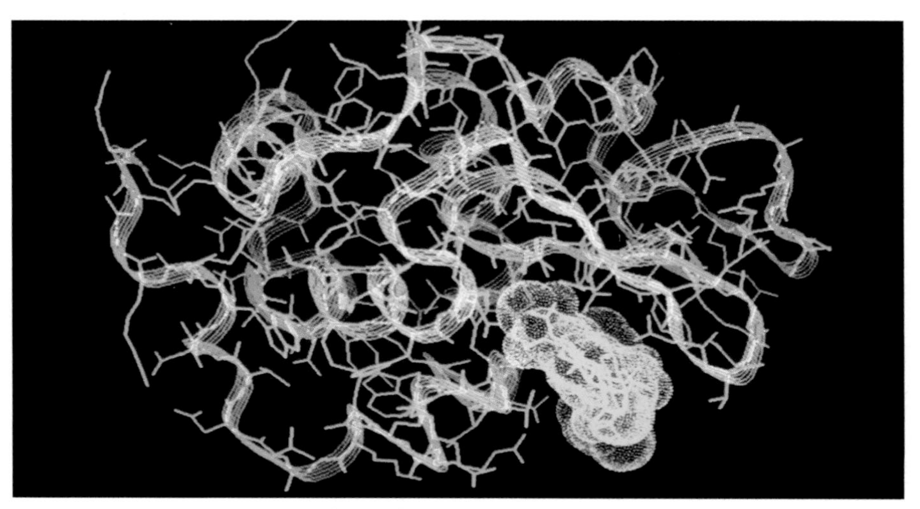

Figure 17.9 ▲
Computer graphic representing the structure of the enzyme lysozyme, which is found in tears and saliva. The skeleton of the enzyme is shown in blue, with the backbone highlighted with a magenta strip. The enzyme has an active site that selectively breaks apart molecules in the cell walls of bacteria. The affected molecules (shown in yellow) fit into the active site of the enzyme.

Most industrial processes involve catalysts. One of the targets of chemists in industry is to develop catalysts that make manufacturing processes more efficient so that they use less energy and produce less waste (see Topic 19).

Catalysts get involved in reactions but they are not reactants and they do not appear in the overall chemical equation. In theory, catalysts can be used over and over again, but there is some loss of catalyst in practice. Sometimes catalysts become contaminated, sometimes they are hard to recover completely from the products and sometimes the catalyst changes its state, such as from lumps to a fine powder, which means that it is no longer useable. Safe disposal of waste catalysts can be a challenge if they are made of toxic or corrosive chemicals.

17.4 Collision theory

Scientists can explain the factors that affect reaction rates. They use a model that gives a picture of what happens to atoms, molecules and ions as they react. This model works best for gases, but it can also be applied to reactions in solution.

Gas molecules in motion

The model that scientists use to explain the behaviour of gases assumes that the molecules in a gas are in rapid random motion and colliding with each other. They call this particles-in-motion model the 'kinetic theory'. The name comes from a Greek word for movement.

The kinetic theory makes a number of assumptions about the molecules of a gas. Applying Newton's laws of motion to the collection of particles leads to equations that can describe the properties of gases very accurately.

The assumptions of the kinetic theory model are that:

- gas pressure results from the collisions of the molecules with the walls of the container
- there is no loss of energy when the molecules collide with the walls of any container
- the molecules are so far apart that the volume of the molecules can be neglected in comparison with the total volume of gas
- the molecules do not attract each other
- the average kinetic energy of the molecules is proportional to their temperature on the Kelvin scale.

A gas that behaves exactly as the model predicts is called an 'ideal gas'. Real gases do not behave exactly like this; however, under laboratory conditions, some gases are close to behaving like an ideal gas. These are the gases that, at room temperature, are well above their boiling points, such as helium, nitrogen, oxygen and hydrogen.

The assumptions built into the model help to explain why real gases approach ideal behaviour at high temperatures and low pressures. At high temperatures, the molecules are moving so fast that any small attractive forces between them can be ignored. At low pressures, the volumes are so big that the space taken up by the molecules is insignificant.

The theory also helps to explain why real gases deviate from ideal gas behaviour as they get nearer to becoming a liquid. As a gas liquefies, the molecules get very close together and the volume of the molecules cannot be ignored. In addition, gases cannot liquefy unless some attractive (intermolecular) forces between the molecules hold them together. The Dutch physicist Johannes Van der Waals (1837–1923) developed his theory of intermolecular forces (see Section 6.13) by studying the behaviour of real gases and their deviations from the behaviour of an ideal gas.

Figure 17.11 ▲
The hot air balloon flights of the Montgolfier brothers in the latter part of the eighteenth century inspired scientists to study the behaviour of gases.

Maxwell–Boltzmann distribution

Two physicists used the kinetic theory to explore the distribution of energies among the molecules in gases. They worked out the proportion of molecules with a given energy at a particular temperature. The two physicists were James Maxwell (1831–1879), in Great Britain, and Ludwig Boltzmann (1844–1906), in Austria.

Figure 17.13 shows the distribution of energies for the molecules of a gas under two sets of conditions. This Maxwell–Boltzmann distribution helps to explain the effects of temperature changes and catalysts on the rates of reactions.

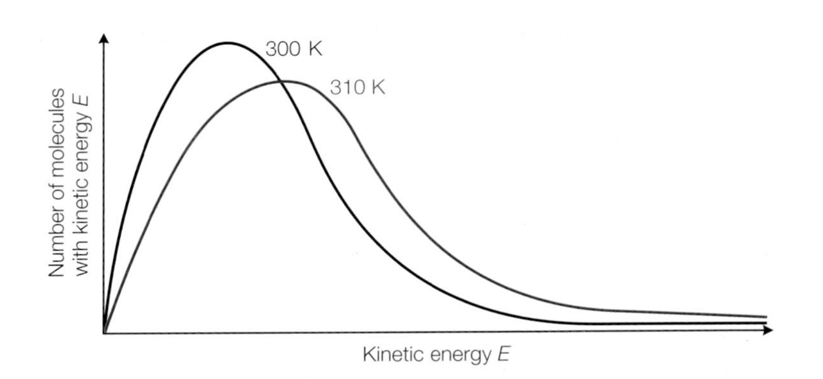

Figure 17.12 ◄
The Maxwell–Boltzmann distribution of the kinetic energies of the molecules of a gas at two temperatures. The area under the curve gives the total number of molecules. This area does not change as the temperature increases, so the peak height falls as the temperature rises and the curve spreads to the right.

Explaining the effects of concentration, pressure and surface area on reaction rates

In any reaction mixture, the atoms, molecules or ions are forever bumping into each other. When they collide, there is a chance that they will react. Increasing the pressure of a gas means that the reacting particles are closer together. There are more collisions and therefore the reaction goes faster. Increasing the concentration of reactants in solution has a similar effect.

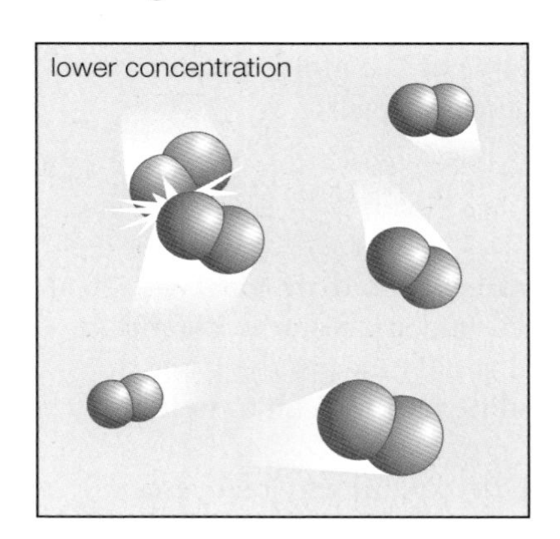

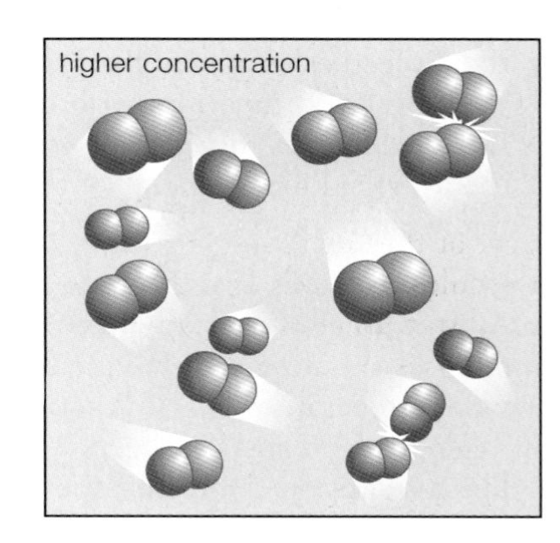

Figure 17.13 ▶ Increasing the pressure, or concentration, means that the reacting atoms, molecules or ions are closer together. There are more collisions and the reactions are faster.

In a reaction of a solid with a liquid or a gas, the reaction is faster if the solid is broken up into smaller pieces. Crushing the solid increases its surface area, and this means that collisions can be more frequent and the rate of reaction is faster.

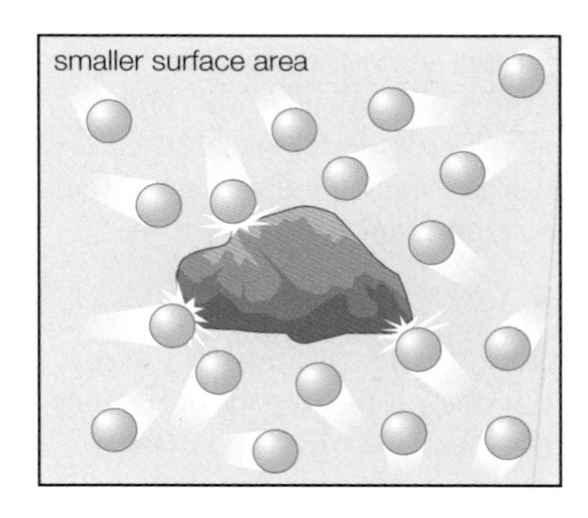

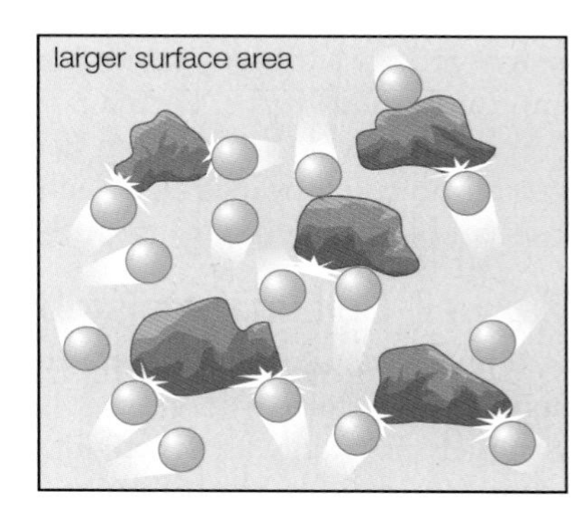

Figure 17.14 ▶ Breaking a solid into smaller pieces increases the surface area exposed to reacting chemicals in a gas or solution.

Explaining the effects of temperature on reaction rates

It is not enough simply for the molecules to collide. Some collisions do not result in a reaction. In soft collisions, the molecules simply bounce off each other. Molecules are in such rapid motion that if every collision led to a reaction, most reactions would be explosive. Only those molecules that collide with enough energy to stretch and break chemical bonds can lead to new products.

Definition

A **reaction profile** is a graph that shows how the total enthalpy (energy) of the atoms, molecules or ions changes during the progress of a reaction from reactants to products.

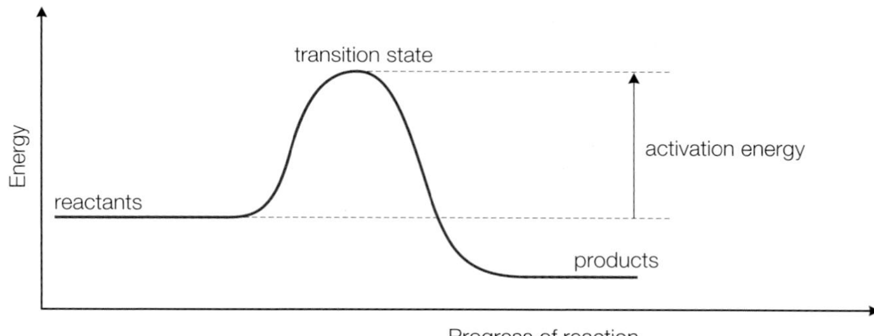

Figure 17.15 ▶ Reaction profile showing the activation energy for a reaction.

Chemists use the term activation energy to describe the minimum energy needed in a collision between molecules if they are to react. Activation energies account for the fact that reactions go much more slowly than would be expected if every collision between atoms and molecules led to a reaction. Only a very small proportion of collisions bring about chemical change. Molecules can react only if they collide with enough energy for bonds to stretch and then break so that new bonds can form. At around room temperature, only a small proportion of molecules have enough energy to react.

The Maxwell–Boltzmann distribution shown in Figure 17.16 shows that the proportion of molecules with energies greater than the activation energy is small at a temperature around 300 K.

The shaded areas in Figure 17.16 show the proportions of molecules with at least the activation energy for a reaction at 300 K and 310 K. This area is larger at the higher temperature. At a higher temperature, therefore, more molecules have enough energy to react when they collide and the reaction goes faster. In addition, when the molecules are moving faster, they collide more often.

Definition

The **activation energy** is the height of the energy barrier that separates reactants and products during a chemical reaction. It is the minimum energy needed for a reaction between the amounts in moles shown in the equation for the reaction.

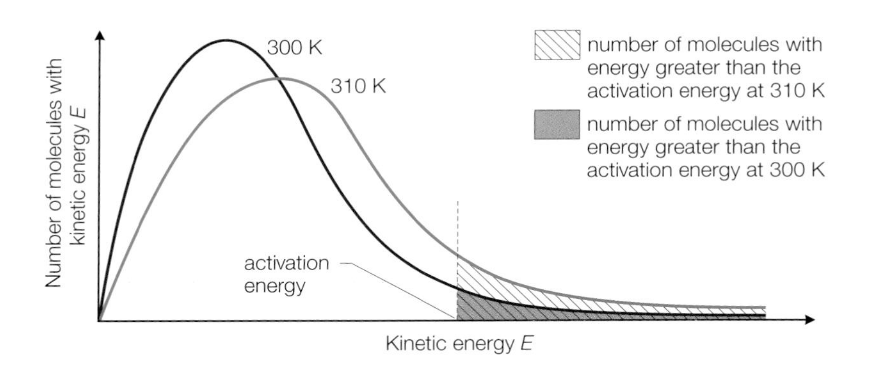

Figure 17.16 ◀ Maxwell–Boltzmann distribution of kinetic energies in the molecules of a gas at 300 K and 310 K. The area under each curve is a measure of the number of molecules. At 310 K, more molecules have enough energy to react when they collide with other molecules.

Test yourself

5 Two factors explain why reactions go faster when the temperature increases. Identify these two factors in terms of the speed and energy of atoms, ions and molecules.

Explaining the effects of catalysts on reaction rates

A catalyst works by providing an alternative pathway for the reaction with a lower activation energy. Lowering the activation energy increases the proportion of molecules with enough energy to react.

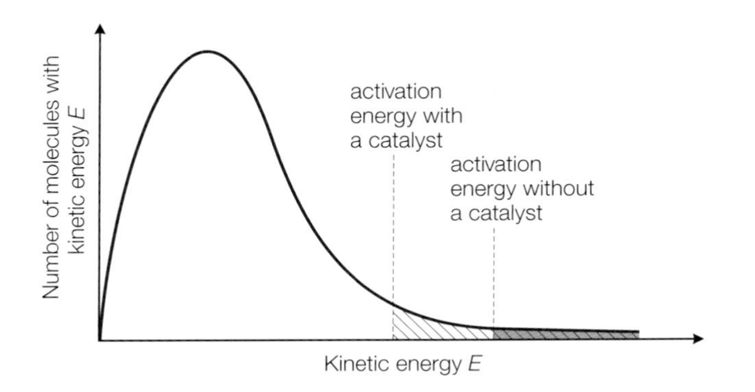

Figure 17.17 ◀ Distribution of molecular energies in a gas showing how the proportion of molecules able to react increases when a catalyst lowers the activation energy.

Often a catalyst changes the mechanism of a reaction and makes a reaction more productive by increasing the yield of the desired product and reducing waste.

One of the ways in which a catalyst can change the mechanism of a reaction is to combine with the reactants to form an intermediate. The intermediate is a stage in the transition from reactants to products. The intermediate breaks down to give the products and the catalyst is released. This frees up the catalyst to interact with further reactant molecules and the reaction continues.

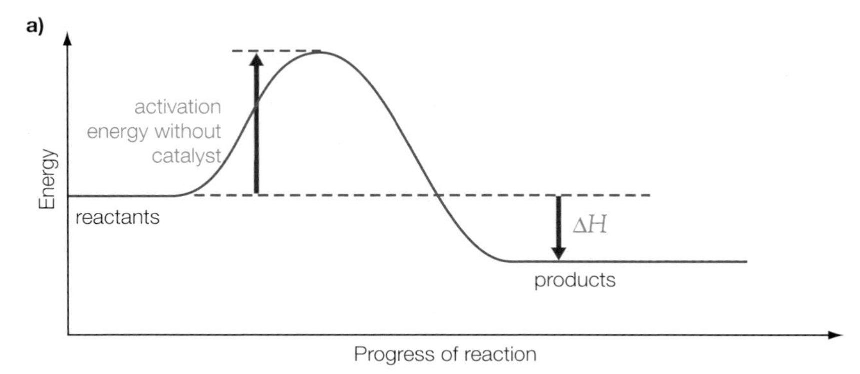

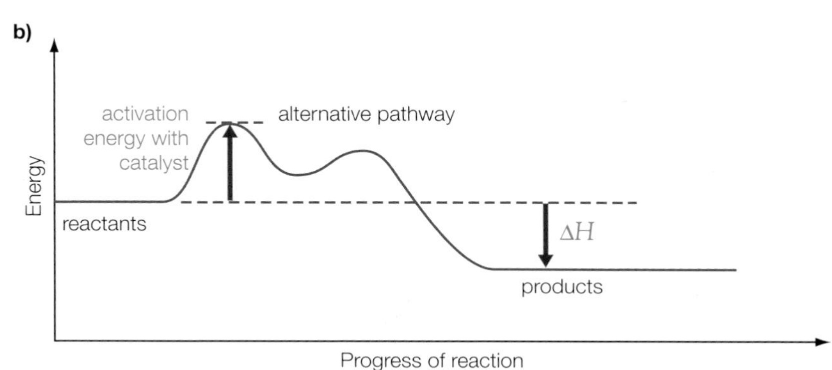

Figure 17.18 ▶ Reaction profiles for a reaction **a)** without a catalyst and **b)** with a catalyst. The dip in the curve of the pathway with a catalyst shows where an unstable intermediate forms.

Test yourself

6 Which part of Figure 17.18b) shows the formation of an intermediate?

7 a) Why is a match or spark needed to light a Bunsen burner?

b) Why does the gas keep burning once it has been lit?

8 Suggest a reason why catalysts are often specific for a particular reaction.

17.5 Stability

The study of energetics (thermochemistry) and rates of reaction (kinetics) helps to explain why some chemicals are stable while others react rapidly.

Compounds are stable if they always stay the same and do not decompose into their elements or into other compounds. Sometimes there is no tendency for a reaction to happen, because the reactants are stable relative to the products. This is usually the case if the enthalpy change for the reaction is positive. Magnesium oxide, for example, has no tendency to split up into magnesium and oxygen. In this case, the reactants are described as thermodynamically stable.

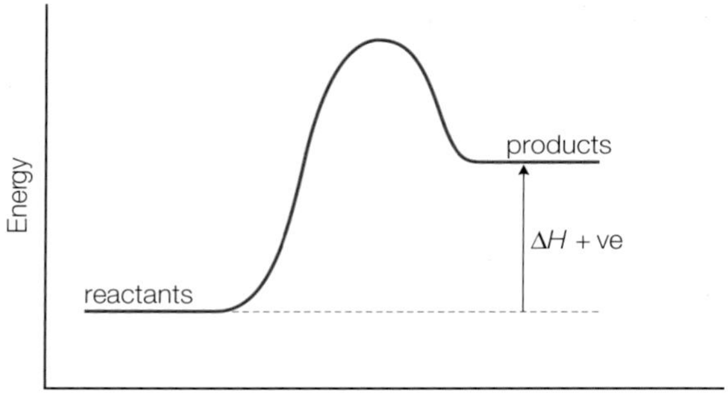

Figure 17.19 ◀
Reaction profile for a change that does not tend to happen because the reactants are more stable than the products.

Sometimes a compound seems stable at lower temperatures even though the enthalpy change suggests that the reaction should happen. An example is the gas dinitrogen oxide, N_2O, which is unstable relative to its elements. As expected, its enthalpy change of formation is positive ($\Delta H_f^\ominus = +82\ \text{kJ mol}^{-1}$). The decomposition reaction from the compound into its elements is exothermic. The compound tends to decompose into its elements, but the rate is very slow under normal conditions because of the large activation energy of the reaction. In this case, chemists say that N_2O is kinetically inert at room temperature.

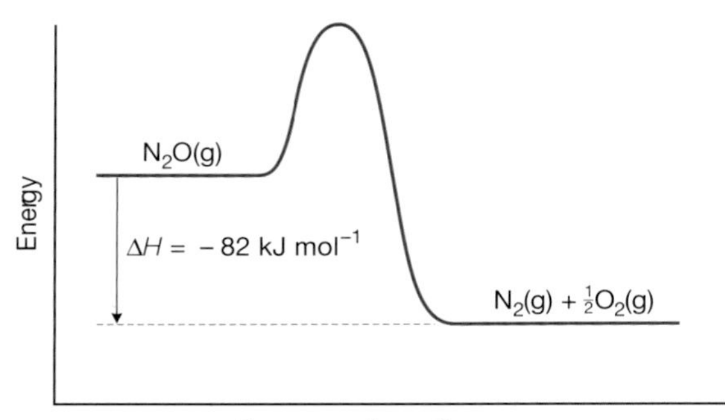

Figure 17.20 ◀
Reaction profile for the decomposition of dinitrogen oxide (kinetic inertness).

Dinitrogen oxide does decompose on heating, however. It can even relight a glowing splint because the red hot wood increases the proportion of molecules with enough energy to overcome the activation energy and decompose into nitrogen and oxygen. Enough oxygen is produced for a glowing splint to burst into flames.

Test yourself

9 Explain the following observations.
  a) Methane gas does not catch fire at room temperature when mixed with oxygen.
  b) Water does not decompose into hydrogen and oxygen when it boils and turns to steam.
  c) Diamonds do not change into graphite.

10 The following seem stable at room temperature. In each case, what is the possible reaction? Which are examples of thermodynamic stability and which are examples of kinetic inertness?
  a) Gunpowder in a firework
  b) Crystal of calcium carbonate
  c) Unlit candle
  d) Piece of copper metal in dilute hydrochloric acid

Definitions

A chemical or mixture of chemicals is **thermodynamically stable** if there is no tendency for a reaction. Usually, but not always, a positive enthalpy change (ΔH) indicates that the reaction does not tend to occur.

A chemical or mixture of chemicals is **kinetically inert** if nothing happens even when the enthalpy change for the reaction is negative. Kinetic inertness is a consequence of a high activation energy for the change, which means that the reaction rate is very slow.

Figure 17.21 ▲
Frying an egg is not reversible. Once cooked, the egg cannot be uncooked.

Figure 17.22 ▲
Stalactites and stalagmites form in limestone caves because the reaction of calcium carbonate with carbon dioxide and water is reversible.

Figure 17.23 ▲
Using cobalt chloride paper to test for water.

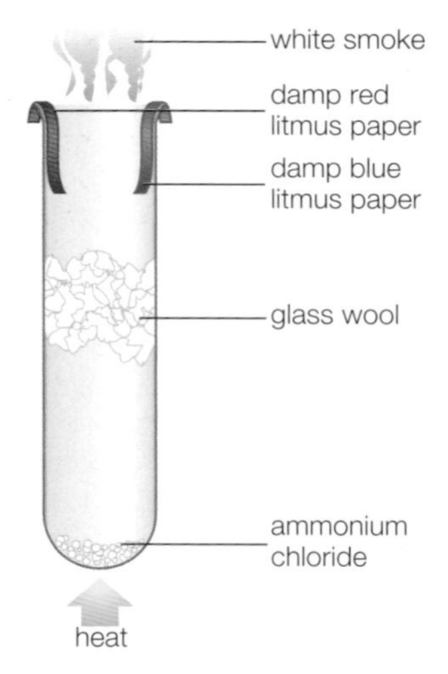

Figure 17.24 ▲
Investigating the effect of heating ammonium chloride.

17.6 Reversible changes

Some changes happen in only one direction. Baking bread is an example. Once bread is baked in an oven, there is no way to reverse the process and split a loaf back into flour, water and yeast. Some chemical reactions are like this.

Many other changes are reversible. Haemoglobin, for example, combines with oxygen as red blood cells flow through the lungs, but it then releases the oxygen to be used in respiration as blood flows in the capillaries throughout the rest of the body.

The study of reversible reactions helps chemists answer the questions 'How far?' and 'Which direction?' These are questions that need to be answered when chemists try to make new chemicals in laboratories and in industry.

Burning of fuels, such as natural gas and petrol, is an example of a one-way processes. Once these fuels have burned in air to make carbon dioxide and water, it is impossible to turn the products back to natural gas and petrol. The combustion of the fuels is an irreversible process.

Many other chemical reactions are reversible. One example is the basis of a simple laboratory test for water. Hydrated cobalt(II) chloride is pink and so is a solution of the salt in water. Heating filter paper soaked in the solution in an oven makes the paper turn blue, because water is driven off from the solution to leave anhydrous cobalt(II) chloride on the paper.

$$\underset{\text{pink}}{CoCl_2.6H_2O(s)} \rightarrow \underset{\text{blue}}{CoCl_2(s)} + 6H_2O(g)$$

The blue paper provides a sensitive test for water, because it turns pink again if it is exposed to water or water vapour. At room temperature, water rehydrates the blue salt:

$$\underset{\text{blue}}{CoCl_2(s)} + 6H_2O(l) \rightarrow \underset{\text{pink}}{CoCl_2.6H_2O(s)}$$

The reaction of ammonia with hydrogen chloride is another reaction in which the direction of change depends on the temperature. At room temperature, the two gases combine to make a white smoke of ammonium chloride.

$$NH_3(g) + HCl(g) \rightarrow NH_4Cl(s)$$

Heating reverses the reaction and ammonium chloride decomposes at high temperatures to give hydrogen chloride and ammonia.

$$NH_4Cl(s) \rightarrow NH_3(g) + HCl(g)$$

The demonstration illustrated by Figure 17.24 shows that ammonium chloride decomposes into two gases on heating. Ammonia gas diffuses through the glass wool faster than hydrogen chloride. After a short time, the alkaline ammonia rises above the plug of glass wool and turns the red litmus blue. A while later, both strips of litmus paper turn red as the acid hydrogen chloride arrives. A smoke of ammonium chloride appears above the tube when both gases meet and cool.

Changing the temperature is not the only way to alter the direction of change. Hot iron, for example, reacts with steam to make iron(III) oxide and hydrogen. Supplying plenty of steam and 'sweeping away' the hydrogen means that the reaction continues until all of the iron changes to its oxide.

$$3Fe(s) + 4H_2O(g) \rightarrow Fe_3O_4(s) + 4H_2(g)$$

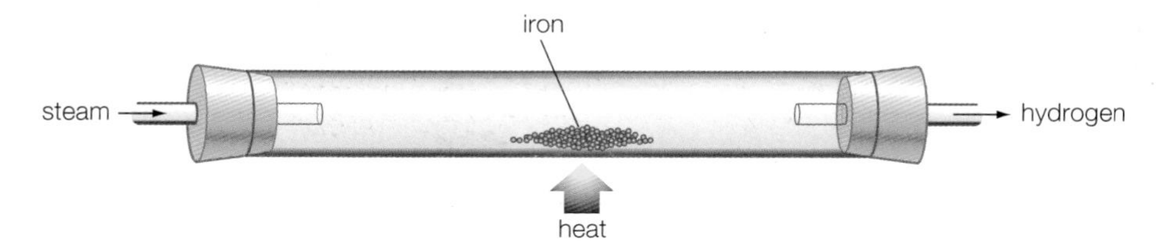

Figure 17.25 ▲
The forward reaction happens when the concentration of steam is high and the hydrogen is swept away, which keeps its concentration low.

Altering the conditions brings about the reverse reaction. A stream of hydrogen reduces all the iron(III) oxide to iron as long as the flow of hydrogen sweeps away the steam that has formed.

$$Fe_3O_4(s) + 4H_2(g) \rightarrow 3Fe(s) + 4H_2O(g)$$

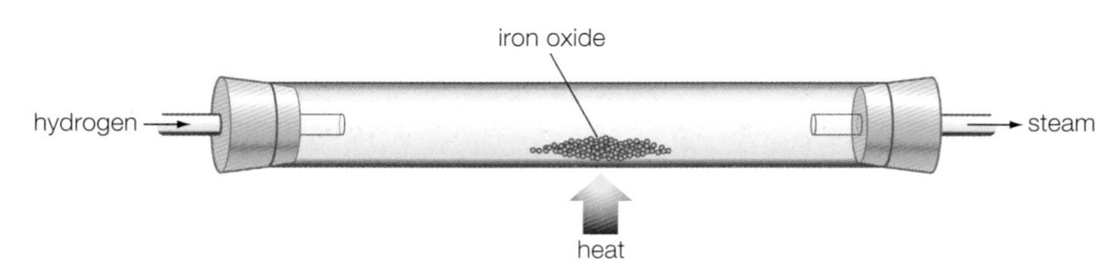

Figure 17.26 ▲
The backward reaction happens when the concentration of hydrogen is high and the steam is swept away, which keeps its concentration low.

Test yourself

11 Can the following changes be reversed either by changing the temperature or by changing the concentration of a reactant or product?
   a) Freezing water to ice
   b) Changing blue litmus to its red form
   c) Converting blue copper(II) sulfate to its white form by heating

12 Explain why wet washing does not dry if it is kept in a plastic laundry bag but it does dry if it is hung out on a line.

17.7 Reaching an equilibrium state

Reversible changes often reach a state of balance or equilibrium. What is special about chemical equilibria is that nothing seems to be happening, but there is ceaseless change at a molecular level.

When chemists ask the question 'How far?', they want to know what the state of a reaction will be when it reaches equilibrium. At equilibrium, the reaction shown by an equation may be well to the right (mostly new products), well to the left (mostly unchanged reactants) or somewhere in between.

Balance points exist in most reversible reactions when neither the forward nor the reverse reaction is complete. Reactants and products are present together and the reactions seem to have stopped. This is the state of chemical equilibrium.

One way to study the approach to equilibrium is to watch what happens when a small crystal of iodine is shaken in a test tube with hexane and a solution of potassium iodide, KI(aq). The liquid hexane and the aqueous solution do not mix.

Iodine freely dissolves in hexane, which is a non-polar solvent (see Section 6.15). The non-polar iodine molecules mix with the hexane

Note

In an equation, the chemicals on the left-hand side are the reactants. Those on the right are the products. The 'left-to-right' reaction is the 'forward' reaction and the 'right-to-left' reaction is the backward reaction.

molecules. There is no reaction. The solution is a purply-violet colour – the same colour as iodine vapour. Iodine hardly dissolves in water, but it does dissolve in a solution of potassium iodide. The solution is yellow, orange or brown depending on the concentration. In the solution, iodine molecules, I_2, react with iodide ions, I^-, to form tri-iodide ions, I_3^-.

Figure 17.27 is a study of changes that can be summed up by this equilibrium:

$$I_2\text{(in hexane)} + I^-\text{(aq)} \rightleftharpoons I_3^-\text{(aq)}$$

Note

The symbol $\rightleftharpoons$ represents a reversible reaction at equilibrium. In theory it is only possible to achieve a state of equilibrium in a closed system (see Section 16.1).

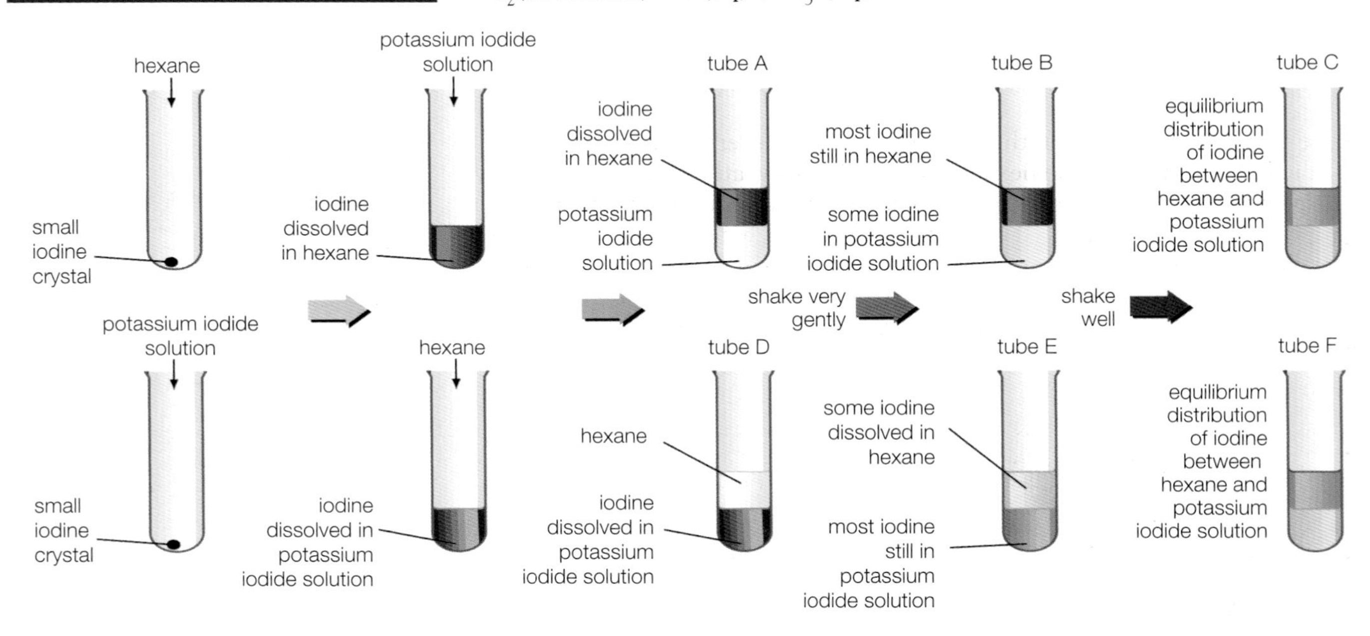

Figure 17.27 ▲
Two approaches to the same equilibrium state.

The graphs in Figure 17.28 show how the iodine concentration in the two layers changes with shaking. After a little while, no further change seems to take place. Tubes C and F look just the same, and both contain the same equilibrium system. This demonstration shows two important features of equilibrium processes:

- at equilibrium, the concentration of reactants and products does not change
- the same equilibrium state can be reached from either the 'reactant side' or the 'product side' of the equation.

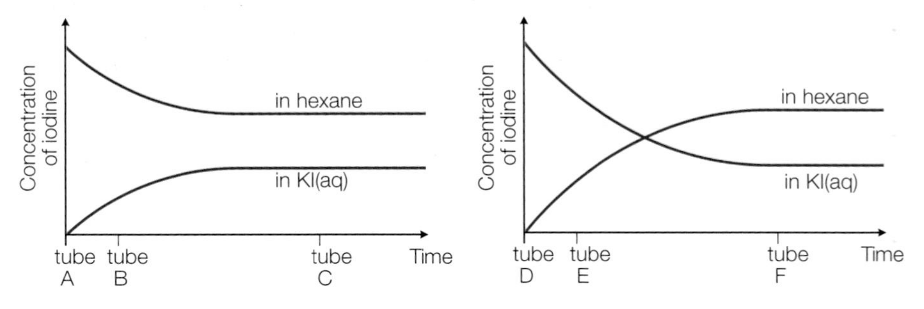

Figure 17.28 ▶
Change in concentration of iodine with time in the mixtures shown in Figure 17.27.

17.8 Dynamic equilibrium

Figure 17.29 shows what is happening at a molecular level in the equilibrium involving iodine moving between hexane and a solution of potassium iodide. Consider tube A in Figure 17.27. All of the iodine molecules start in the upper hexane layer. On shaking, some move into the aqueous layer. At first, molecules can move only in one direction (the forward reaction). The forward reaction begins to slow down as the concentration in the upper layer falls.

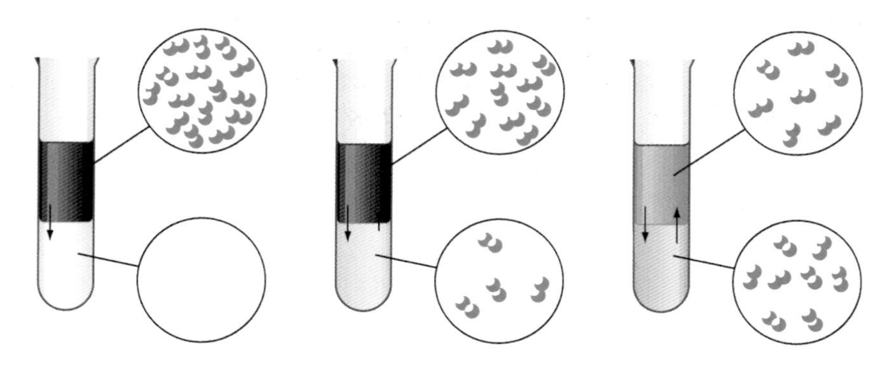

Figure 17.29 ◀
Iodine molecules reaching dynamic equilibrium between hexane and a solution of potassium iodide. (The formation of I_3^- ions in the aqueous layer is ignored in this diagram.)

Once there is some iodine in the aqueous layer, the reverse process can begin, with iodine returning to the hexane layer. This backward reaction starts slowly but speeds up as the concentration of iodine in the aqueous layer increases.

In time, both forward and backward reactions will happen at the same rate. Movement of iodine between the two layers continues, but overall there is no change. In tube C in Figure 17.27, each layer is gaining and losing iodine molecules at the same rate. This is an example of dynamic equilibrium.

Definition

In **dynamic equilibrium**, the forward and backward reactions continue but at equal rates, so the overall effect is no change. At a molecular level, there is continuous movement. At the macroscopic level, nothing seems to be happening.

Test yourself

13 Under what conditions are the following in equilibrium?
   a) Water and ice
   b) Water and steam
   c) Copper(II) sulfate crystals and copper(II) sulfate solution

14 Draw a diagram to represent the movement of particles between a crystal and a saturated solution of the solid in a solvent.

17.9 Factors affecting equilibria

Changing the conditions can disturb a system at equilibrium. At equilibrium, the rate of the forward and backward reactions is the same, and anything that changes the rates can shift the balance.

Predicting the direction of change

Le Chatelier's principle is a qualitative guide to the effect of changes in concentration, pressure or temperature on a system at equilibrium. The principle was suggested as a general rule by the French physical chemist Henri le Chatelier (1850–1936). The principle states that when the conditions of a system at equilibrium change, the system responds by trying to counteract the change.

Changing the concentration

Table 17.2 shows the effects of changing the concentration in the generalised equilibrium system:

$$A + B \rightleftharpoons C + D$$

Table 17.2 ◀

Disturbance	How the equilibrium mixture responds	The result
Concentration of A increases	System moves to the right Some A is removed by reaction with B	More C and D form
Concentration of D increases	System moves to the left Some of the added D is removed by reaction with C	More A and B form
Concentration of D decreases	System moves to the right to make up for the loss of D	There is more C and less A and B in the new equilibrium

The reaction of bromine with water provides an example of the predictions based on le Chatelier's principle. A solution of bromine in water is a yellow-orange colour because it contains bromine molecules in the following equilibrium:

$$\underbrace{Br_2(aq)}_{\text{orange}} + H_2O(l) \rightleftharpoons \underbrace{OBr^-(aq) + Br^-(aq) + 2H^+(aq)}_{\text{colourless}}$$

Adding alkali turns the solution almost colourless. This is because hydroxide ions in the alkali react with hydrogen ions and remove them from the equilibrium. As the hydrogen ion concentration decreases, the equilibrium shifts to the right, and orange bromine molecules are converted to colourless ions. Lowering the hydrogen ion concentration slows down the backward reaction while the forward reaction continues as before. The position of equilibrium shifts until the rates of the forward and backward reactions are the same again.

Adding acid increases the concentration of hydrogen ions. This speeds up the backward reaction and makes the solution turn orange-yellow again. The equilibrium shifts to the left, which reduces the hydrogen ion concentration and increases the bromine concentration until the forward and backward reactions are in balance again.

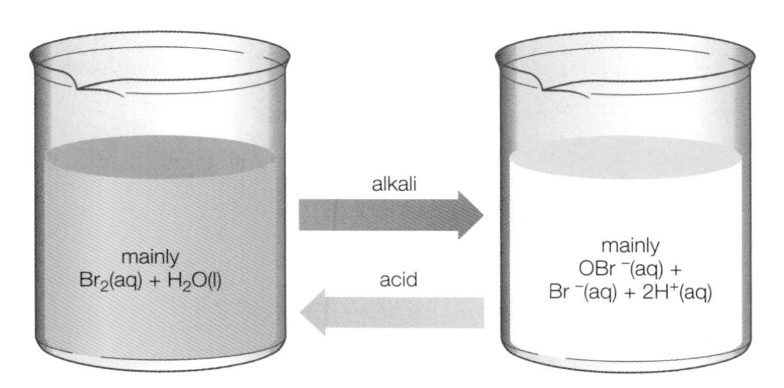

Figure 17.30 ▶ Visible effects of adding alkali and acid to a solution of bromine in water.

Test yourself

15 Write an ionic equation for the reversible reaction of silver(I) ions with iron(II) ions to form silver atoms and iron(III) ions. Make a table similar to Table 17.2 to show how le Chatelier's principle applies to this equilibrium.

16 Yellow chromate(VI) ions, $CrO_4^{2-}(aq)$, react with aqueous hydrogen ions, $H^+(aq)$, to form orange dichromate(VI) ions, $Cr_2O_7^{2-}(aq)$, and water molecules. The reaction is reversible. Write an equation for the system at equilibrium. Predict how the colour of a solution of chromate(VI) ions would change:
a) on adding acid
b) on then adding hydroxide ions that neutralise hydrogen ions (see Section 3.6).

17 Heating limestone, $CaCO_3$, in a closed furnace produces an equilibrium mixture of calcium carbonate with calcium oxide, CaO, and carbon dioxide gas. Heating the solid in an open furnace decomposes the solid completely into the oxide. How do you account for this difference?

Changing the pressure and temperature

Many industrial processes happen in the gas phase. High pressures and high temperatures are often needed, even when there is a catalyst. One important example of this is the Haber process that is used to make ammonia. The equilibrium system involved is:

$$N_2(g)+3H_2(g) \rightleftharpoons 2NH_3(g) \qquad \Delta H = -92.4\,\text{kJ mol}$$

The reaction takes place in a reactor packed with an iron catalyst. Adding a catalyst does not affect the position of equilibrium. Instead, it speeds up both the forward and backward reactions so it shortens the time taken to reach equilibrium.

Le Chatelier's principle helps to explain the conditions chosen for the Haber process. There are 4 moles of gases on the left-hand side of the equation but only 2 moles on the right. Increasing the pressure makes the equilibrium shift from left to right, because this reduces the number of molecules and tends to reduce the pressure. Increasing the pressure thus increases the proportion of ammonia at equilibrium.

The reaction is exothermic from left to right and therefore endothermic from right to left. Le Chatelier's principle predicts that increasing the temperature makes the system shift in the direction that takes in energy (tending to lower the temperature). Increasing the temperature thus decreases the proportion of ammonia at equilibrium.

Tutorial

Activity

Ethanol from ethene in industry

In the UK, ethanol is manufactured by the reaction of ethane with water in the presence of a catalyst. The reaction is reversible and exothermic, with an enthalpy change of reaction of $-46\ \text{kJ mol}^{-1}$.

The catalyst is phosphoric acid. It is supported on the surface of an inert solid (silica). The proportion of steam in the reaction has to be controlled to prevent the catalyst taking up water so that it is diluted and runs off the surface of the solid.

The process is carried out at 300 °C at 60–70 times atmospheric pressure. The water:ethene ratio is around 0.6:1.

1 Which alternative method of making ethanol is mainly used in other parts of the world?

2 How can ethene for this process be made?

3 **a)** Write an equation for the reaction of ethene with steam and show the enthalpy change.
b) What type of reaction is this?

4 Suggest a reason for supporting the phosphoric acid on the surface of silica.

5 **a)** State three conditions that favour the formation of ethene at equilibrium according to le Chatelier's principle.
b) Suggest reasons why the working conditions are not those suggested by your answer to part **a)**.

6 About 5% of the ethene is converted to ethanol each time the reaction mixture passes through the catalyst bed. Suggest how a yield of 95% conversion is achieved.

REVIEW QUESTIONS

Extension questions

1 Hydrogen peroxide solution, $H_2O_2(aq)$, decomposes slowly and releases oxygen. The reaction is catalysed by manganese(IV) oxide. Table 17.3 shows the volume of oxygen collected at regular intervals when one measure of MnO_2 powder is added to 50 cm³ of a hydrogen peroxide solution at 20 °C.

Time/s	0	20	40	60	80	100	120	140	160	180
Volume of oxygen/cm³	0	10	20	26	32	35	38	39	40	40

Table 17.3 ▲

a) Write an equation for the reaction. (1)

b) Draw a graph of the results on axes with a vertical scale that shows the volume of oxygen up to 100 cm³. (2)

c) Explain the shape of your graph. (2)

d) On the same axes, sketch the graphs you would expect if, in separate experiments, all the conditions are the same except that:

i) the temperature is increased to 40 °C
ii) the volume of hydrogen peroxide solution is 100 cm³
iii) manganese(IV) oxide granules are used in place of powder
iv) the concentration of the hydrogen peroxide solution is halved. (8)

2 a) Explain why most collisions in a reaction mixture do not result in a reaction. (2)

b) How can the collision frequency between molecules in a gas be increased without changing the temperature? (1)

c) Why can a small increase in temperature lead to a large increase in the rate of a reaction? (3)

3 Sketch the reaction profile for a reaction that takes place with a catalyst given that:

- the reaction is endothermic
- the activation energy for the formation of an intermediate is higher than the activation energy for the conversion of the intermediate to the products. (3)

4 a) Sketch a graph, with labelled axes, to show the Maxwell–Boltzmann distribution of the energies of the reactant molecules. (3)

b) On the energy axis, mark plausible values for the activation energy of a reaction without a catalyst and the activation energy with a catalyst. (2)

c) Use your graph to explain why a catalyst speeds up the rate of a reaction. (3)

5 Carbon dioxide is dissolved in water under pressure to make sparkling mineral water. In a bottle of sparkling mineral water, there is an equilibrium between carbon dioxide dissolved in the drink, $CO_2(aq)$, and carbon dioxide in the gas above the drink, $CO_2(g)$.

a) Write an equation to represent the equilibrium between carbon dioxide gas and carbon dioxide in solution. (1)

b) Use this example to explain the term 'dynamic equilibrium'. (2)

c) Explain why lots of bubbles of gas form when a bottle of sparkling mineral water is opened. (2)

d) Less than 1% of the dissolved carbon dioxide reacts with water to form hydrogencarbonate ions:

$$CO_2(g) + H_2O(l) \rightleftharpoons HCO_3^-(aq) + H^+(aq)$$

Use this equation to explain why carbon dioxide is much more soluble in sodium hydroxide solution than in water. (2)

6 Haemoglobin is a large molecule in red blood cells that can be represented by the symbol Hb. Haemoglobin carries oxygen in the blood from our lungs to the cells in our bodies.

$$Hb(aq) + 4O_2(g) \rightleftharpoons HbO_8(aq)$$

a) Suggest a reason why haemoglobin takes up oxygen as blood passes through the blood vessels in the lungs. (2)

b) Suggest a reason why haemoglobin releases oxygen as blood passes through the blood vessels between cells in muscles. (2)

c) Haemoglobin molecules are affected by the presence of carbon dioxide. The molecules hold onto oxygen less strongly if the carbon dioxide concentration is higher. Why does this help the blood to deliver oxygen to cells in muscles? (2)

7 Methanol is manufactured from carbon monoxide and hydrogen by the following reaction:

$$CO(g) + 2H_2(g) \rightleftharpoons CH_3OH(g) \quad \Delta H = -91 \text{ kJ mol}^{-1}$$

The reaction is carried out in the presence of aluminium oxide pellets coated with a mixture of copper and zinc oxides. The process runs at a pressure that is about 100 times atmospheric pressure and at a temperature of 575 K.

a) Explain why the process operates at a high pressure. (2)

b) What is the purpose of the pellets coated with metal oxides? (1)

c) i) What is the effect of increasing the temperature on the yield of methanol at equilibrium. Explain your answer. (2)

ii) Suggest why a temperature as high as 575 K is used. (2)

18 Chemistry of the air

From the chemist's viewpoint, the atmosphere can be imagined as a giant chemical reactor with chemical inputs from land, sea, living things and human activities. The mixture of chemicals is irradiated with energy from the Sun.

18.1 The atmosphere

The atmosphere is in layers. We live in the lowest layer, which is called the troposphere. The air gets cooler with increasing height in the troposphere. There is then an important point at which the temperature begins to increase again. This temperature inversion marks the beginning of the second layer, which is called the stratosphere (Figure 18.1).

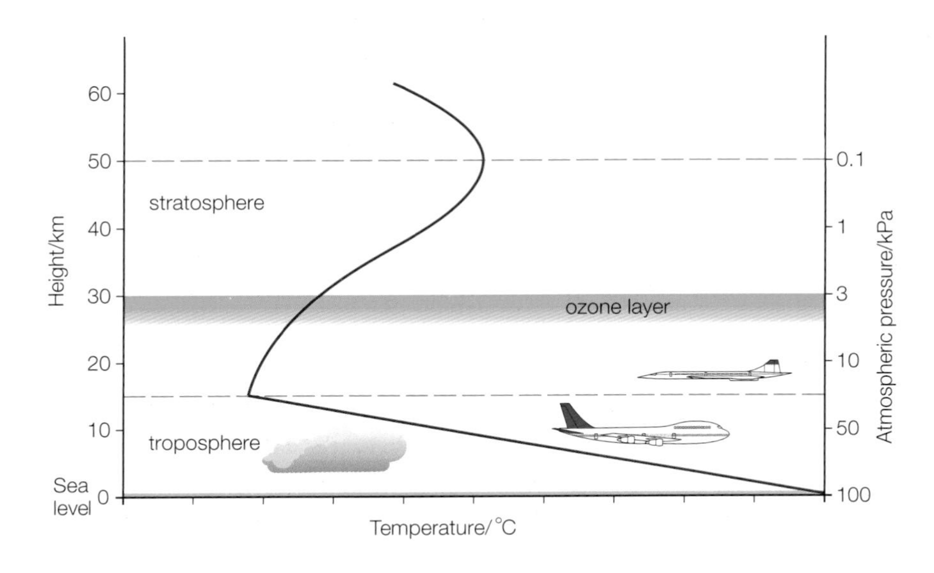

Figure 18.1 ◄ Structure of the Earth's atmosphere.

The thin cold air at the top of the troposphere is trapped between the warmer air of the stratosphere and the warmer denser air in the troposphere below. The effect is to create a barrier that makes it difficult for molecules to diffuse upwards into the stratosphere.

Human activities change the composition of the atmosphere. These human, or anthropogenic, inputs to the troposphere include pollutant gases from motor vehicles, power stations and industrial processes. Most of these pollutants do not remain unchanged for long enough to reach the stratosphere. An important exception to this is the group of compounds called CFCs (see Section 14.5) – very stable compounds that do not break down in the lower atmosphere. In time, these compounds move up into the stratosphere, where they can destroy the ozone that protects us from damaging radiation from the Sun.

Definition

Anthropogenic effects are changes brought about by human activities. They include factors that affect the atmosphere, such as burning fossil fuels, deforestation and intensive agriculture.

18.2 Evidence for atmospheric change

Chemists, and other scientists, provide the evidence that shows that global warming and climate change has happened in the past and is happening now. Sensitive techniques of chemical analysis, such as the analysis of gases trapped in glacial ice, have played an important part in obtaining data about changes in the atmosphere that happened hundreds of thousands of years ago.

Scientists have been measuring the concentration of carbon dioxide in the air since 1958. The main site for making measurements is the Mauna Loa Observatory in Hawaii. Mauna Loa is the highest mountain in Hawaii, and the observatory is 3400 metres above sea level. The air samples are collected

Test yourself

1 Suggest reasons why the area round the Mauna Loa observatory is a good place to measure the concentration of carbon dioxide in the air.

through inlet tubes placed several metres above the ground in the barren region of volcanic rock. The air is analysed using an infrared gas analyser, which continuously records the concentration of carbon dioxide in the stream of gas passing through it.

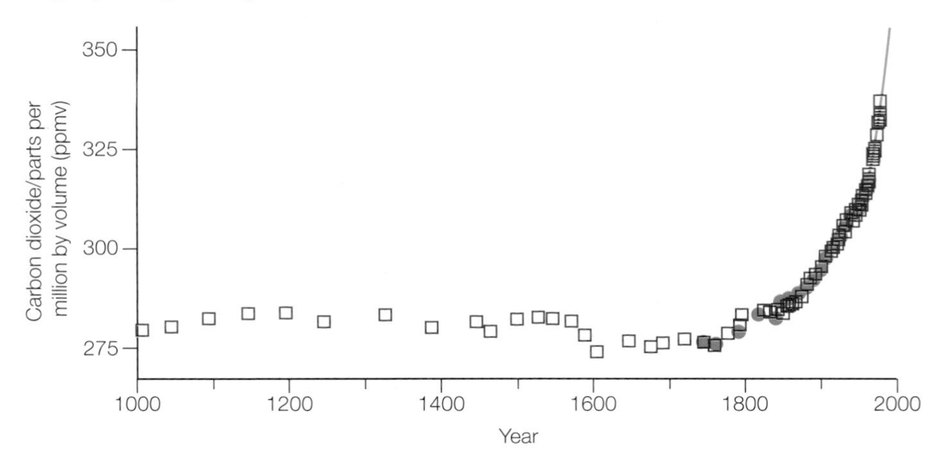

Figure 18.2 ▶
Concentrations of carbon dioxide in the atmosphere over the last 1000 years. The red squares and green dots are the results from air trapped in glacial ice. The blue line is based on direct measurements of the gas in the atmosphere.

Figure 18.3 ▲
Dr Eric Wolff, principal investigator at the British Antarctic Survey, studying an ice core in Antarctica.

Accurate information about the concentration of gases in the air before 1958 comes from studies of ice in the Antarctic and other cold regions of the Earth. Snow falling in the Antarctic never melts. Instead, it is compressed by more snow falling on top of it and turns it to ice. Over hundreds of thousands of years, the Antarctic ice caps have built up to a depth of more than 4 kilometres.

The falling snow traps pollutants and particles. The snow crystals also trap bubbles of air, which remain in the ice forever. Even the water in the snow carries clues to past climates. Most of the water is $H_2{}^{16}O$, but there is also a small proportion of $H_2{}^{18}O$. The proportion of the two isotopes depends on the Earth's temperature at the time the snow falls. Water with the heavier isotope has a slightly greater tendency to condense out as air moves from warmer regions to the Antarctic. The colder the climate, the more water condenses as clouds and rain before it reaches the Antarctic where snow falls. So, the colder the climate, the higher the ratio of $^{16}O:^{18}O$ in the snow.

Scientists drill down into the ice to extract cylindrical cores about 10 cm in diameter, which they store at −20 °C. They cut the cores into sections, some of which are analysed straight away in the field. Other samples are sent to laboratories for analysis by techniques that are not possible in the Antarctic, such as mass spectrometry.

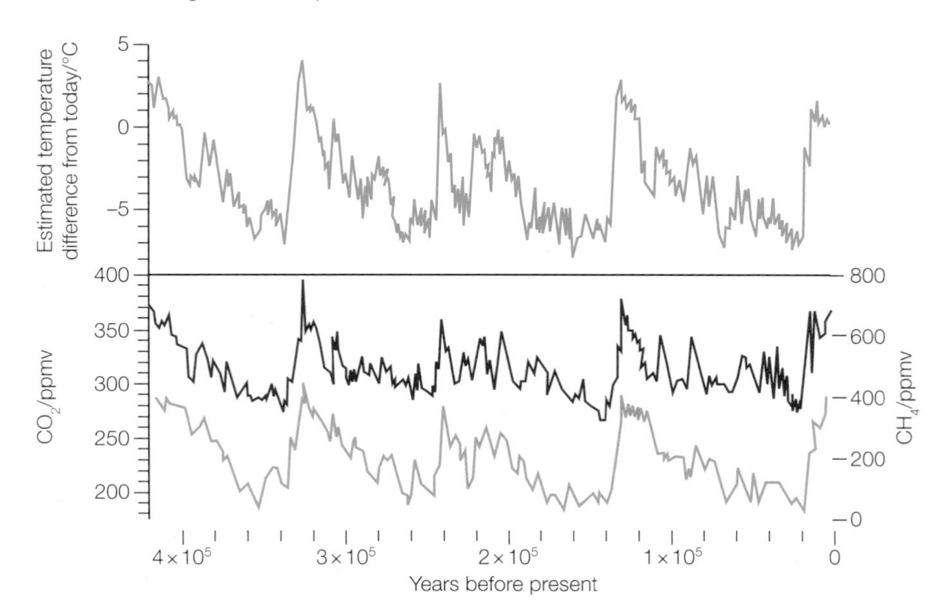

Figure 18.4 ▶
Changes to temperature and concentrations of carbon dioxide and methane in the atmosphere over the last 420 000 years from analysis of ice cores in Antarctica. The red line shows data for methane, CH_4 (right axis). The orange line shows data for carbon dioxide, CO_2 (left axis).

Test yourself

2 Two of the techniques used to analyse ice samples are mass spectrometry and infrared spectroscopy (see Topic 15).
   a) Which technique is suitable for measuring the ratio of oxygen isotopes in ice?
   b) Which technique is suitable for measuring the concentration of carbon dioxide in air samples from ice bubbles?
3 Ice samples contain only tiny traces of some pollutants. Suggest two precautions necessary to achieve accurate results when samples are sent to laboratories for analysis.
4 What can you conclude about changes in the atmosphere of CO_2 from the information in Figure 18.4?
5 Analysis of ice in Greenland has shown that the level of lead pollution in 1969 was 200 times the level in 1700. In 1990, the level of lead was just 10 times its natural level. Suggest explanations for these changes.

18.3 Greenhouse gases and climate change

Without the atmosphere, the whole Earth would be colder than Antarctica. The greenhouse effect keeps the surface of the Earth about 30 °C warmer than it would be if there were no atmosphere. With no greenhouse effect, there would be no life on Earth.

Most of the energy in the Sun's radiation is concentrated in the visible and ultraviolet regions of the spectrum. When this radiation reaches the Earth's atmosphere, about 30% is reflected into space, 20% is absorbed by gases in the air and about 50% reaches the surface of the Earth.

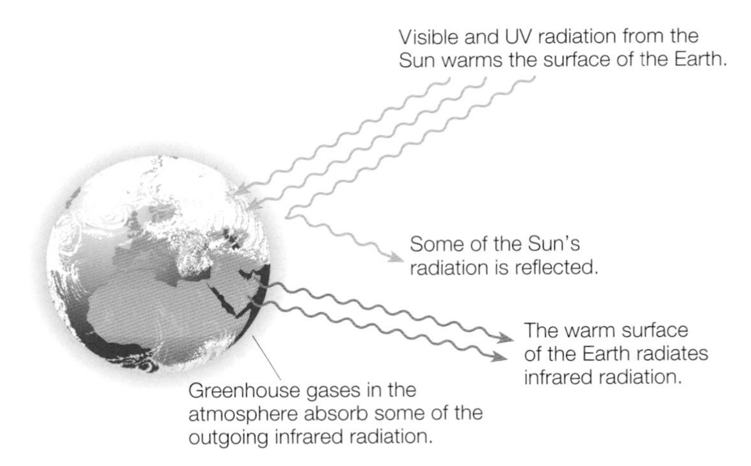

Figure 18.5 ◀ Greenhouse effect.

The surface of the Earth is very much cooler than the Sun. This means that it radiates most of its energy back into space at longer infrared wavelengths. Some of this infrared radiation is absorbed by molecules in the air and warms up the atmosphere. This is the greenhouse effect.

The Earth's average temperature stays the same if the energy input from the Sun is balanced by the re-radiation of energy back into space. Anything that changes this steady state gradually leads to global warming or global cooling.

Nitrogen and oxygen make up most of the air, but they do not absorb infrared radiation. This is because they are not polar. They are never polar, even when they vibrate, which means they cannot absorb infrared radiation (see Section 15.1).

The gases in the air that do absorb infrared radiation are called greenhouse gases. The natural greenhouse gases that help keep the Earth warm are water vapour, carbon dioxide and methane. Water vapour makes the biggest contribution.

Since the Industrial Revolution, human activity has led to a marked increase in the concentrations of greenhouse gases in the atmosphere. More recently, new synthetic chemicals such as CFCs have joined the mix. The CFCs and their related compounds are very powerful greenhouse gases.

Adding greenhouse gases to the air means that more infrared radiation from the surface of the Earth is trapped. This enhances the greenhouse effect. As a result, the Earth warms up more than it would naturally, and as the Earth warms up, its climate changes.

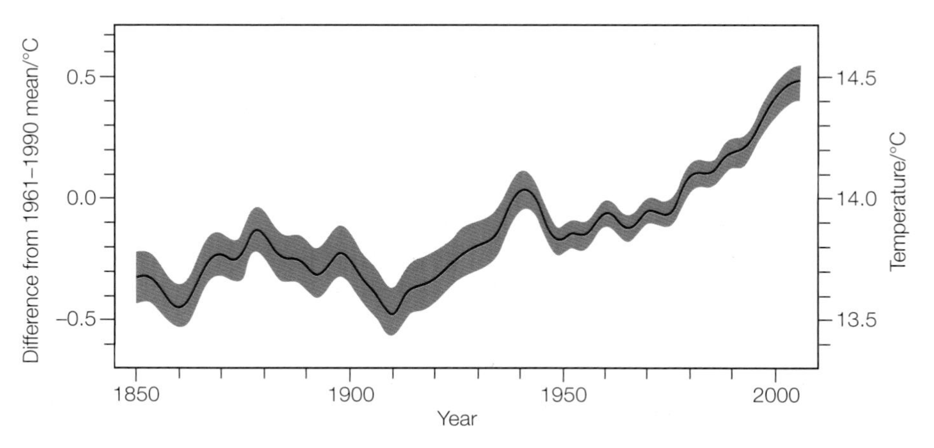

Figure 18.6 ▶
Variations in the global mean temperature since 1850 compared with the average value from 1961–1990. The blue area indicates the range of uncertainty around the red line that shows the mean value. In 2005, the concentration of CO_2 by far exceeded the natural range over the last 650 000 years.

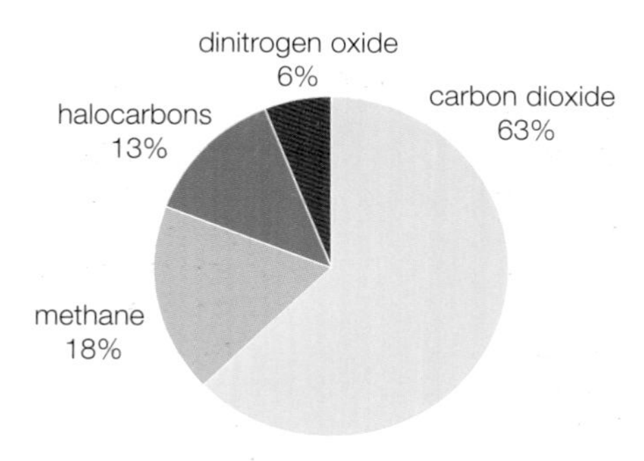

Figure 18.7 ▲
Contributions of the main greenhouse gases to the enhanced greenhouse effect.

The contribution that a greenhouse gas makes to the enhanced greenhouse effect is determined by several factors. These include how strongly the gas absorbs infrared energy, how much of it is already in the atmosphere, how much is being added to the atmosphere and how long it survives in the atmosphere.

Test yourself

6 Methane has non-polar molecules, yet it can absorb infrared radiation.
 a) Draw the structure of methane and show the polarity of the bonds.
 b) Why is methane non-polar overall?
 c) Using a simple diagram, suggest a pattern of vibrations for C–H bonds in methane that can lead to a varying dipole.

7 Human activities produce a lot of water vapour. Why does this have little effect on the concentration of water in the atmosphere?

8 Give three examples of human activities that have led to a marked increase in the concentration of carbon dioxide in the atmosphere.

9 The concentration of all CFCs in the atmosphere is about one million times less than the concentration of CO_2, yet these compounds make a significant contribution to the enhanced greenhouse effect. Suggest an explanation for this.

10 Why do you think most public debate about climate change and greenhouse gases focuses on carbon dioxide?

11 The blue area on either side of the graph line in Figure 18.6 is narrower for the year 2000 than for the year 1900. Suggest reasons for this.

12 How does the data in Figure 18.6 suggest that recent global warming is the result of human activities and not just the result of natural climate variations?

18.4 Impacts of climate change

In 2007, the Intergovernmental Panel for Climate Change (IPCC) issued its fourth assessment of scientific evidence. The IPCC report showed a strong scientific consensus that climate change is happening and is the result of human activity.

Since 2005, the concentration of carbon dioxide in the atmosphere has been higher than it has ever been for the last 650 000 years. Some of the consequences are clear from the evidence. Eleven of the warmest years observed since records began were in the 12 years leading up to the report in 2007. Average sea levels rose by 17 cm in the twentieth century and will continue to rise. Large quantities of snow and ice have vanished from the northern hemisphere, and this threatens water supplies for hundreds of millions of people.

The weather is altering in many regions. Extreme events such as hurricanes, droughts and heat waves are becoming more frequent. At the same time, patterns of rainfall and snowfall are changing, with more precipitation in the temperate parts of the world but less in the tropics and around the Mediterranean. These changes are a threat to both food crops and wildlife. A quarter of plant and animal species will be at risk of extinction if the increase in mean temperatures exceeds 1.5–2.5 °C. These changes will also have consequences for human health and disease.

Most governments now accept that we must make huge changes to the way we supply and use energy and in all other activities that release greenhouse gases. The government in the UK has a strategy for dealing with climate change. Each year, by law, it must report to Parliament on progress towards meeting the targets aimed at reducing emissions of CO_2. Organisations and individuals are being encouraged to consider their own carbon footprint and to find ways of reducing it.

A carbon footprint measures the total amount of carbon dioxide, and other greenhouse gases, released into the air over the full lifecycle of a process or product. On average, the carbon footprint for one person in the UK is just under 11 tonnes of carbon dioxide per year from all activities. Of this amount, more than 40% comes directly from our individual activities, such as heating and lighting our homes and driving vehicles. The average driver in the UK produces 4 tonnes of CO_2 per year just from motoring.

Figure 18.8 ▲
These two satellite images show the retreat of the Helheim glacier in Greenland between May 2001 and June 2005. The glacier is the grey region at the left of each image. To its right is a narrow fjord filled with icebergs that have broken from the glacier's front. The glacier length was fairly consistent until 2001 but then it retreated around four kilometres in four years. The retreat of the glacier is a consequence of climate change.

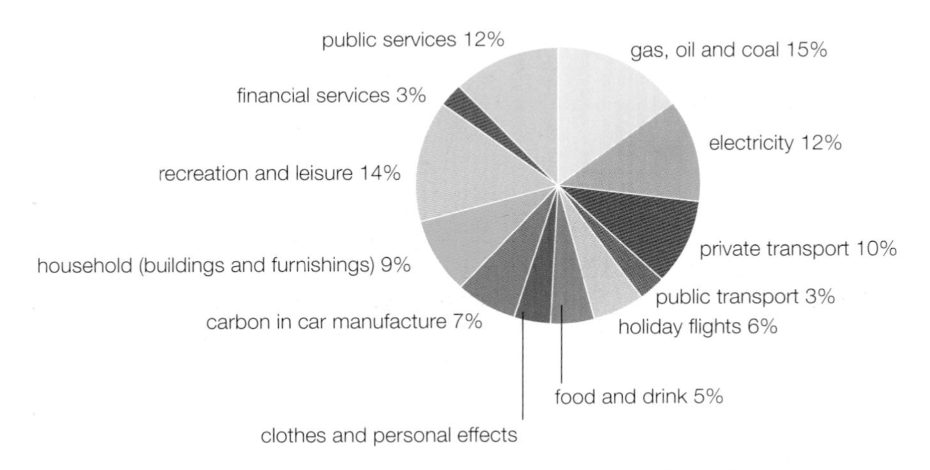

Figure 18.9 ◄
Part of a carbon footprint comes directly from activities such as home heating and travel, which involve burning fuels. Other parts of the footprint are indirect and arise from the whole lifecycle of the products and services we use.

Test yourself

13 Identify possible changes to your lifestyle that could significantly reduce your carbon footprint.

18.5 Carbon capture and storage

Carbon capture and storage (CSS) is one approach being explored for cutting emissions of carbon dioxide on a large scale. This type of approach is only appropriate when there is a concentrated source of carbon dioxide in a flow of waste gas from a power station or industrial site. The source also has to be fairly close to where the carbon dioxide is to be stored. It is then possible to capture between 85% and 90% of the gas.

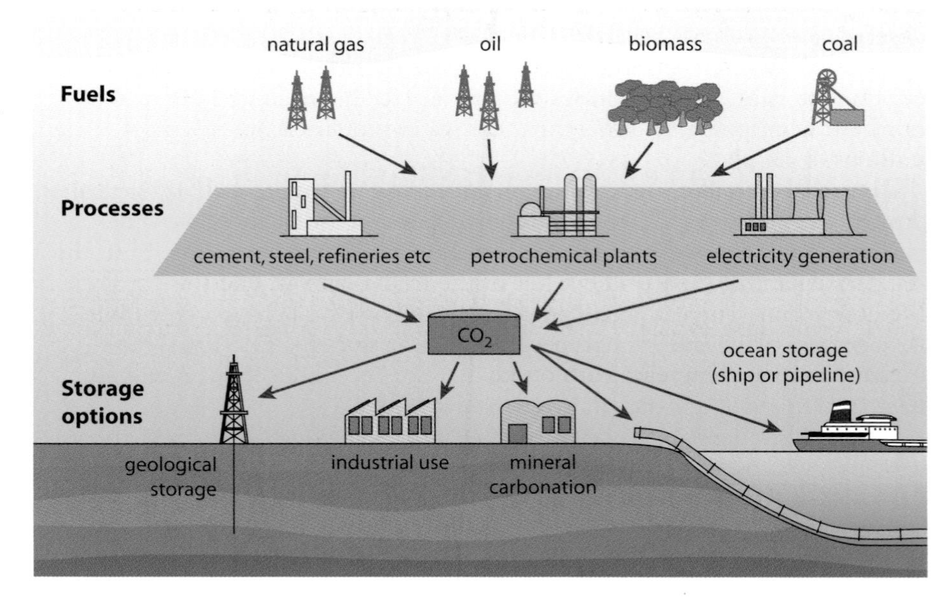

Figure 18.10 ▶ Options for carbon dioxide capture and storage: storage in deep geological formation, injection deep in the oceans or reaction with metal oxides to form stable carbonate minerals.

Test yourself

14 Identify three contributions that chemists can make to investigating solutions to climate change or its impacts.

15 Suggest three health, safety or environmental hazards that need to be considered when planning carbon capture and storage in the deep oceans or by mineral carbonation.

Injection of carbon dioxide into the deep oceans is still in the research phase; so too are the methods being investigated to combine carbon dioxide with metal oxides to form carbonate minerals. Injecting carbon dioxide into oil and gas fields is now feasible and economically viable in some circumstances.

The largest study of the injection of carbon dioxide into an oil field has been taking place at Weyburn in Canada since 2000. This involves injecting underground between 3000 and 5000 tonnes per day of carbon dioxide. Over the lifetime of the project, around 20 million tonnes of carbon dioxide will have been stored instead of being released into the atmosphere. At the same time, the injection of gas has increased the yield of oil from the field.

18.6 Ozone layer

The ozone layer is a concentration of the gas in the stratosphere at about 10–50 km above sea level. At this altitude, ultraviolet (UV) light from the Sun splits oxygen molecules into oxygen atoms:

$$O_2 + \text{UV light} \rightarrow O\bullet + O\bullet$$

These are free radicals that then convert oxygen molecules to ozone:

$$O\bullet + O_2 \rightarrow O_3$$

The ozone formed also absorbs UV radiation. This splits ozone molecules back into oxygen molecules and oxygen atoms and destroys ozone. This is the reaction that protects us from the worst effects of sunburn:

$$O_3 + \text{UV light} \rightarrow O_2 + O\bullet$$

In the absence of pollutants, a steady state is reached, with ozone being formed at the same rate as it is destroyed. Normally, the steady-state concentration of ozone is sufficient to absorb most of the dangerous UV light from the Sun.

The problem with CFCs and some other halogen compounds is that they are chemically very unreactive. They escape into the atmosphere where they are so stable that they last for many years – long enough for them to diffuse up to the stratosphere. In the stratosphere, the intense UV light from the Sun splits up CFCs to produce compounds such as hydrogen chloride, HCl.

Test yourself

16 Why is the presence of ozone in the stratosphere important to life on Earth?

These chlorine compounds then undergo a series of reactions that only happen in winter in the stratosphere above Antarctica. Once chemists understood these reactions, they could explain why the 'hole' in the ozone layer is most extreme over the South Pole.

In the dark winter over Antarctica, the stratosphere is so cold that clouds of ice crystals form. These clouds circle the pole in a spinning vortex of air that traps the ice crystals. On the surface of the ice crystals, a whole series of reactions take place that cannot happen anywhere else in the atmosphere. The reactions are fast. The end result of the reactions produces chemicals, including chlorine, Cl_2, that can destroy ozone in sunlight. When sunlight returns to the Antarctic in the spring of the southern hemisphere, the chlorine molecules are quickly split into chlorine atoms:

$$Cl_2 + \text{UV light} \rightarrow Cl\bullet + Cl\bullet$$

Chlorine atoms then react with ozone:

$$Cl\bullet + O_3 \rightarrow ClO\bullet + O_2$$
$$ClO\bullet + O\bullet \rightarrow Cl\bullet + O_2$$
$$\text{Overall: } O\bullet + O_3 \rightarrow 2O_2$$

The second reaction involves oxygen atoms, which are common in the stratosphere, and it reforms the chlorine atoms. In effect, therefore, one chlorine atom can rapidly destroy many ozone molecules. This effect was noticed in the early 1980s, when scientists in the Antarctic found that the ozone concentration in the stratosphere was much lower than expected. Since then, the ozone layer has been monitored by satellites, and this has confirmed that there is a 'hole' in the ozone layer not only over the Antarctic but also in other regions.

Figure 18.11 ▲
Polar stratospheric clouds form between 20 and 30 km above the ground. They only form at very low temperatures. That is why they only appear in winter and mainly over Scandinavia, Scotland, Alaska and Antarctica.

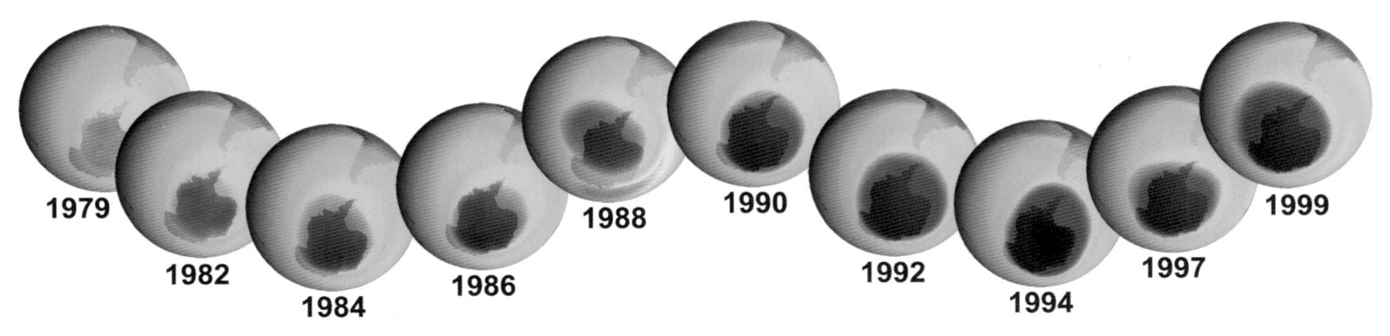

Figure 18.12 ▲
Growth of the ozone hole over Antarctica over 20 years. The dark blue end of the spectrum indicates the region of maximum ozone depletion. The red end of the spectrum shows the area of least depletion.

Test yourself

17 a) State the type of bond breaking when chlorine molecules split to form chlorine atoms.
b) Why are chlorine atoms in the atmosphere examples of free radicals?

18 a) Identify two reactions that remove ozone from the stratosphere.
b) Why can a single chlorine atom destroy 100 000 ozone molecules?

19 Why is UV spectroscopy a suitable technique for measuring concentrations of ozone in the stratosphere?

20 Why is the depletion of the ozone layer most severe over Antarctica?

Worldwide production of CFCs has fallen sharply since their impact on the ozone layer was discovered. At a conference of the United Nations in 1987, governments agreed on large reductions in the production of CFCs leading to an eventual ban. This was the Montreal Protocol. The protocol has been regularly revised to keep it up to date. It is often quoted as the most successful international treaty about the environment to date.

Concentrations in the atmosphere of the key CFCs have levelled off or started to fall since the Montreal Protocol came into effect. Meanwhile, the concentrations of the HCFCs have increased, because they were the early replacements for CFCs as solvents and refrigerants (see Section 14.5).

The effect of oxides of nitrogen on the ozone layer was discovered before the effects of CFCs were understood. Like CFCs, the nitrogen oxides NO and NO_2 react catalytically with ozone. They upset the steady state in the stratosphere and speed up the rate of breakdown of ozone:

$$NO + O_3 \rightarrow NO_2 + O_2$$
$$NO_2 + O\bullet \rightarrow NO + O_2$$

$$\text{Overall: } O\bullet + O_3 \rightarrow 2O_2$$

This happens to some extent naturally. Micro-organisms in the soil and oceans reduce nitrogen compounds to dinitrogen oxide, N_2O. Some of this gas is carried up into the stratosphere, where it reacts with oxygen atoms to form NO. High-flying aircraft can significantly increase the role of oxides of nitrogen in destroying ozone. This is especially so for supersonic aircraft that fly in the stratosphere. They can release nitrogen oxides from their engines directly into the ozone layer at altitudes of 20 km.

Test yourself

21 Explain why concentrations of CFCs in the atmosphere have started to fall, while concentrations of HCFCs have increased.

22 Write an equation to show the formation of NO from N_2O.

23 Suggest reasons why the effects of CFCs and related compounds on the ozone layer are much more significant than the effect of nitrogen oxides from aircraft.

18.7 Controlling air pollution

Monitoring air pollutions

For many people, a more urgent issue than climate change is the quality of the air they breathe from day to day. This is a particular concern for people who have asthma or heart disease. In the UK, 1500 monitoring stations measure the levels of pollutants in the air from motor vehicles, power stations, agriculture and industry. Some of these are in urban areas and others are in the countryside.

Many of the monitoring sites are automatic, and they record the concentrations of pollutants such as carbon monoxide, nitrogen oxides, hydrocarbons and ozone. Computers control the programme of monitoring and store the data. They feed the information to a central station. In the UK, many of the networks are connected to the internet, so that up-to-date information is available to the public hour by hour.

Figure 18.13 ▶ This automatic monitoring station monitors air quality in the City of Norwich. Air samples are drawn in through the inlet above the container of the instruments.

The monitoring techniques are based on spectroscopy. Infrared spectroscopy can be used to measure concentrations of carbon monoxide, while ozone concentrations are measured by ultraviolet spectroscopy.

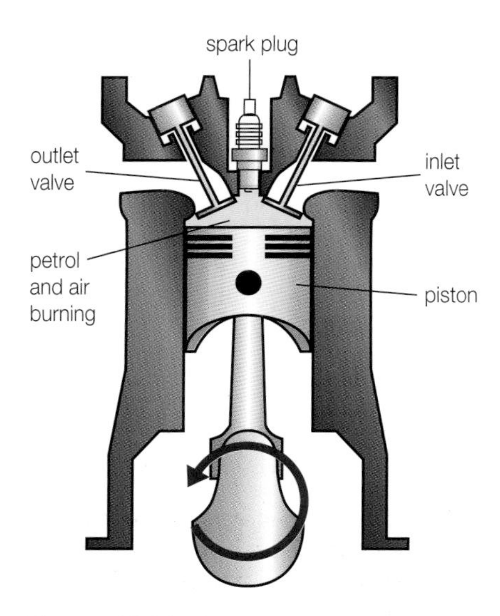

Figure 18.14 ▲
Petrol burns at a high temperature, under pressure, in the cylinder of a car engine.

Air pollution from motor vehicles

Motor vehicles are generally the main source of air pollutants in towns and near motorways. Engines that burn petrol or diesel fuels can pollute the air for three main reasons:

- they do not burn the fuel completely
- the fuel contains impurities
- they run at such a high temperature that nitrogen and oxygen in the air can react.

A controlled quality of air and fuel enters the cylinder of an engine before being compressed and burned. This means that some of the fuel may not burn completely and that some of the fuel may not burn at all. The pollutant gases that leave the engine are the primary pollutants. Some of these gases take part in reactions in the air that produce secondary pollutants.

Pollutant	Sources	Properties
Carbon monoxide	Incomplete combustion of fuel	A toxic gas that combines strongly with haemoglobin, which means that the blood can carry less oxygen. In low doses, this can put a strain on the heart. In higher doses, it kills
Oxides of nitrogen – NO and NO_2 – that are sometimes referred to as NO_x	Reaction of nitrogen with oxygen at the high temperature in the cylinder of an engine	Reacts with moist air to make nitric acid in acid rain. Affects the lungs and can worsen the symptoms of asthma
Unburned hydrocarbons	Incomplete combustion and evaporation of fuel	Some of the hydrocarbons, such as benzene, are carcinogenic. Others react with ozone to form harmful pollutants
Ozone	Reaction of oxygen with oxygen atoms formed from NO_2 in sunlight	Irritates the eyes, nose and throat
Oxidation products of hydrocarbons	Oxidation of unsaturated hydrocarbons by ozone	Toxic chemicals in photochemical smog, which irritate the eyes, nose and throat

Table 18.1 ◄
Gaseous air pollutants from motor vehicles.

One source of secondary pollutants is the decomposition of NO_2 in sunlight to form NO and free oxygen atoms. The oxygen atoms combine with oxygen molecules to form ozone. Ozone alone is a serious pollutant, but it can lead to further harm in still, sunny weather near cities, when it mixes with unburned hydrocarbons. The ozone reacts with the hydrocarbons to form a complex mixture of irritant chemicals that, in the absence of any wind, build up to create photochemical smog.

Test yourself

24 Classify the pollutants shown in Table 18.1 as primary or secondary pollutants.

25 In cities affected by photochemical smog, the levels of nitrogen oxides and unburned hydrocarbons generally peak in the early morning, but the peak levels for photochemical smog are around midday. Suggest a reason for this.

26 Explain why ozone in the stratosphere is good for our health while ozone in the troposphere is bad for our health.

27 Explain why the carbon dioxide from burning fuels is a serious pollutant of the atmosphere but is not included in Table 18.1.

Figure 18.15 ▲
Photochemical smog over Hong Kong, China.

Activity

Catalytic converters

Catalytic converters have done a great deal to improve air quality in our towns and cities. The catalyst speeds up reactions that remove pollutants from motor vehicle exhausts. The reactions convert oxides of nitrogen to nitrogen and oxygen. They also convert carbon monoxide to carbon dioxide and unburned hydrocarbons to carbon dioxide and steam.

The catalyst in a catalytic converter is made from a combination of platinum, palladium and rhodium. The pollutants are adsorbed onto the surface of the catalyst, where they react. The products are then desorbed into the stream of exhaust gases.

The catalyst must not adsorb molecules so strongly that the reactive sites on the surface of the metal are inactivated. However, the interaction between pollutant molecules and the metal surface has to be strong enough to weaken bonds and provide a reaction mechanism that is fast enough under the conditions in the exhaust system. The reaction products then have to be so weakly attracted that they are quickly released into the gas stream.

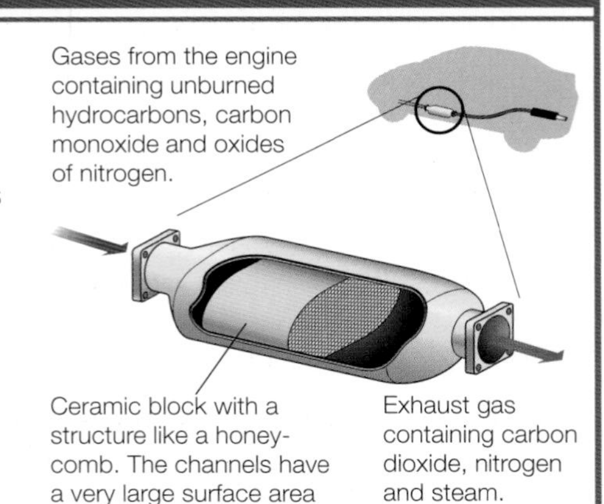

Figure 18.16 ▲ Diagram to show a catalytic converter.

1 Why do you think the catalyst in a catalytic converter is present as a very thin layer on the surface of many fine holes running through a block of inert ceramic?

2 Suggest reasons why the catalyst in a catalytic converter is only fully effective:
   a) after the engine has been running for some time
   b) if the engine is properly maintained so that it runs with the right mixture of air and fuel.

Figure 18.17 ▲ Surface of the metal catalyst in a catalytic converter absorbs the pollutants NO and CO, where they react to form CO_2 and N_2. Note that carbon atoms are shown as green in this computer graphic.

3 Write equations for the reactions catalysed by a catalytic converter that remove carbon monoxide and nitrogen monoxide from exhaust gases.

4 What is the effect of the catalyst that means it speeds up the reactions that destroy pollutants?

5 Why would the catalyst be ineffective if the bonding between the catalyst surface and the reactants was:
   a) too strong
   b) too weak?

6 Identify two ways, other than fitting catalytic converters, to reduce air pollution from motor vehicles in cities.

7 What contribution, if any, do catalytic converters make to solving the problem of climate change?

Definitions

Adsorption is a process in which atoms, molecules or ions are held onto the surface of a solid.
Desorption is the opposite of adsorption, when atoms, molecules or ions are released from a solid surface.

REVIEW QUESTIONS

1 a) What is the origin of the infrared radiation absorbed by gases in the air that leads to global warming? (3)

b) Explain, with examples, how the model of polar covalent bonds allows chemists to understand why some gases in the atmosphere contribute to global warming while others do not. (3)

2 a) Use the example of CFCs to show how an understanding of the mechanisms of reactions can help scientists understand the causes of environmental problems. (4)

b) Why have scientists recommended that CFCs should no longer be used in aerosols, foams and refrigerants? (2)

c) How have governments, industry and individuals responded to scientific information about ozone-depleting chemicals? (3)

4 a) Give examples to explain the difference between greenhouse gases and gases that help to deplete the ozone layer. (4)

b) Give two examples of compounds that are both greenhouse gases and ozone-depleting chemicals. Explain why this is possible. (3)

5 Three of the pollutants produced in petrol engines are carbon monoxide, oxides of nitrogen and unburned hydrocarbons.

a) Explain how these three types of pollutant are formed. (3)

b) Outline the harmful effects of each of these types of pollutant. (3)

c) Explain how a catalytic converter reduces the emissions of these pollutants from motor vehicles. (4)

19 Green chemistry

The aim of green chemistry is to meet all our needs for chemicals without damaging our health and the environment. Green chemistry makes the chemical industry more sustainable by favouring renewable feedstocks, reducing the use of energy resources and cutting down on waste. This is in line with the Rio Declaration on Environment and Development, which was agreed by governments in 1992. The declaration asserts that states should reduce and eliminate unsustainable patterns of production and consumption.

19.1 The chemical industry

The chemical industry converts raw materials into useful products. The industry manufactures some chemicals on a scale of thousands or even millions of tonnes per year. These are the bulk chemicals such as sodium hydroxide, sulfuric acid, chlorine and ethene.

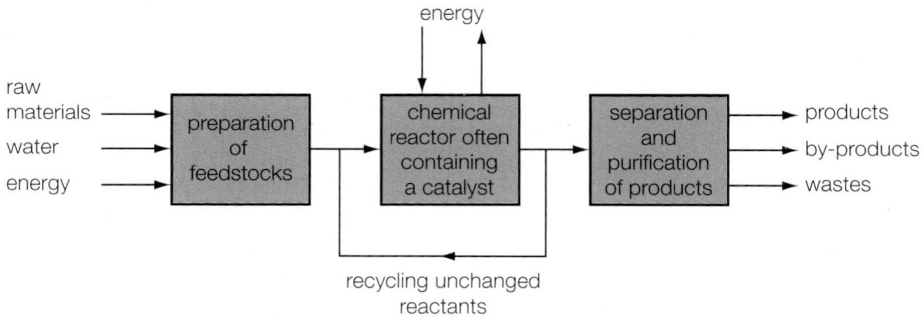

Figure 19.1 ▶
Principles of green chemistry can be applied to all aspects of a chemical process.

raw materials
water
energy
preparation of feedstocks
energy
chemical reactor often containing a catalyst
separation and purification of products
products
by-products
wastes
recycling unchanged reactants

Figure 19.2 ▲
The chemical industry helps to provide many of the products we rely on day by day, such as the chemicals used to make clothing fibres, building materials, food preservatives, drugs and medicines – however, some of these synthetic chemicals can be harmful.

The industry makes chemicals such as drugs and pesticides on a much smaller scale. It also makes a wide range of speciality chemicals that are needed by other manufacturers for specific purposes. These include products such as flame retardants, food additives and liquid crystals for flat-screen computer displays.

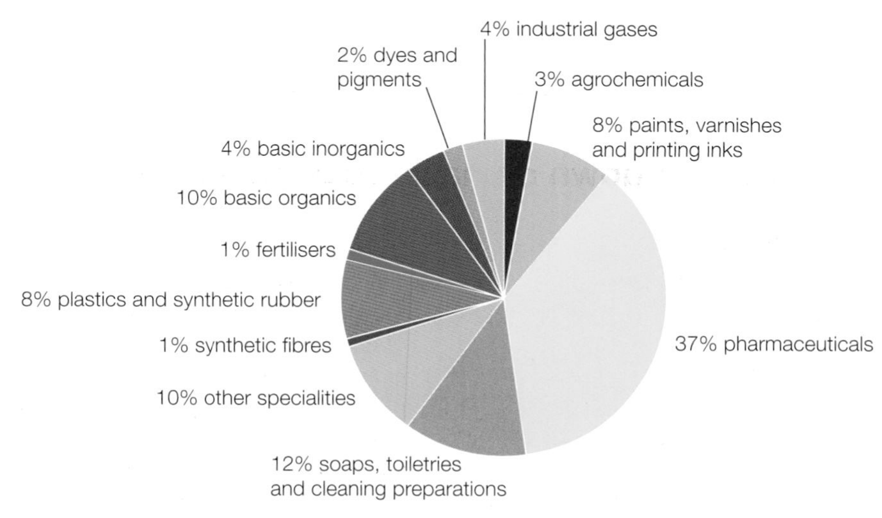

Figure 19.3 ▶
The range of products made by the chemical industry in the UK with their relative values.

Test yourself

1 Give the name and formula of a bulk chemical.
2 Pharmaceuticals are speciality chemicals not bulk chemicals. Suggest reasons why the pharmaceutical sector is so large in Figure 19.3.

19.2 Principles of green chemistry

In 1990, the US Pollution Prevention Act led to new thinking about ways to protect human health and the environment – a major area of interest was the chemical industry. In response to the act, scientists at the US Environmental Protection Agency were the first to describe the main features of green chemistry.

The key principles of green chemistry cover five broad areas of the production and the use of chemicals. These principles aim to make the chemical industry and our use of chemicals more sustainable:

- **Finding alternatives to hazardous chemicals** – this includes producing chemical products that are effective but not harmful and developing manufacturing processes that avoid toxic intermediates and solvents that are hazardous to people and the environment.
- **Designing more efficient processes with high atom economies** – this includes the development of catalysts to make possible processes with high atom economies and catalysts that are highly selective so that only the desired product is formed, thus cutting down wastes.
- **Changing to renewable resources** – this means choosing raw materials or chemical feedstocks that can be renewed instead of relying on crude oil and natural gas.
- **Seeking alternative sources of energy and improving energy efficiency** – this involves developing alternatives to fossil fuels and devising processes that run at low temperatures and pressures, as well as making good use of the energy released by chemical changes.
- **Reducing waste, recycling waste products and preventing pollution** – this includes minimising waste so that there is less to treat or dispose of, recycling materials and creating biodegradable products that do not persist in the environment but break down to harmless chemicals. Chemists can contribute through the increasing use of sensitive methods of analysis to detect pollution.

Many of the changes brought about by applying these principles not only make the chemical industry safer but also make it more sustainable.

Definition

Green chemistry is the design of chemical processes and products that reduce or eliminate the production and use of hazardous chemicals.

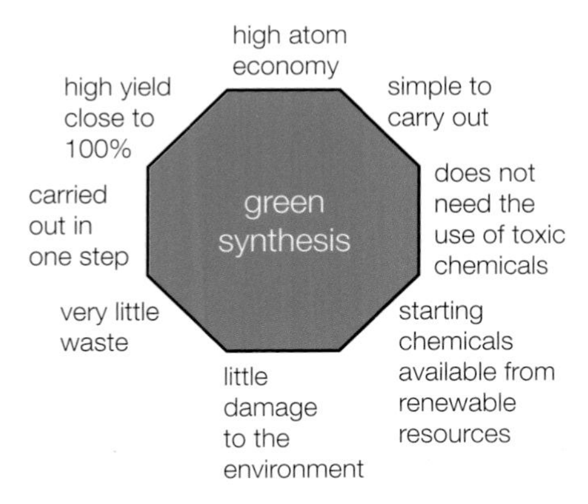

Figure 19.4 ▲
Key features of green synthesis.

Definition

Sustainable development meets the needs of the present without compromising the ability of future generations to meet their own needs.

Test yourself

3 Give examples to explain why the traditional chemical industry is not sustainable.

19.3 Cutting down on hazardous chemicals

Safer processes

The chemical industry uses some highly toxic and corrosive chemicals. One example is hydrogen cyanide, which is acutely toxic and requires special handling to minimise the risks to workers, the community and the environment. The industry has learned to work safely with such chemicals, but accidents do happen, and chemicals such as hydrogen cyanide are a threat to people who work in chemical plants and those who live in neighbouring communities.

The chemical company Monsanto, for example, manufactures a herbicide called Roundup®. The original process for making this product required both methanal and hydrogen cyanide – two unpleasant and toxic chemicals. The key reactions in the original process were exothermic and produced intermediates that became unstable if the temperature was too high. The overall process also created 1 kg of waste for every 7 kg of product, and the waste needed treating before it could be disposed of safely.

Fortunately, Monsanto has been able to develop a new process that relies on a copper catalyst and uses raw materials that are much less toxic and do not include hydrogen cyanide or methanal. The process is much safer and easier to control because the key reaction is endothermic. There is no waste because the catalyst can be filtered off for reuse and any chemicals not converted to the product are pure enough to be recycled directly into the reactor. There are fewer steps in the new process, and the yield is higher.

Figure 19.5 ▶
This herbicide is now made by a greener process that avoids toxic chemicals such as hydrogen cyanide and methanal.

Test yourself

4 a) Explain the danger of a runaway reaction in the older method for making Roundup®.
b) Explain why this danger is absent with the newer method.
5 Why is it an advantage to reduce the number of reaction steps involved in the manufacture of a chemical?

Safer solvents

Most of the organic chemicals used in industry are insoluble in water, which means that many important reactions must take place in other solvents. Traditionally, the chemical industry used a range of toxic liquids such as benzene and tetrachloromethane. The most hazardous solvents have largely been phased out, but the industry still uses other volatile organic compounds (VOCs) as solvents. The VOCs are pollutants when they evaporate into the air. Research by green chemists aims to replace harmful solvents with safer alternatives or even to use one of the reactants as the solvent.

The pharmaceutical company Pfizer has applied the principles of green chemistry to the manufacture of the drug Zoloft®, which is prescribed to people with depression. Zoloft® was first made in a multi-stage process that involved the use and recovery of four solvents: dichloromethane, tetrahydrofuran, methylbenzene and hexane. The new process takes place in two key steps, the solvent is ethanol and the use of a more selective palladium catalyst reduces waste.

As well as being a user of solvents, the chemical industry supplies solvents to customers, including people in the dry-cleaning business. For many years, the main solvent for dry cleaning has been 1,1,2,2-tetrachloroethene. This is used to remove grease and dirt from fabrics that would be damaged by water with detergent or soap. However, continued exposure to tetrachloroethene can cause liver and kidney damage in humans. The solvent may also be a carcinogen. As a result, alternative cleaning processes based on other solvents are being developed. One new system is based on 'supercritical' CO_2, which is carbon dioxide under high pressure. Supercritical CO_2 has properties intermediate between those of a liquid and a gas. The need to operate under pressure makes the equipment for dry cleaning with supercritical CO_2 expensive.

Definitions

A gas cannot be liquefied by pressure alone above its **critical temperature**. Carbon dioxide exists in the **supercritical state** above its critical temperature of 31.1 °C and its critical pressure, which is 73 times atmospheric pressure.

Test yourself

6 How has Pfizer improved both worker and environmental safety by redeveloping the process for making Zoloft®?

7 The uptake of dry-cleaning processes based on supercritical CO_2 has been slow. Suggest factors that make a system based on supercritical CO_2:
   a) attractive
   b) unattractive.

8 Supercritical CO_2 is now used to remove caffeine from coffee beans instead of solvents such as dichloromethane. Suggest two advantages of using supercritical CO_2 for this purpose.

Safer products

As well as avoiding hazardous chemicals as reactants and solvents, green chemists aim to reduce the risks from chemicals in the final products. A notable example has been the elimination of lead or lead compounds from petrol, paint and electrical components.

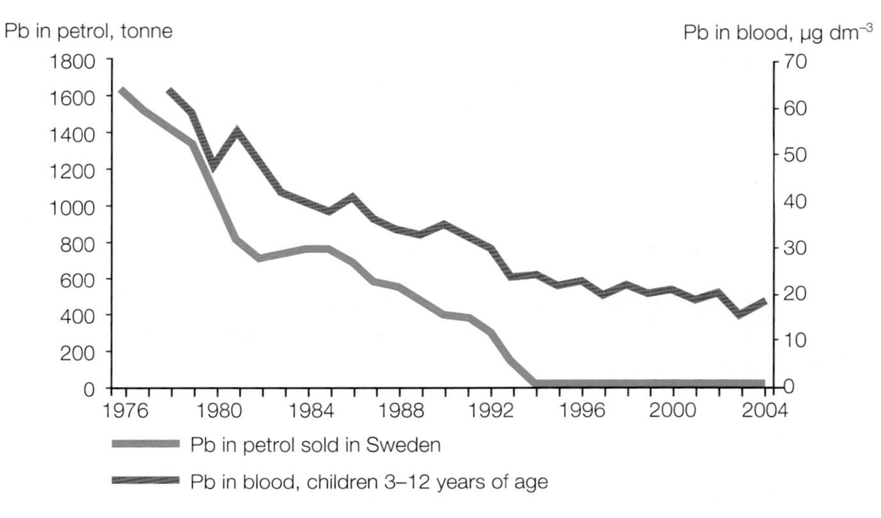

Figure 19.6 Changes in the levels of lead in the blood of children aged 3–12 years in two towns in southern Sweden compared with the mass of lead in petrol sold in Sweden per year.

Concern has been growing about 12 chemicals, or types of chemicals, that do not break down in the environment for a very long time. These chemicals spread widely around the world and accumulate in the fatty tissues of animals. They are toxic to people and wildlife. This set of chemicals – the so-called 'dirty dozen' – includes eight organochlorine pesticides (including DDT), two chlorinated compounds used in industry (including polychlorinated biphenyls, PCBs) and two by-products of industrial activity (including dioxins).

All of the chemicals in the 'dirty dozen' list are classified as persistent organic pollutants (POPs). The characteristics of POPs are that they:

- are highly toxic
- are persistent, lasting for years or even decades before degrading into less-dangerous forms
- evaporate and travel long distances through the air and through water
- accumulate in fatty tissue.

Test yourself

9 a) What can you conclude from the data presented in Figure 19.6?
   b) Suggest explanations for the fact that levels of lead in the blood of children have not fallen to zero since lead in petrol was phased out in Sweden.

10 a) Explain why POPs are particularly hazardous.
   b) Suggest reasons why chemicals of this kind were developed and put to use.
   c) Why was an international agreement needed to control POPs?

As evidence grew about the possible damage done by POPs, the United Nations Environment Programme started negotiations in 1995. Later, in 2001, a convention was drawn up in Stockholm. The 'Convention on Persistent Organic Pollutants' became effective in 2004, and about 100 countries have agreed to outlaw nine of the 'dirty dozen' chemicals. Those who have signed also agree to do everything possible to cut down on the industrial by-products. The use of the insecticide DDT to control mosquitoes that spread malaria is allowed in parts of the world where malaria is found.

Activity

Fire-extinguishing foam

In the 1960s, the US Navy devised foam fire-extinguishing systems to deal with burning hydrocarbons. They were effective, but they released hydrofluoric acid and fluorocarbons at the temperature of a fire. The mixture used to make foams included surface-active agents mixed with water to form stable foams. Unfortunately, the surface-active agents used by the US Navy washed into the ground. They were not biodegradable, so they contaminated water supplies.

In 1993, Pyrocool Technologies developed a fire-extinguishing foam (FEF) that was designed to be much less harmful to the environment. The special foam contains no fluorine compounds and is based on biodegradable surfactants. Very low concentrations of the new foams are needed, and they are very effective.

Figure 19.7 ◀ Fighting a fire with foam.

1. Suggest reasons why a foam of air and water is much more effective at putting out burning hydrocarbons than water alone.
2. Explain why it is hazardous to release hydrogen fluoride into the air.
3. In the presence of electrical equipment, hydrocarbon fires can be put out with halons such as CF_3Br. What is the environmental damage caused by halons?
4. Explain why Pyrocool Technologies' FEF is a 'greener' product than other chemicals used to put out fires.

19.4 Designing more efficient processes

Many of the reactions traditionally used to make drugs and speciality chemicals have involved reagents such as metals, metal hydrides, acids and alkalis. These reagents are used up in the reaction and turn into chemicals that have to be separated from the main product and then treated as wastes. These processes have low atom economies (see Section 10.7).

One of the challenges for green chemists is to develop catalytic methods to create new chemicals from simple raw materials such as hydrogen, steam, ammonia and carbon dioxide. In an efficient process, the catalyst is constantly recycled and not used up. These methods have high atom economies.

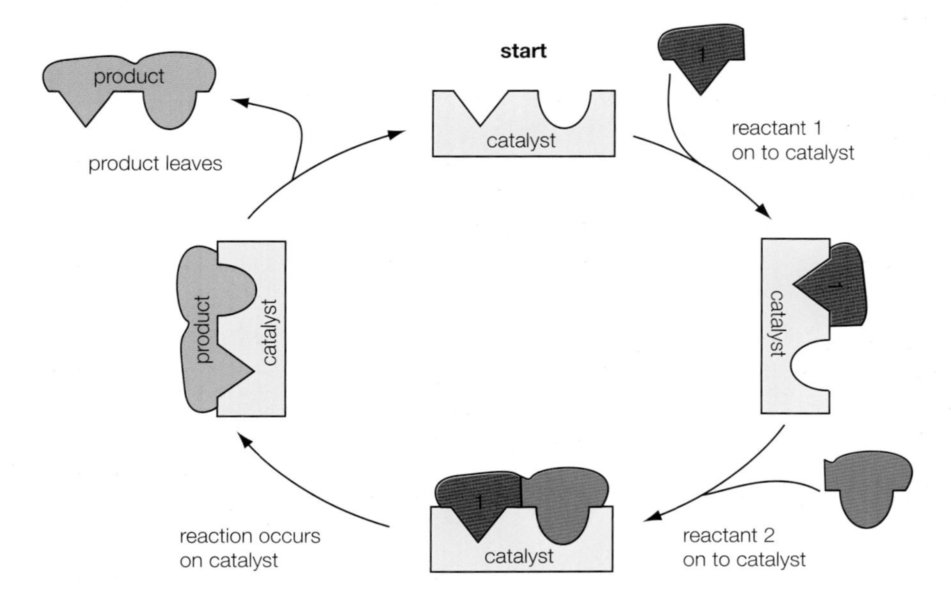

Figure 19.8 ◀ A catalytic cycle. The catalyst is not used up. A long-lasting catalysts can convert a very large amount of reactants into product before it has to be replaced.

Over 40 years, developments in catalysts have made the manufacture of ethanoic acid very much more efficient (see pages 232–3). Worldwide, more than six million tonnes of the acid is made each year. Ethanoic acid is an important bulk chemical as it is a key intermediate on the way to making a very wide range of other chemicals, including polymers, drugs, herbicides, dyes, adhesives and cosmetics.

Test yourself

11 a) Sketch a graph to suggest how the energy profile of the system might vary as the reaction proceeds from reactants to products for the catalysed reaction shown in Figure 19.8.

b) Add to the same graph a curve to represent the change in energy for the equivalent uncatalysed reaction.

19.5 Using renewable resources

The main raw materials for making organic chemicals are crude oil and natural gas. These are non-renewable resources.

The use of renewable resources is still at an early stage in the chemical industry. There are two main approaches:

- One method is simply to extract useful chemicals directly from plant materials. This is what the perfume industry has done for centuries when obtaining ingredients from rose petals, lavender and other plant materials.
- The other method is to break down plant materials into simple chemicals for further use. Worldwide, for example, this is the way in which ethanol is made by fermentation of starch and sugars.

Activity

Greening the manufacture of ethanoic acid

Until the 1970s, the main method of making ethanoic acid was to oxidise hydrocarbons from crude oil with oxygen in the presence of a cobalt(II) ethanoate catalyst. This process ran at 180–200 °C and at 40–50 times atmospheric pressure. The reaction produced a mixture of products that had to be separated by fractional distillation. The economics of the process depended on finding markets for all of the products.

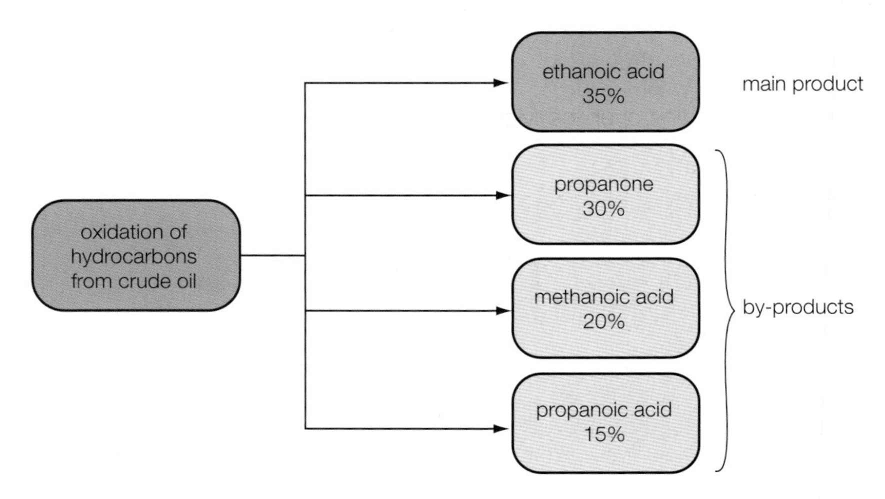

Figure 19.9 ◄ Percentages of the product and by-products from the manufacture of ethanoic acid by oxidising hydrocarbons. The atom economy of the process is about 35%.

In 1970, Monsanto introduced a new process. In this process, methanol and carbon monoxide combine to make ethanoic acid as the only product in the presence of a catalyst made from rhodium and iodide ions. This process runs at 150–200 °C and 30–60 times atmospheric pressure. This is a very efficient way to make the acid with high yields and a high atom economy. Much less energy is needed than in the older method based on hydrocarbons. The reaction is fast and the catalyst has a long life.

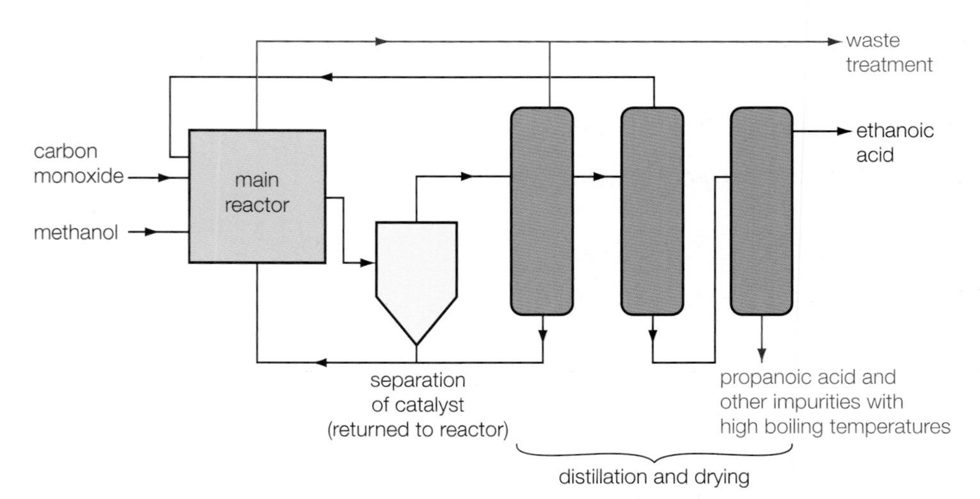

Figure 19.10 ◄ Flow diagram for the Monsanto process for making ethanoic acid.

The Monsanto process does have some disadvantages. Rhodium is very expensive. There also has to be a high enough concentration of water present in the reactor to prevent the rhodium and iodide parts of the catalyst combining to form insoluble salts. Such salts are inactive and have to be recovered and converted back to the catalyst. Water has to be separated from the product by distillation. Furthermore, rhodium catalyses side reactions such as the reaction of carbon monoxide with water to form carbon dioxide and hydrogen.

In 1986, the oil company BP bought all the rights to the process from Monsanto. The company now runs a variant of the process called the Cativa process. The reaction of methanol with carbon monoxide is the same, but the Cativa process uses an iridium catalyst instead of a rhodium catalyst. The catalyst is made even more effective by adding ruthenium compounds as catalyst promoters that speed up the reaction by a factor of three.

The Cativa process has several advantages over the Monsanto process:

- Iridium is much cheaper than rhodium.
- The process is much faster, so more product can be made without increasing the size of the plant.
- Less carbon monoxide reacts with water, so the yield from CO is more than 94%.
- The catalyst is more selective in producing ethanoic acid, so less energy is needed to purify the product.
- The catalyst is more stable, so it lasts longer.
- Water concentrations in the plant can be lower, so less energy is needed to dry the product.

1 a) Write an equation for the reaction used to make ethanoic acid in the Monsanto process.
b) What is the atom economy for the Monsanto process?

2 The yield of the Monsanto process is 98% based on methanol but 90% based on carbon monoxide. Explain why this discrepancy arises.

3 Suggest reasons why the Monsanto process uses much less energy than the older process based on oxidising hydrocarbons.

4 Why is the Monsanto process more economical in meeting the varying demand for ethanoic acid than the older process?

5 What is the atom economy for the Cativa process?

6 In what ways is the Cativa process greener than the Monsanto process?

7 Metals are used in the catalysts involved in all three processes for making ethanoic acid. Where do these metals appear in the periodic table?

A good example of a green plastic made from renewable resources is poly(lactic acid), PLA. The feedstock is lactic acid, which is made by using bacteria to break down sugars or starch. In the presence of a catalyst, lactic acid molecules join up in pairs to form a cyclic compound that polymerises in the presence of a second catalyst.

Poly(lactic acid) can be recycled, but it is also biodegradable. It has a range of medical applications, such as stitches for wounds, and is also used for packaging, waste sacks and disposable eating utensils.

Test yourself

12 a) Explain why crude oil is not renewable.
b) Explain why starch and sugars are renewable.

13 Draw the structure of lactic acid, which has the systematic name 2-hydroxypropanoic acid.

Figure 19.11 ▶ Spring water from Colorado in bottles made with poly(lactic acid).

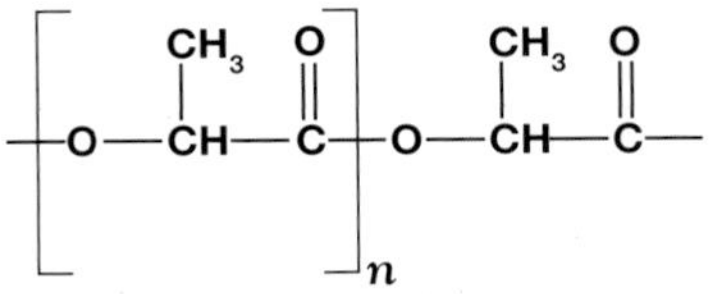

$$-\left[O-CH(CH_3)-C(=O)\right]_n-O-CH(CH_3)-C(=O)-$$

Figure 19.12 ▲ Structure of poly(lactic acid).

19.6 Seeking alternative energy resources

Energy efficiency

New catalysts can make a big contribution to reducing the energy needed for a particular process. For this reason, as well as other reasons, the chemical industry now makes much more efficient use of energy. The average energy needed per tonne of chemical product is less than half of that needed 50 years ago.

An efficient chemical plant harnesses the energy from exothermic processes to produce steam for heating and for generating electricity. A notable example of this is the manufacture of sulfuric acid. The first stage of this process involves burning sulfur to make sulfur dioxide. This is so exothermic that a sulfuric acid plant can produce enough steam not only to generate all the electricity needed to run the plant but also to provide a surplus of electric power that can be sold to the National Grid.

Test yourself

14 Identify four reasons why the chemical industry needs energy to convert raw materials into useful products.

Definitions

A process is **carbon neutral** if the carbon dioxide (or other greenhouse gases) it releases is balanced by actions that remove an equivalent amount of carbon dioxide (or other gas) from the atmosphere.

Bioethanol is ethanol made from plant crops or biomass.

Alternative sources of energy

The move to greater sustainability includes the shift to alternative energy resources to replace finite resources such as fossil fuels. The chemical industry is helping to supply alternative fuels such as bioethanol, biodiesel and hydrogen.

In recent years, there has been a rapid expansion in the production of bioethanol as a fuel throughout the world, particularly in Brazil and the United States. The apparent advantage of bioethanol is that it is carbon neutral. The crops grown to make ethanol take in carbon dioxide during photosynthesis as they grow. The same quantity of carbon dioxide is then given out as the bioethanol burns.

The problem with bioethanol in the United States is that its production uses a lot of energy from oil. This includes fuel to make and run farm machinery and energy and chemicals to manufacture pesticides and fertilisers, as well as energy to process the crop and produce ethanol.

Studies of the overall impact of biofuels have produced conflicting results. Some have even suggested that more energy is expended in making bioethanol than is available from burning the fuel. A more optimistic study has suggested that ethanol from maize in the United States does reduce greenhouse gas emissions but only by about 12% compared with petrol.

The reduction in CO_2 emissions is more favourable for biodiesel obtained from vegetable oil. Biodiesel from soyabeans can reduce emissions by 41% compared with conventional diesel fuel. This is mainly because energy is not needed for distillation during the production of the fuel. In addition, far fewer fertilisers and pesticides are used in growing soyabeans.

Instead of making bioethanol from maize, a better solution is to produce ethanol from non-food sources, such as woody plants, and agricultural wastes, including straw. The use of micro-organisms to break down cellulose to sugars does not compete with food supplies and makes it possible to process large amounts of waste. However, this approach is still at the stage of research and development.

The conditions for producing bioethanol in Brazil are more favourable. Ethanol is manufactured there by fermenting sugars from sugar cane. The refineries that make bioethanol in Brazil have the advantage that they can meet all their energy needs for heating and electricity by burning the sugar-cane waste. However, there are several negative aspects of this industry in Brazil, where, among other things, large-scale deforestation has been carried out to make way for sugar-cane plantations.

Figure 19.13 ▲
Refinery in the United States, which converts starch from maize into bioethanol.

Test yourself

15 a) Why, in theory, is the use of bioethanol as a fuel carbon neutral?
b) Why, in practice, is the use of bioethanol as a fuel not carbon neutral?
16 Why is the growing interest in biofuels a threat to food supplies?

19.7 Reducing waste and cutting pollution

Cutting out toxic wastes

One of the key principles of green chemistry is that it is better to prevent waste than to treat or clean up waste after it has formed. This has led to an emphasis on catalytic processes with high atom efficiencies.

The synthesis and manufacture of new drugs often produces relatively large volumes of waste. Syntheses of this kind sometimes involve several steps, each with its own reactants and solvents, as well as chemicals to purify the product. Lilly Research Laboratories has applied the principles of green chemistry to come up with a new process designed to make a drug that stops convulsions.

The original process used large volumes of solvent and several toxic chemicals, including chromium oxide, which can cause cancer. For every 100 kg of product, the new process cuts out the need for 35 000 litres of solvent and eliminates the production of 300 kg of waste chromium compounds.

Figure 19.14 ▲
Toxic waste is hazardous and expensive to dispose of safely.

Test yourself

17 Suggest three reasons why the chemical industry can become more efficient and more economic by reducing waste from its processes.

Recycling

Recycling waste products helps to conserve resources of materials and energy. Recycling helps to slow down the rate at which valuable resources are converted to waste (Figure 19.15).

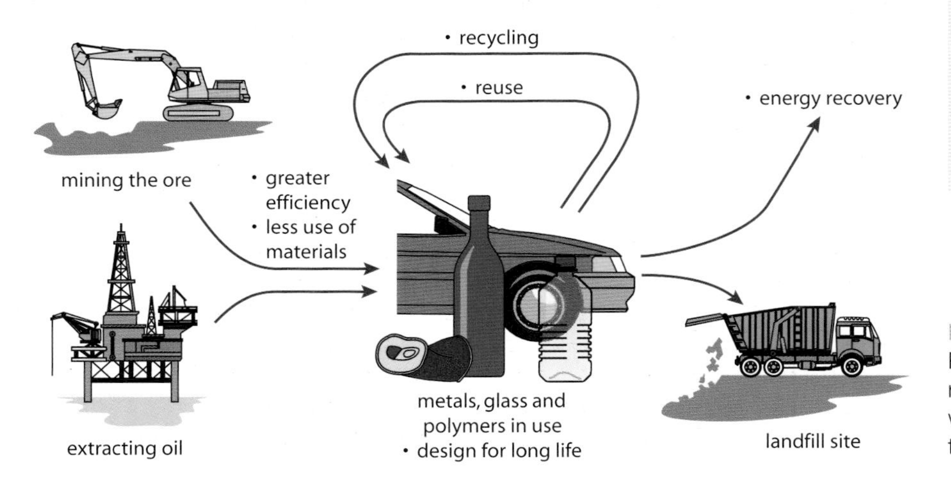

Figure 19.15 ◀
Processes and products can be redesigned to slow down the rate at which valuable raw materials are transformed into waste.

Test yourself

18 a) Give an example to illustrate each of the six bullet points in Figure 19.15.
b) For each example, explain why it contributes to making our use of materials and energy resources more sustainable.

Activity

Buses running on hydrogen

Hydrogen has advantages as a fuel because it produces only water when it reacts with oxygen. Transport for London ran a three-year trial of buses fuelled by hydrogen. The hydrogen combined with oxygen in fuel cells to provide electricity for the motors. The three buses used in the trial proved so reliable that the trial period was extended. The main aim was to cut local pollution of the air in London streets by avoiding the use of diesel fuel.

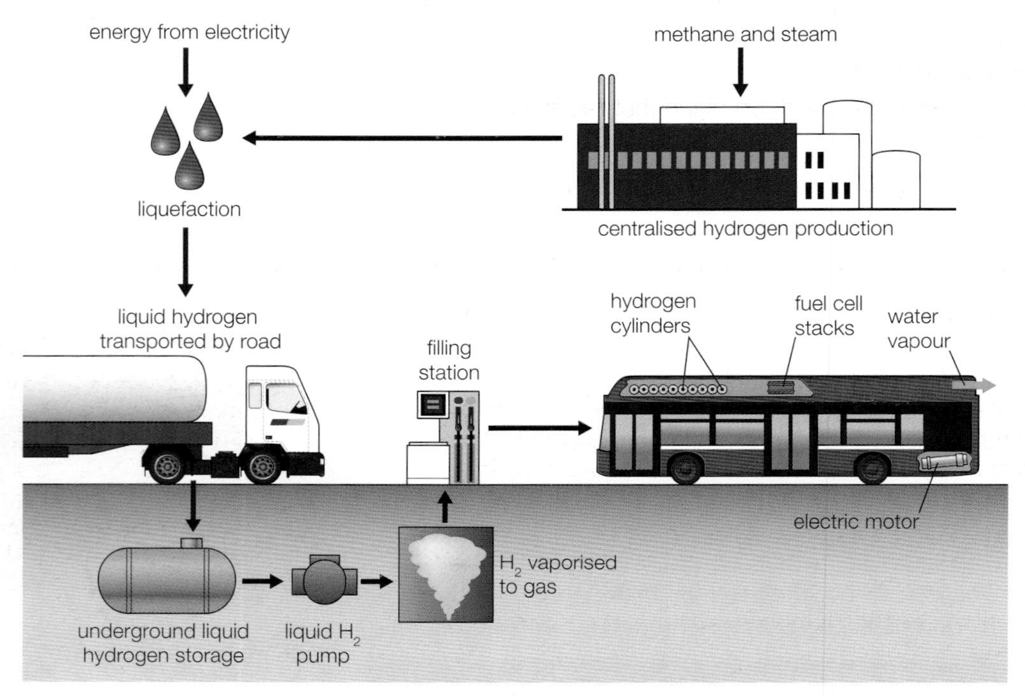

Figure 19.16 ◀ Infrastructure needed to support the trial of fuel-cell buses in London.

There are no natural reserves of hydrogen. The gas must be produced from compounds such as water or methane. In effect, the hydrogen is a store of energy.

Hydrogen for the London bus trial was supplied by a petrochemical company that makes the gas from natural gas (methane) and steam. This is a catalytic process. The first stage is an endothermic reaction that produces hydrogen and carbon monoxide. In the exothermic second stage, the carbon monoxide reacts with more steam to make hydrogen and carbon dioxide.

The carbon dioxide and hydrogen are separated with the help of a molecular sieve that traps carbon dioxide molecules but not the smaller hydrogen molecules. A molecular sieve is usually a zeolite: a crystalline material with channels in the giant structure through which smaller molecules can pass freely. From time to time, the zeolite has to be regenerated by flushing out the carbon dioxide with a stream of hydrogen.

1 a) Write an equation for the reversible reaction of methane with steam.
b) What conditions favour the formation of products in this equilibrium?

2 a) Write the equation for the reaction in the second stage.
b) Overall, in the two stages, how many moles of hydrogen are formed from each mole of methane?

3 Explain why hydrogen is described as a 'store of energy'.

4 Was the use of buses with hydrogen fuel cells in the London trial a carbon-neutral system? Give your reasons.

5 Which of the following methods of making hydrogen have the potential to be carbon neutral and why?
a) Electrolysis of water
b) Growing genetically-modified algae that release hydrogen during photosynthesis.

6 Comment on the extent to which the bus trial in London reduced air pollution.

Recycling is well established in the metal industries. Scrap metal from all stages of production is recycled routinely. Much metal is also recycled at the end of the useful life of products made of metal.

Recycled steel can be as good as new after reprocessing. The scrap is fed into a furnace, where it melts and mixes with fresh metal to make new steel. For every tonne of steel recycled, there is a saving of 1.5 tonnes of iron ore and half a tonne of coal.

Glass is made from raw materials that are relatively cheap. The basic raw materials are sand, limestone and salt. There are plenty of these materials in the UK. However, making glass uses a great deal of energy. Recycling waste glass can save up to a quarter of the energy needed to make glass from its raw materials. Energy is saved:

- by reducing the energy needed to extract and process the raw materials
- because it takes less energy to melt recycled glass than its raw materials.

Recycling polymers is not so easy for a number of reasons:

- There are many different types of polymers, and it is difficult to sort them from wastes.
- Polymer products are often made by combining several polymers and ticking them together in ways that are hard to separate.
- Polymers are often mixed with fillers, pigments, plasticisers and other ingredients.

Instead of recycling a plastic or fibre directly, it can be better to convert the polymer back to simple chemicals. This produces raw materials for the chemical industry, which can replace some of the chemicals from oil. Under favourable circumstances, it is possible to recover more than 80% while burning the remainder to provide the energy for the process.

Sometimes the best solution is to burn polymer waste and to use the energy to generate electricity. It is important to burn the waste at high temperatures under controlled conditions to avoid the formation of toxic chemicals, such as dioxins, which can form if PVC is burned under the wrong conditions.

Some polymer waste ends up being tipped into landfill sites, while other waste litters the environment. Under these circumstances, it can be better if the waste is biodegradable so that it gradually rots away.

Test yourself

19 Suggest reasons why it is easier to recycle metals such as iron or aluminium than polymers such as poly(ethene) or PVC.
20 Suggest three ways by which recycling metals, glass or polymers helps to limit harmful impacts on the environment.

Definition

A material is **biodegradable** if it can be broken down by micro-organisms in the environment to products that are not harmful

REVIEW QUESTIONS

Extension questions

1 Calculate the atom economies for the two methods of producing $C_6H_5COCH_3$ in the processes a) and b) below.

a) $3C_6H_5CH(OH)CH_3 + 2CrO_3 + 3H_2SO_4 \rightarrow 3C_6H_5COCH_3 + Cr_2(SO_4)_3 + 6H_2O$ (2)

b) $2C_6H_5CH(OH)CH_3 + O_2 \xrightarrow{\text{catalyst}} 2C_6H_5COCH_3 + 2H_2O$ (2)

c) Identify features of the two reactions that mean that process b) is greener than process a). (2)

2 The main method of making ethanol worldwide is by anaerobic fermentation of sugars with yeast. After fermentation, the solution contains about 14% ethanol, which is separated from the solution by distillation.

a) Write an equation for the fermentation of glucose, $C_6H_{12}O_6$, to make ethanol and carbon dioxide. (1)

b) Work out the atom economy for this process. (2)

In the UK, the main method of manufacturing ethanol is by hydrating ethene with steam in the presence of a phosphoric acid catalyst. Some side reactions occur and produce other chemicals such as methanol and ethanal. The process runs at 300 °C and a pressure of 60–70 times atmospheric pressure. The product is purified by distillation.

c) Write an equation for the hydration of ethene to make ethanol. (1)

d) Work out the theoretical atom economy for the process shown in the equation. (1)

e) Why, in practice, is the efficiency of the process less than your answer to part d) would suggest? (1)

f) Compare the sustainability of the two processes. (6)

3 a) State what you understand by the term 'sustainable development'. (2)

b) Explain, with at least two examples, why applying the principles of green chemistry helps to make the chemical industry more sustainable. (8)

The periodic table of elements

Key

relative atomic mass **atomic symbol** name atomic (proton) number

1	2											3	4	5	6	7	0(8)
(1)	(2)	(3)	(4)	(5)	(6)	(7)	(8)	(9)	(10)	(11)	(12)	(13)	(14)	(15)	(16)	(17)	(18)
							1.0 **H** hydrogen 1										4.0 **He** helium 2
6.9 **Li** lithium 3	9.0 **Be** beryllium 4											10.8 **B** boron 5	12.0 **C** carbon 6	14.0 **N** nitrogen 7	16.0 **O** oxygen 8	19.0 **F** fluorine 9	20.2 **Ne** neon 10
23.0 **Na** sodium 11	24.3 **Mg** magnesium 12											27.0 **Al** aluminium 13	28.1 **Si** silicon 14	31.0 **P** phosphorus 15	32.1 **S** sulfur 16	35.5 **Cl** chlorine 17	39.9 **Ar** argon 18
39.1 **K** potassium 19	40.1 **Ca** calcium 20	45.0 **Sc** scandium 21	47.9 **Ti** titanium 22	50.9 **V** vanadium 23	52.0 **Cr** chromium 24	54.9 **Mn** manganese 25	55.8 **Fe** iron 26	58.9 **Co** cobalt 27	58.7 **Ni** nickel 28	63.5 **Cu** copper 29	65.4 **Zn** zinc 30	69.7 **Ga** gallium 31	72.6 **Ge** germanium 32	74.9 **As** arsenic 33	79.0 **Se** selenium 34	79.9 **Br** bromine 35	83.8 **Kr** krypton 36
85.5 **Rb** rubidium 37	87.6 **Sr** strontium 38	88.9 **Y** yttrium 39	91.2 **Zr** zirconium 40	92.9 **Nb** niobium 41	95.9 **Mo** molybdenum 42	[98] **Tc** technetium 43	101.1 **Ru** ruthenium 44	102.9 **Rh** rhodium 45	106.4 **Pd** palladium 46	107.9 **Ag** silver 47	112.4 **Cd** cadmium 48	114.8 **In** indium 49	118.7 **Sn** tin 50	121.8 **Sb** antimony 51	127.6 **Te** tellurium 52	126.9 **I** iodine 53	131.3 **Xe** xenon 54
132.9 **Cs** caesium 55	137.3 **Ba** barium 56	138.9 **La*** lanthanum 57	178.5 **Hf** hafnium 72	180.9 **Ta** tantalum 73	183.8 **W** tungsten 74	186.2 **Re** rhenium 75	190.2 **Os** osmium 76	192.2 **Ir** iridium 77	195.1 **Pt** platinum 78	197.0 **Au** gold 79	200.6 **Hg** mercury 80	204.4 **Tl** thallium 81	207.2 **Pb** lead 82	209.0 **Bi** bismuth 83	[209] **Po** polonium 84	[210] **At** astatine 85	[222] **Rn** radon 86
[223] **Fr** francium 87	[226] **Ra** radium 88	[227] **Ac*** actinium 89	[261] **Rf** rutherfordium 104	[262] **Db** dubnium 105	[266] **Sg** seaborgium 106	[264] **Bh** bohrium 107	[277] **Hs** hassium 108	[268] **Mt** meitnerium 109	[271] **Ds** damstadtium 110	[272] **Rg** roentgenium 111							

Elements with atomic numbers 112–116 have been reported but not fully authenticated

*Lanthanide series	140 **Ce** cerium 58	141 **Pr** praseodymium 59	144 **Nd** neodymium 60	[147] **Pm** promethium 61	150 **Sm** samarium 62	152 **Eu** europium 63	157 **Gd** gadolinium 64	159 **Tb** terbium 65	163 **Dy** dysprosium 66	165 **Ho** holmium 67	167 **Er** erbium 68	169 **Tm** thulium 69	173 **Yb** ytterbium 70	175 **Lu** lutetium 71
*Actinide series	232 **Th** thorium 90	[231] **Pa** protactinium 91	238 **U** uranium 92	[237] **Np** neptunium 93	[242] **Pu** plutonium 94	[243] **Am** americium 95	[247] **Cm** curium 96	[245] **Bk** berkelium 97	[251] **Cf** californium 98	[254] **Es** einsteinium 99	[253] **Fm** fermium 100	[256] **Md** mendelevium 101	[254] **No** nobelium 102	[257] **Lr** lawrencium 103

Index

Page numbers in **bold** refer to illustrations.

A

absorption spectra 174
acceptors 38
accuracy **46**
acid rain **35**
acid-base titrations 45–6
acids 32–40, 55–6, 104, **120**, 155
  carboxylic 160, **161**
  equilibria **212**
actinides 68, 97
activation energies **183**, **205**
active sites **202**
actual yields 130
addition reactions 147–51, **151**
adsorption 224
aerosol propellants 170
Afbau principle 66
agents 55
air chemistry 117, 215–25
air conditioning 169–70
alcohols 117, **127**, 157–63, 175
aldehydes 160, **161**
alicyclic hydrocarbons 133, 134
aliphatic hydrocarbons 133, 134
alkalis **32**, **33**, 34, 36–40
  equilibria **212**
  metals 68, 102–7
  see also bases
alkanes 88, 123–5, **127**, 133–41
  halogens 117, 165–71, **165**, **166**, **167**
alkenes 117, 126, 133–4, 143–55, **143**
  ethane 149–50, 159, 213
  halogens 148
alkyl groups **125**
alpha particles **10**
alternative energy resources 138–9, 227, 234–5
amethyst gemstone **76**
ammonium 38, **78**, **82**, 83
anaerobic respiration 158
analyses 117, 173–9
angles 81–4
anhydrous salts 40, **111**
animal
  stings 36
  testing 114
anions 72, **92**, 113, 128, 151
antacids **36**, 104
Antarctica **216**, 221
anthropogenic effects 215, 218
antihistamines 36
apparatus
  oxidation **161**
  titrations **44**
arenes 133, 134
aromatic hydrocarbons 133
arrangements, molecules 2, 122
arteriosclerosis **133**
aspirin **2**
assumptions 184–5
atmosphere **75**, 215–16, **216**
atom economies 130–1, 227
atomic
  masses 12–16
  numbers 12–13, 68
  orbitals 144
  radii 99, **99**
  structures 1–17
  theory 8
atoms 2, 7–17, **52**, 59–61
  ions 72, 74, 102–3
  structure 1–17
attractions 70–80, 87–90, **92**, 110
automatic monitoring stations **222**
average bond enthalpies 194–5
Avogadro constant 20, 27

B

backward reactions 209–13
balance
  equations 24–5, 56
  states 209–13
ball-and-stick models **6**
barites **105**
barium **105**, **107**
bases 32–40, 104
benzene **133**
beryllium **102**
beta particles 10
biodegradable waste 154–5, 233, 237
biofuels 138, **139**, 158–9, 234–5
bioplastics 155
blowing agents 170
boiling points 74–6, 98, 100
  alkenes 144
  branched alkanes 135
  halogens **110**
  hydrides **90**
  metals 80
  noble gases **88**
  salts 39
  straight chain alkanes 135
Boltzmann, L 203
bomb calorimeters **184**, **185**, 188
bombardment, neutrons 97
bonds 5–7, 24–5, 70–92, **133**, 143–4, 173–5
  enthalpies 194–5
BP *see* British Petroleum
Bragg, L 70
branched chains 133, 135
  see also hydrocarbons
British Antarctic Survey **216**
British Petroleum (BP) 233
bromine **109**, 140–1, 149–50
bromomethane **165**
buckyballs **4**
'build up' principle 66
burettes 41–4
burning 139–40, **183**, 188, 230
  see also combustion
butane **126**

C

calcium **102**, **103**
  carbonates **35**, **39**, 106, **208**
  hydroxides 47
calculations 15, 26–30
  titrations 44
calorimeters **184**, **185**, 188
cancer **4**, 114
capture & storage, carbon (CSS) 219–20
carbon 118–31, 219, **220**
  emissions **140**, 216, 219–20
  footprints **219**
  neutral processes 234
carbon dioxide
  concentrations 216, 219
  injections 220
carbonates 37, **105**
carboxylic acids 160, **161**
catalysts 136–9, 147–8, 202–6, 228, 231
  converters **224**
cations 72, **92**, 128, 150–1
Cativa process 233
Celsius scales 27
CFCs *see* chlorofluorocarbons
Chadwick, J 11, 97
chains 117, 118–31, 133–41, 143–4
  isomers 127
  see also hydrocarbons
charges 12, 61, **87**, 98, 103
Chatelier, H 211
chemicals
  amounts 19–21
  analysis 3
  behaviour 2
  changes 3
  compositions 2–3
  equations 24–6
  equilibria 209–13
  industries **226**, 227
  kinetics 198, 202–6
  quantities 19–30
  structures 2–3
chemists 2, 19
  chlorofluorocarbons 169
  electronegativity 85
  energy levels 61
  global warming 215
  green chemistry 231
  mass spectra 179
  naming compounds 54–5

orbitals 65–8
oxidation numbers 51
chemotherapy **4**
chlorine 104, **109**, **112**, 140–1
chlorofluorocarbons (CFCs) 169, **170**, 171, 215, 222
cis-isomers 145, 146
citric acid 33, **34**
clay **71**
climate change 138, 165, 169–70, 217–19
clingfilm **151**
co-ordinate bonds *see* dative covalent bonds
coils 184
collision theory 202–6
colours 5–6
flames 107
combustion 139–40, 154, 160, 183–4, 188–91
tubes **23**, **111**
comparing atomic masses 13–14
compositions, materials 2–3
composting 153
compounds 2, 5–7, 38–40, 67–8, 178–9, 228
alkaline earth metals 104
alkanes 123
analysis 22
ionic 72–4
naming 54–5, 124–6
concentrations 29–30, 45, 47, 200, **201**, **204**
equilibria **210**, 211–12
conditions 187, 193
conduction 8, 39
conductivity 74, 76, 80, 98
configurations, electrons 66
controlling chemical changes 3
corrosion **50**, 198, 227
covalent
bonds 70, 75–8, **85**, **90**, 134
lattices 76
radii 99
solids 71
cracking 135–9, 143, 153, 190
critical states 228
crops **138**
crude oil 134–7, **136**, 143
crystals 40, 70–3, **92**, **99**, **105**, 236
metals **79**
quartz **76**
see also salts
CSS *see* capture & storage
cutting pollution 235, 237
cycles
catalysts **231**
enthalpies **190**, **192**
cyclic hydrocarbons **133**
cycloalkanes 133
cycloalkenes 143
cyclohexanol **163**
cylinders **223**

D

d-blocks **67**
d sub-shells 63, 65–6
Dalton, J **8**, 94
dative covalent bonds **78**
Davy, H **94**
DDT *see* dichloro-diphenyl-trichloroethane
decomposition 8, **105**, **154**, **202**, **207**, 208
deflection 10, 176
dehydration 162–3
delocalised electrons 79, 100, **133**
delta *H* (enthalpies) 181, 188
Democritus **7**
density plots 65
dental care 36
desorption 224
detecting alcohol 175
diamonds **76**
diatomic molecules 24, 110
dichloro-diphenyl-trichloroethane (DDT) 229, 230
diesel 136
diffraction **71**, 81, 145
digestive systems **107**
dilute solutions 43, **43**, 55
dioxins 154, 229
dipoles 87
directional bonds 81
directions of change 193
'dirty dozen' chemicals 229–30
displayed formulae 122, 123
disposing waste 153–4, 235
disproportionation reactions 113
distillation 233
D¨bereiner, J 94–5
donors 38
'*dot-and-cross*' diagrams **72**, **73**, **77**, **84**, **102**
double bonds *see* multiple bonds
drugs **4**, 198, 228
dyes 109
dynamic equilibrium 210, **211**

E

E-Z stereoisomers 146
earth metals 102–7
economies 227
effective nuclear charges 61, 98, 99, 103
efficiency 227, 231
effluent **154**
eka-silicon 95, **96**
electrical
energy 185
heating coils 184
electrolysis 34, 102
electromagnetic spectra 59, 173
electronic structures 59–68
electrons 8–9, **11**, 12, 59–68, 74–9
delocalised 79, 100, 133
discovery **9**
electronegativity 53, 85–6, 89, 111
halogens 109
mass spectra 178–9
microscopes 169
pairs 78, 81–4, 85, 89
redox 50–7
structures 63–5, 67–8
see also '*dot-and-cross*' diagrams
electrophiles 129, **149**, 150–1
electrostatic attractions 70, 71, 76–7, **92**
electrostatic charge **87**
elements 5–7, 67–8, 94–100, 102–7, 109–15, 176–9
definition 2
periodic tables 139
elimination reactions **162**, **163**
emission spectra 59
empirical formulae 21–2, 121
end points 44, 45
endothermic reactions 60, 62, 182–5, 194
see also energy
energy 117, 153–4, 181–96, 206
ionisations 59–68, **60**, 98, **99**, 103
levels **60**, 61–8, **64**, **65**
resources 227, 234–5
enthalpies 117, 167, 181–96
Environmental Protection Agency 227
enzymes **202**
equations 24–8, 37–8, 44, 51, 56, 147
enthalpies 189
see also reactions
equilibrium 117, 198, 209–13
errors 184–5
esterification 162
ethane **119**
alcohols **118**, **119**, **123**, 158–9, 213
alkenes 149–50, 159, 213
green chemistry **232**, 233
ethers 160
exothermic reactions 181–5, **182**, 188, 193, 194
see also energy
experimental formulae 21–2
experiments 3–4
assumptions 184–5
errors 184–5
explosions 198
extinguishing fires 230
extraction **50**
extreme events 219

F

f-blocks **67**, 68
f sub-shells 63, 66
fast neutrons 97
fats 133, 148
feedstocks 153, 159, 160, 226
FEF *see* fire-extinguishing foam
fermentation 130, **154**, 158–9

fertilisers 33, 36, 138
fingerprints 174
fire-extinguishing foam (FEF) 230
first ionisation energies *see* ionisations
fission 128–9, 140, 150, **168**
flames 107, 109, 184
fluorine **77**
fluorocarbons 109
food manufacturers 19
forces 71, **72**, 87–8, **92**, 110, 203
formation 189, 191–2, **200**
formulae 4, 5–7, 21–3, 44, 121–5
  red copper oxide **23**
forward reactions 209–13
fractional distillation 135, **136**, 143, 159
fragmentation 178–9
free radicals 128, 129, **140**, **141**, 220
free rotation 144, **145**
freezers **170**
freezing points 90
fuel-cells 175, 236
fuels **127**, 134–40, 160, 183–5, **223**, 234–5
functional groups **119**, **120**, 127, 157

G

gases 26–8, 59–60, **72**, **109**, **199**, 202–3
  analysers 216
gasoline 136
Geiger, H 10, 11
gemstones **76**
general formulae 124
geometric isomers 145
giant lattices 71, 73, 76, 79, **90**
glass 237
global mean temperatures **218**
global warming 138, 140, 165, 169–70, 217
goitre 110
Goretex 169
gradients **199**
graduated flasks 42
green chemistry 117, 226–37
greenhouse gases 138, 140, 217–19, **218**
groups 1, 67, 102–7, 109–15, **125**
  see also periodic tables
gypsum **105**

H

Haber process 3, 212, 213
half-equations 51
halides 112–13, 148
halogens 88, 109–15, 140, 220–1
  alkanes 117, 165–71, **165**, **166**, **167**
  alkenes 148
  halogenation 139, 140–1, 154
hazardous elements 109–15, 227–30
HCFCs *see* hydrochlorofluorocarbons
HDPE *see* high-density polyethene
heart disease 148
heat 166, **167**, 208
  coils 184
  loss 184
Helheim glacier **219**
helium **11**, **12**
herbicides 227, **228**
Hess's Law 189–93, **190**
heterolytic fission **128**, **129**, 140, 150, **168**
hexane **124**
HFCs *see* hydrofluorocarbons
high-density polyethene (HDPE) **153**
Hodgkin, D 4
holes, ozone 221
homologous series 119, **120**, 143
homolytic fission 128, 140
hormones 109
household waste **153**
human impacts 215, 218
hydrated salts 40, **92**, **105**
hydrates *see* hydrated salts
hydrides **90**
hydrocarbons 118–31, 133–41, 143–55, 157–63, 165–71
hydrochlorofluorocarbons (HCFCs) 169
hydrofluorocarbons (HFCs) 169
hydrogen **11**, **28**, 118–31, 147–8, 236
  bonding **89**, **90**
  bromine 149–50
  halides 113, 148
  hydrogenation 147–8
  ions 32, 34, 38
hydrolysis 166–7, **167**
hydrosphere **75**
hydroxides 102–7

I

Ibuprofen **131**
ice 89, **90**
ideal gases 203
incinerators 154
incomplete combustion 154, 184
indicators 32, **33**, 34, **35**, 40–1, **45**
indigestion 36
indirect determination 189
'indivisible' particles 7–8
industrial feedstocks 159
Industrial Revolution 218
industries **226**, 227
infrared gas analysers 216
infrared spectroscopy 161, 173–5, **174**, 223
initiation 140, 141
injections, carbon dioxide 220
inlet tubes 216
inner shells *see* shells
inner transition elements 68
inorganic materials 75
insecticides 229, 230
insolubility 91
instrumental analysis 117, 173–9
Intergovernmental Panel for Climate Change (IPCC) 218
intermediates 168, **206**
intermolecular forces 71, 75, 87–8, 91–2, 100, 203
  alkanes 134, 135
internal combustion engines **136**
International Union of Pure and Applied Chemistry (IUPAC) 124, 125, 157
internuclear distances **99**
intramolecular forces 71
iodine **76**, 109–10, **210**
ionic
  bonds 70, 72–4, 81, **85**
  compounds 6–7, 38, 39–40
  equations 51
  halides 111
  precipitations 198
  radii 103
  solids 71
ionisations 178–9
  energies 59–68, **60**, 98, **99**, 103
ions **12**, 50–3, **92**, 102–3, 178–9
  acids & bases 32, 34, 37–8
  bonding 72–4, **78**, 81–4
  definition 2
  equilibria 212
  halides 112, 113
  mass spectrometry 176
IPCC *see* Intergovernmental Panel for Climate Change
iridium 233
irreversible changes 208
isomers 126–7, **126**, **127**, 145
isotopes 15–16, 97
IUPAC *see* International Union of Pure and Applied Chemistry

K

Karlsruhe conference 95
Kelvin scales 27, 203
kerosene 136
ketones **161**
kinetics 198, 202–7
kJ (kilojoules) *see* energy
Kroto, H **4**

L

laboratory animals 114
lactic acid 33, **120**
landfill 154
lanthanides 68
lattices 71–4, 76, 79, **90**, **92**
laws
  motion 203
  octaves' 95
  periodic 95
laxatives 104, 105

Le Chatelier's principle 211–13
lengths, bonds 81, 99
lifestyles 138
Lilly Research Laboratories 235
lime kilns **130**
lime water 47
limestone **35**, 106, **106**, **130**, **208**
limiting reactants 130
liquids **72**
  chromatography 4
  liquification 203
lithosphere **75**
litmus 32, **33**, 34, **35**, **45**, **208**
lone pairs 78, 83–4, 89
loss, heat 184
low yields 129–30
lysis 166–7, **167**
lysozyme **71**, **202**

M

magnesium **24**
management, waste 153–4
manufacturing alcohols 158–61
maps 65
Marsden, E 10, 11
mass **11**, 12–14, 19–21, 26
  numbers 12–13
  spectra **15**, **16**, 178–9, **179**
  spectrometry **4**, **13**, 15, 59, 173, 176–9
mass-to-charge ratios 176
materials 2–3
  properties 70
Mauna Loa Observatory 215–16
Maxwell-Boltzmann distribution 203, **205**
mean bond enthalpies 194–5
mean temperatures **218**
measurements
  chemical amounts 19
  volumes **27**
medicine production 109
melting points 74–6, 98, 100
  metals 80
  salts 39
Mendeleev, D 4, **95**, 98
metals 6–7, 39–40, **52**, 55–6, 100–7, 237
  atomic radii 99
  carbonates 37
  halogen reactions 111
  ionic bonding 72–4
  metallic bonds 70, 79–81
  metallic radii 99
  oxides 37
  periodic tables 67–8
methane **5**, **25**, **77**, **81**, 83
  alcohols **160**, 161
methylated spirit (meths)158, **183**
methylpropane **126**
Midgeley, T 170
'Milk of Magnesia' 36, 104
modelling equations 25
models **6**, 70
molar masses 19–21
molecules 2, 81–2, 83–4, 118–31
  arrangements 2
  forces 71, **72**
  formulae 21–2, 122
  orbital theory 144
  structures 71, 75–6
moles 19–21
Mongolfier brothers **203**
monitoring air pollution 222–3
mono-atomic elements 100
Monsanto 227–8, **232**
Montreal Protocol 169
Moseley, H **97**
motor vehicles **223**
MTBE 160
Mulliken, R 85
multiple bonds 77–8, 83–4, 126, 133–4, 143–4, 148

N

naming compounds 54–5, 124, 125, 126
nanotechnology 3
neon **59**, 60
neutralisation 35–8, **37**
neutrons 8, **11**, 12, 97, **102**
Newlands, J **95**
Newton, I 203
Nitram 36
nitrates 107
nitrogen **57**
  oxides 222, 223
noble gases 63, 67, **88**
non-biodegradable waste 154
non-bonded electrons 83–4
non-linear molecules 83
non-metals 5–7, 39–40, **52**, 72–8, 100, 109–15
  atomic radii 99
non-renewable resources 231
non-stick polymers 109
nuclear charges 61, 98, 103
nuclear model **11**
nuclei 10, 60–8, 84, **85**, 99
  hydrogen bonding 89
  ionisation energies 98
nucleophiles 129, **167**, **168**

O

oil 134–7, 143
olive oil **143**
open systems 181
orbitals 65–8, 144
organic chemistry 75, 117, 118–31, 143–55, 157–63, 165–71
organochlorine pesticides 229
outermost shells *see* shells
oxidation 50–1, 53, 55, 160, **161**, 184
  agents 55
  numbers 51–5, **52**, **53**, **54**
  states 54, 103, 112–13
  see also redox
oxides 37, **57**, **102**, 104, 222, 223
oxoanions 113
oxygen 103
  see also combustion
ozone layer 220–2

P

p-blocks **67**
p orbitals **66**
p sub-shells 63–6
pairs 78, 81–4, 85, 89
palladium catalysts 228
paraffin 136
parent ions 179
patterns
  chemical behaviour 2
  fragmentation 178
  ionisation energies 61–2
  periodic tables 94–100
  solubility 91–2
Pauling, L 85
PCBs *see* polychlorinated biphenyls
peaks **174**
percentage uncertainties **187**
percentage yields 129–31
periodic tables 4, 59–62, 64, 67–8, 74, 94–100, 239
  electronegativity **85**
periodicity 94–9, **99**, **100**
periods 67–8, 74, 99
  see also periodic tables
permanent dipoles 87, **88**, 89–90
  see also hydrogen bonding
persistent organic pollutants (POPs) 229–30
pest control 165, 229
PET *see* polyethene terephthalate
petrol 134–7, **223**
Pfizer 228
pH 33, **36**, **45**, **115**
pharmaceutical companies 19
phenolphthalein indicator 40–1
Phoenix lander 176, **177**
photochemical smog **223**
photosynthesis 181, **182**
physical properties, periodicity 98
pi bonds **144**
pipetting **19**, **40**, 41–4
PLA *see* poly(lactic acid)
plant stings 36
plastics 153–5, 233
plots 65
plum pudding model **9**
plutonium **97**
polar covalent bonds **85**
polar stratosphere clouds **221**
polarity 85–7, 89, 91–2, **129**, **173**

alkanes 134
poles 87
pollution 153, 165, 169–70, 215–16, 222–3, 235–7
prevention 227
see also air chemistry; green chemistry
polychlorinated biphenyls (PCBs) 229
polyethene terephthalate (PET) **153**
poly(lactic acid) (PLA) 155, 233, **234**
polymers **3**, 109, 119, 151–2, 153–5, 237
polystyrene (PS) **153**
polythene **119**
polyvinylchloride (PVC) 109, **151**, **152**, 168
POPs *see* persistent organic pollutants
position isomers 127
positive nuclear charges 61, 98
potassium dichromate(VI) 160
precision **46**
predictions, Mendeleev 95, 96
pressure 204
equilibria 211, 212–13
preventing pollution 227
primary alcohols 157
see also alcohols
primary pollutants 223
primary standards 42
products 127, 147, 182, 194, **199**
safety 229
stability 206
see also equations; reactions
profile diagrams **182**
propagation 141
propane **123**, **134**, **140**
propellants 170
properties, materials 70
proteins **71**, **121**
protons 8, 9, 10, **11**, 12
PS *see* polystyrene
PVC *see* polyvinylchloride
Pyrocol Technologies 230

Q

quadrupole instruments 176
quantities 19–30
quantum shells 60–8
quartz **76**
quicklime 104, 106, **130**

R

radiation **173**, 217, 220
radicals **128**, 129, **140**, **141**, 220
radii 99, 103
radioactivity **97**
random errors 46
rapeseed **138**
rates 117, 198–213
reactions 166, 167
reactants 127, 130, 147, 182, 194, **199**
stability 206
see also equations; reactions
reactions 26–8, 35–8, 103–4, 112–13, 127–31, **183**
addition 147–51
alkanes 139–41
elimination 162–3
endothermic 60, 62, 182–5, 194
enthalpies **185**, **186**, 190–2, 194
exothermic 181–5, **182**, 188, 193, 194
halogenation 140–1
halogens 111–12
profiles **204**, **207**
rates 166, 167, 198–213
redox 50–7
reversible 208–13
substitution 166–8
volumes 26–8, 40
see also equations
reactivity 109–15, 139
reagents 147
see also reactants
recycling 153, 227, 235, 237
red copper oxide, formulae **23**
redox 50–7
see also oxidation; reduction
reducing waste 206, 227, 228, 235, 237
reduction **23**, 50–1, 53, 55
see also redox
refineries, bioethanol **235**
reflux condensers 161
reforming 135, 136, **137**
refrigeration 169–70
rehydration 208
relative charges 12
relative masses 11–16
renewable resources 226, 227, 231, 233–4
resolutions 176
resources 117, 227, 234–5
retardants 109
reversible changes 208–9
rhodium 232
ring structures **133**
see also hydrocarbons
Rio Declaration on Environment and Development 226
roadside testing 175
rotation 144, **145**
Roundup (herbicide) 227
rules, oxidation numbers **52**
ruthenium compounds 233
Rutherford, E 10, 11, 59

S

s-blocks **67**
s orbitals **66**
s sub-shells 63–6
safety 227–30
salts 38–40, **72**, 92, 104, **111**, **208**
green chemistry 232
see also halogens
saturated
alkenes 143
fats 133, 148
hydrocarbons 133
solutions 91
scales, electronegativity 85
scanning tunneling microscopes 3
scientists, global warming 215
Seaborg, G 97–8
secondary alcohols 157
see also alcohols
secondary pollutants 223
shapes 81–2, **83**, **84**, **134**
shells 60–8, **64**, **65**, 103
shielding effect 61, **98**, 111
SI *see* standard units
sigma bonds **144**
simple molecular structures 71, 75–6
single-beam spectrometer **173**
skeletal formulae 123
slaked lime 106
'smeared out' electrons 65
society, science values 3
sodium aluminium silicate 136
sodium chloride **72**, **73**
solar
energy 181
radiation 217
solids 201, **204**
solubility 33, 91–2
hydroxides 104
solutes 91–2
solutions 29–30, 42–7, 91–2, 110–11, **185**
solvents 91–2, **112**, 228
sources, alternative energy 227, 234–5
space research 176–7
space-filling models **6**
specific heat capacities 184
spectator ions 37, 38
spectra 15–16, 59, 85, **97**, 173, 178–9
spectrometry **4**, **13**, 15, 59, 173, 176–9
spectroscopy 173–5, 223
stability 63, 105, 188, 189, 206–7
standard
conditions 187
enthalpies 187–9, **192**
solutions 42–3
states 188, 189
units (SI) 20, 27
starch 155
states of balance 209–13
steam 148–9
stereoisomers **145**, 146
sting treatments 36
straight chains 133, 135
see also hydrocarbons
stratosphere 215, **221**
structural
formulae 122–3
isomers 126

structures 59–68, 70–92
alkanes **143**
atomic 1–17
electrons 63–5, 67–8, 74
materials 2–3
three-dimensional 81
sub-shells 61–8, **64**
substitution reactions **140**, **141**, 166–8
successive ionisation energies *see* ionisations
sulfates 105, 107
sunflower oil **143**
supercritical states 228
surface areas 201, **204**
sustainable lifestyles 138
symbols 4
synthesis intermediates 168
systematic errors 46
systems 181

T

temperatures 201, **202**, 204–5, **218**, 228
equilibria 211, 212–13
temporary dipoles 87, 110
see also hydrogen bonding
termination 141
tertiary alcohols 157
see also alcohols
testing 32–3, 114, 148, 208
theoretical yields 129–31
theories 3–4
thermochemistry 184–5, 187, 206
see also enthalpies
thermodynamic stability 207
Thomson, JJ 9
thyroxine 109
titanium **131**
titrations 41–6, **42**
formulae 44
toxicity 227, 229, 235
trans-fats **148**
trans-isomers 145, **146**
transfers **50**
transition elements 67, 68
transmission quadrupole instruments 176
transmittance 174
transuranian elements 97
trays 135
treatments 36
trends 94, 98, 99, **103**
carbonates 105
electronegativity **85**
halogens 110, 111
hydroxides 104
sulfates 105
triads 94
trihalomethanes 114
triple bonds *see* multiple bonds
troposphere 215

U

u-shaped tubes **178**
ultraviolet (UV) radiation 220
uncertainties **187**
United Nations Environment Programme 230
Universal indicator 32, **33**, 45
unsaturated
alkenes **143**
fats 133
hydrocarbons 133, 148
unsymmetrical alkenes 150–1
UV *see* ultraviolet

V

V-shaped molecules 83
values, society 3
van der Waals' forces 87–8, 92, 110
vegetable oil 148
Venn diagrams **33**
vibrations 173–5, **173**, 174
vinegars **40–1**
volatile organic compounds (VOCs) 228
volumes 26–7, **28**, 40
volumetric analysis 42

W

Waals, J 87, 203
waste 153–4, 159, 206–8, 235, 237
water **6**, 34–5, 89–92, 104, 148–9, 162–3
molecular shapes 83
molecules **71**
testing 208
treatment 114–15
water of crystallisation 40
wavelengths **173**, 217
see also infrared spectroscopy
wavenumbers 174
Winkler, C 96
witherites **105**
Wolff, E **216**

X

X-rays **70**, 79, **97**, 99, **107**, **133**
diffraction **71**, 81, 145

Y

yeast 158–9
yields 129–31, 206

Z

zeolites 236
Zoloft (drug) 228